BASIC MOLECULAR BIOLOGY

BASIC MOLECULAR BIOLOGY

FRED W. PRICE

PROFESSOR OF BIOLOGY
STATE UNIVERSITY COLLEGE,
BUFFALO, NEW YORK

JOHN WILEY & SONS

NEW YORK · SANTA BARBARA · CHICHESTER · BRISBANE · TORONTO

Library of Congress Cataloging in Publication Data

Price, Fred William.
  Basic molecular biology.

  Includes bibliographical references and index.
  1. Molecular biology.   I. Title.
QH506.P74   574.8'8   78–525
ISBN 0–471–69729–X

Printed in the United States of America

10 9 8 7 6 5 4 3 2 1

Preface

This is an introductory textbook in molecular biology. It is based on the one-semester course that I teach at the State University College at Buffalo, New York, where it is offered as an elective subject for undergraduate students majoring in biology. This does not mean that the book is directed toward any particular college or university level; it is suitable for freshmen with a good background in general biology and chemistry. In compiling the text, I have assumed that anyone wishing to acquire a knowledge of molecular biology will have already studied general chemistry and biology. I have not, therefore, included the customary chapters on the elementary facts of atomic and molecular structure, but have included chapters on cell division and Mendelian genetics for the sake of completeness in developing the themes of biological replication and information transfer.

There are several good books on molecular biology currently available, but many of these seem to be either too advanced or too elementary as beginning textbooks. Others, although of the right level of difficulty, are too detailed. This textbook will satisfy a need felt by other teachers in the field for a textbook of a specific level.

My approach is structural. The introductory chapter on selected relevant physicochemical principles is followed by four main sections. First, protein molecules and assemblies of protein molecules are discussed under the broad theme of the relation of structure to function. Here, the structural and mechanical roles of fibrous proteins are described and the influence of amino acid composition in determining their distinctive properties is discussed. The role of the globular proteins as the carriers and catalysts of cells and organisms is presented next. Distinction is made between *enzymes*, which are globular proteins with specific combining capacity and catalytic activity, and *emphores*, such as antibodies and the oxygen-binding proteins, which have specific binding capacity but no catalytic function. The importance of quaternary structure in physiological regulation is discussed.

Biomembranes as protein-lipid assemblies are considered under the general area of bioenergetics and energy transduction. In this section, the complex structure of biological energy transducers is described with emphasis on the common features—the intricate structure of stacked parallel phospholipid membranes, compartmentalization, and spatial organization.

Biological replication and the storage and transmission of genetic information is the subject of the third section, which deals with protein-nucleic acid assemblies. Recent ideas on the ultrastructural organization of the interphase cell nucleus and of metaphase chromosomes are introduced, and ribosomes and their role in the structural basis of protein synthesis are discussed.

The preceding sections may be described as analytic, because the cell is systematically dissected into its macromolecular components. The fourth section can be thought of as synthetic; it demonstrates how these macromolecular components are put together to form a living cell. The underlying theme is that despite the seeming diversity of cell types and structures, cells really are composed of only a relatively few "standard parts." These are found in fairly simple form in procaryotic cells, but they are elaborated into more complex structures in eucaryotic cells.

The temptation to delve deeply into certain exciting aspects of molecular biology has been resisted. In an introductory text, the clear exposition of fundamentals with a minimum of detailed information is of greater importance. There are ample suggestions for further reading if the student wishes to pursue a given topic in greater depth.

The description of the genetic control of protein synthesis and the treatment of molecular genetics are not as detailed or as thorough as they are in many other molecular biology textbooks. This is intentional. The modern science of molecular biology was ushered in with the elucidation of the role of nucleic acids in determining the structure of cellular proteins. Exciting and fundamental as this topic is, I have tried to show that molecular biology is something more than this, and that equally exciting fields in areas other than molecular genetics are opening up and expanding all the time. In other words, my aim is a balanced approach, which some of the otherwise good textbooks in this field do not provide.

Fred W. Price

Acknowledgments

I extend sincere thanks to my publishers, John Wiley & Sons, Inc., for their help during the writing and preparation of the manuscript, and to the anonymous reviewers who offered many valuable suggestions and constructive criticism.

I also thank the various authors and investigators who have allowed me to reproduce photographs and drawings from their published works.

I greatly appreciate the help of Michelle Brucato, secretary in the Biology Department, State University College, Buffalo, New York, who gave unstintingly of her time in the final typing of most of the chapters.

Special heartfelt thanks go to my parents and to my friend Joseph Provato for their encouragement and moral support during the writing of this textbook.

F.W.P.

To my beloved parents,
Corona A. and William G. Price,
this book is affectionately dedicated

Contents

BASIC MOLECULAR BIOLOGY

PART ONE
INTRODUCTORY

PART ONE

INTRODUCTORY

CHAPTER ONE
The science of molecular biology

Molecular biology is the study of structure, organization and function in living matter at the molecular level. It attempts to explain biological phenomena in terms of the laws of chemistry and physics that govern the behavior of the simpler nonliving physical and chemical systems that have been successfully analyzed and explained in terms of them. The success of this approach depends, of course, on the validity of the assumption that living organisms, despite their complexity, are subject to these same laws.

NATURE AND AIMS OF MOLECULAR BIOLOGY

Students often ask, "What is the difference between biochemistry and molecular biology?" I usually answer this by asking them to think of two features of living matter at the molecular level—metabolism and structure. Regarding metabolism, consider the stages in the glycolytic breakdown of the sugar molecule in cellular respiration. All of the changes undergone by the glucose molecule, the enzymes and the cofactors that catalyze the process, and the energy yield are examples of the subject matter of biochemistry, which is principally concerned with the metabolic transformations of relatively small molecules in the cell. Regarding structure, the elucidation of how protein molecules, phospholipids, and cholesterol are organized within cell membranes or how the protein molecules in the contractile machinery of muscle are organized provide examples of the subject matter of structural molecular biology (i.e., we are concerned here with macromolecular organization of cellular structures; molecular biology is therefore mainly concerned with proteins, nucleic acids, and other macromolecules and their interactions).

In the following chapters, four main aspects of biological macromolecules are discussed.

1. Their nature; that is, their chemical structure, size, and shape and three-dimensional organization.
2. Their functions and how these are related to structure. One of the main aims of molecular biology is to attempt to correlate gross biological function with events that occur at the molecular level.
3. Their organization in cellular components.
4. The manner in which they are built up from their smaller monomeric subunits and how living cells overcome the "energy barriers" and entropy problems in synthesizing them.

As we have already mentioned, molecular biology is concerned not only with molecular structure and organization—it would indeed be a sterile subject if this were so—but also with the correlation of function with structure, such as the elucidation of how the structure and organization of the protein filaments in muscle fibers are related to muscular contraction and how they interact to bring this contraction about; how proteins and lipids are organized in cell membranes in relation to the permeability and other properties of cell membranes; and how chromosome and ribosome structure lead to specificity in protein synthesis.

One of the most fascinating aspects of biomolecular structure–function relationships is that when biomolecules associate in certain ways, new and unique functions emerge that were not inherent in the individual molecules; in a much simpler case, the water molecule possesses unique properties not found in its constituent atoms. Molecular aggregates come together to form more highly organized structures with their own characteristic properties, and these—the functional units of the cell—exist together in a complex, regulated, functional interdependence from which emerges the unique and mysterious phenomenon called "life." The recognition of these interrelations between "supramolecular structures" shows that we cannot expect to be able to "explain" life phenomena in terms of isolated molecules or even in terms of systems of interacting molecules alone.

There is another sense in which molecular biology may be defined; this dates roughly from the elucidation by J. D. Watson and F. H. C. Crick in 1953 of the structure of the genetic material (DNA), based on the X-ray diffraction data of M. F. H. Wilkins and R. Franklin. This tremendous achievement was followed by one spectacular discovery after another in various laboratories, until finally the mechanism of cellular information transfer and protein synthesis and its specificity was worked out and summarized in what has become known as the "central dogma" of molecular biology (see Chapter Eighteen); this is a high-sounding name for what is really a rule-of-thumb principle. Closely related to this has been the discovery of various kinds of metabolic regulatory processes in cells; some are genetic and others are nongenetic. Molecular biology is therefore often thought of in this narrower sense as primarily the study of information transfer and metabolic regulation in cells.

J. C. Kendrew pointed out that there are, in fact, two kinds of molecular biologists, structurists and informationists. At one time these were almost completely separate schools. There was the "phage molecular biologists" of M. Delbrück and S. E. Luria, who were principally interested in analyzing and explaining the genetics of viruses and bacteria in terms of a one-dimensional, submicroscopic, informational entity that ultimately turned out to be the DNA molecule. The other school, represented by W. T. Astbury, J. D. Bernal, and their pupils, was interested in developing techniques for determining the three-dimensional structure of biological macromolecules, especially proteins. This group was not concerned with or especially interested in genetics. Today, happily, there is far more integration of the two schools because, after all, genetic phenomena and protein structure and function are inseparably linked, as is recognized in the central dogma, perhaps even more so now that dissenting voices have been heard questioning the strictly one-way direction of cellular information transfer (Chapter Eighteen) implied by the dogma in its original form.

ORIGIN AND BRIEF HISTORY OF MOLECULAR BIOLOGY

The origin of the term molecular biology dates from Astbury's use of it in his 1950 Harvey Lecture. However, the molecular approach to the study of life is much older than this, as the following brief review of ideas and findings about protein structure will show.

That proteins were linear sequences of various smaller molecular units linked by covalent forces was vaguely realized even before 1900. However, until 1925, a more popular concept was that proteins were association complexes in which secondary forces were responsible for maintaining their structural integrity. Gradually, however, the idea that proteins were macromolecular (i.e., high molecular weight linear polymeric chains of covalently linked monomers) gradually displaced the association complex view.

As early as 1840, hemoglobin was obtained in crystalline form; many other proteins, including enzymes, were prepared as crystals by the turn of the century. In 1909 E. T. Reichert and A. P. Brown published their monograph, "The

Crystallography of the Hemoglobins," which contains crystallographic data of hemoglobins from more than 100 vertebrate species. Despite the obvious implication (i.e., since proteins could be crystallized, their molecules must have a definite three-dimensional shape), most biologists did not realize the significance of this until about the middle of the 1900s. Perhaps this was a result of deep-rooted, unconscious, psychological conditioning, implying that protoplasm and proteins were far too complex for there to be any hope that mortals would ever be able to analyze them by methods that worked for simple organic and inorganic molecules. This spell was broken by F. Sanger's elucidation in 1955 of the covalent structure of the insulin molecule (Chapter Three), which stimulated other groups and led to the determination of the primary structures of several other proteins relatively soon afterward.

As long ago as 1871, R. Lankester expressed the belief that the discovery and analysis of the chemical and molecular differences between different groups of organisms may be more important for determining phylogenetic relationships that the comparative study of gross morphology. The immunological, chromatographic, electrophoretic, and other separative technics and powerful methods for protein primary structure determination have made such comparative studies of molecular morphology fairly commonplace and have fully borne out Lankester's remarkable early insight. It is well illustrated by the species variations in the molecules of hemoglobins (Chapter Nine) and cytochromes (Chapter Three).

A GLIMPSE INTO THE FUTURE

Although the history of molecular biology has been fascinating, the future could be even more exciting. Techniques for synthesis of nonnatural nucleic acids and proteins with strange and unexpected properties have been developed, and macromolecules can now be made that outperform their natural counterparts in specific tasks. Experiments in which missing or defective genes have been artificially introduced into the defective organism's genome may hold hope for achieving a permanent cure of genetically determined diseases in humans. The puzzle of the origin and nature of cancer and its eventual cure will almost certainly be solved from studies of normal and abnormal cells at the molecular level. The same approach will also, no doubt, ultimately enable us to explain and understand the miracle of growth and differentiation of multicellular organisms from the fertilized egg cell.

Most of all, advances in the approach to cell and organismal function, which involves not only the study of the properties of biomolecules, but also seeks to elucidate and to analyze the laws and forces governing the complex interactions of molecular and supramolecular complexes and aggregates, provide the best hope for the eventual analysis of the life process. This, to me, is the special fascination and aesthetic appeal of molecular biology.

CHAPTER TWO
Some physicochemical principles

As mentioned in the preface, I assumed that in addition to a knowledge of general biology, the reader has a thorough understanding of basic chemical concepts and topics. These should include:

1. The definitions of matter, element, compound, atom, molecule, and ion.
2. Covalent and ionic bonding.
3. Structure and properties of simple aliphatic organic compounds: hydrocarbons, alcohols, aldehydes, ketones, ethers, carboxylic acids, esters, amides, and α-amino acids.
4. Structures of simple aromatic organic compounds: benzene and five- and six-membered heterocyclic ring compounds.
5. Isomerism and asymmetry in organic compounds.

Knowledge of the structures and properties of the four main classes of biological macromolecules—proteins, polysaccharides, polynucleotides, and lipids—is not assumed and will be described in the relevant chapters.

In addition to this elementary chemical background, a familiarity with certain physicochemical principles and phenomena is desirable for the proper understanding of certain cell and molecular biological phenomena and concepts. Some of the more important of these are the law of mass action, chemical equilibrium, dissociation of water, pH and buffer action, Le Chatelier's principle, hydrogen bonding, the "law of increasing disorder," diffusion, osmosis, and dialysis. These will be simply explained in the following sections.

LAW OF MASS ACTION

Suppose we have two substances, A and B, that react together to give the products C and D. For the sake of argument, we may imagine that A and B are dissolved in water and react to give C and D as soluble products.

$$A+B \longrightarrow C+D \tag{1}$$

The rate at which A and B react to give C and D is determined by several things, such as temperature, pressure (if one or both were gases), and the concentrations of the reactants. Our aim is to determine the effect of concentration, so we will assume that temperature, pressure, and other factors that are likely to affect reaction rate are kept constant.

Common sense and experience tell us that increasing the concentration (moles per liter) of A or B or both will increase the rate at which they will react, but what is not immediately obvious is how the rate of reaction is related to the two concentrations simultaneously. One might guess that perhaps the rate will be proportional to the sum of the concentrations of A and B (written [A] and [B] where the square brackets stand for "concentration of"). Actually, the reaction rate is proportional to the product of the concentrations. This is written:

$$R \propto [A][B]$$

where

$$R = \text{rate of reaction}$$

$$\propto = \text{"proportional to"}.$$

At first sight it may seem strange that the rate is proportional not to the sum, but to the product of the concentrations; however, application of a little probability theory should straighten things out.

Probability theory has a bearing on chemical reaction rates because we are dealing, in this case, with atoms, molecules, and ions in solution and these are constantly dashing about in a completely random manner. For reaction to occur, two or more of the particles in solution must collide. If they did not collide it is difficult to know how else reaction could occur. Reaction rate therefore depends on collision frequency; if we want to calculate this we need a measure of the probability of two particles colliding.

In the above example we have A and B in solution at known concentrations at the beginning of the experiment. The particles of A and B (whether they are atoms, molecules, or ions) are evenly distributed in the solution and are in a state of constant random motion. Let us consider the probability that a given particle of A will be at any one particular point anywhere in the volume of the solution at a given moment. A moment's thought will correctly suggest that this probability will be directly proportional to the concentration of A; the more A particles are flying around, the more likely it is that any one particle is going to be at a

particular point at a given moment. If the concentration is doubled, the probability is doubled. The same reasoning applies to the particles of B. For reaction
to occur, an atom of A must collide with an atom of B (i.e., both must be in the
same place at the same time). We have just found that the probability of either A
or B particles being at a given place depends on how many of both there are in a
given volume (i.e., on their concentrations). What we now need to calculate is
the probability that a particle of A and a particle of B will be in the same place at
the same time. Probability theory tells us that the chance of two separate and
isolated independent events occurring simultaneously is equal to the product of
their separate probabilities. This is the law that governs the results of tossing two
coins or dice and the probability of any sequential combination of heads and tails
or of numbers occurring.

The separate probabilities of A and B particles being at any one point at any
moment is proportional to the separate concentrations of A and B, respectively.
Therefore, it follows that the probability of an A and a B colliding and reacting
and, hence, the reaction rate is proportional to the product of the concentrations. This may be written:

$$R = k[A][B]$$

where $k =$ the proportionality constant. This is the mathematical expression of
the law of mass action, which states that the rate of a chemical reaction is directly
proportional to the products of the *active masses* of the reactants. Active mass is
here interpreted as concentration that is sufficiently accurate for our purpose.

CHEMICAL EQUILIBRIUM

As A and B continue to react, the concentrations of both, of course, are
decreasing, and so the rate of reaction gets slower. At the same time, C and D
are accumulating. In principle, every chemical reaction is reversible, that Equation 1 is more correctly written:

$$A + B \rightleftharpoons C + D \tag{2}$$

Admittedly, most reactions go so far in one or the other direction as to appear
irreversible, but there is always some back reaction, although this may be
infinitesimally small in extent. Therefore, as C and D accumulate, they will react
to give A and B at an increasing rate as their concentrations increase, and this
rate (R') will be proportional to their concentrations.

$$R' \propto [C][D]$$

or

$$R' = k'[C][D]$$

Sooner or later, the rates of the forward-and-back reaction will be equal, and equilibrium will be established with no further net change in the composition of the reaction mixture. Therefore, at equilibrium:

$$R = R'$$

that is,

$$k[A][B] = k'[C][D]$$

Rearranging:

$$\frac{[C][D]}{[A][B]} = \frac{k}{k'} = K_E$$

where

$$K_E = \text{the } \textit{equilibrium constant} \text{ for the reaction}$$

LE CHATELIER'S PRINCIPLE

Imagine, now, what would happen if the concentration of any one of the reactants, say A, in the above equilibrium was increased (i.e., more was added). The rate of the forward reaction would increase because the product [A][B] would now be greater than before the extra A was added. More C and D would be formed, and the rate of the back reaction will therefore increase. Eventually a new equilibrium would be established.

The important thing to notice here is that by adding more A, we have disturbed the equilibrium and, in so doing, the system has reacted in such a way as to tend to remove that which is disturbing the equilibrium; the forward reaction goes faster, and this tends to remove the added A that disturbed the original equilibrium. The same applies to any of the other reactants, either singly or if the concentration of any two were increased or decreased. Thus, when an equilibrium is disturbed, the system reacts in such a way as to try to remove the disturbance. This is what is stated in Le Chatelier's principle:

> *If to a system in equilibrium a constraint is applied, that change takes place that tends to remove the constraint.*

The word "constraint" means anything that disturbs the equilibrium. Note that Le Chatelier's principle states that a system in equilibrium reacts to a constraint in such a way that it *tends* to remove or to neutralize it; it may not succeed in doing this completely. Le Chatelier's principle enables us to predict how a system in equilibrium will react to any changes produced in it. A bowl of water in an airtight bell jar is in equilibrium with its own vapor and the atmosphere within the bell jar, each of which is exerting its own partial pressure. If the bell jar is evacuated (i.e., if the equilibrium is disturbed by reducing the pressure), the system reacts by attempting to restore the original pressure—more water evaporates. A piston in a cylinder containing air is in equilibrium with the outside atmospheric pressure. If the equilibrium is disturbed by the piston being pushed in, the internal gas pressure increases, and a push is developed upward on the piston that tends to oppose the constraint.

Le Chatelier's principle can be widely applied in many spheres of activity, as the exercise of a little imagination will show. Stable societies, for instance, are like systems in equilibrium, and any attempt to upset the status quo and to foist change is invariably resisted by people; this is well known today and throughout history. Many more examples could be quoted. In every case, change applied to a system in equilibrium tends to be resisted. Le Chatelier's principle might well be nicknamed the "law of maintenance of the status quo."

Many examples of the application of this principle may be found in biological systems. The living cell, which is a highly complex system in dynamic, ever-changing equilibrium with its environment, reacts to disturbing forces by attempting to neutralize or remove them. A good example is the efficient buffering systems in cells that resist changes in hydrogen ion concentration, since so many vital cellular reactions proceed at an optimal rate only within narrow limits of hydrogen ion concentration. This important topic will now be considered.

DISSOCIATION OF WATER AND pH

All cellular activities occur in an aqueous medium, and the fact that the water molecule is even only slightly dissociated into hydrogen and hydroxyl ions is of considerable importance. The dissociation of the water molecule may be represented thus

$$H_2O \rightleftharpoons H^+ + OH'$$

A more correct version is:

$$H_2O + H_2O \rightleftharpoons H_3O^+ + OH'$$

Two water molecules react to give a hydrated H^+ ion (hydronium ion) and a hydroxyl ion. However, the first, simpler version will be used in the following discussion, as it is sufficiently accurate. Since the dissociation is reversible, the law of mass action may be applied to the reaction, and we can write:

$$\frac{[H^+][OH']}{[H_2O]} = K_a \tag{3}$$

where K_a is the *acid dissociation constant* of water, *which has been experimentally* found to be equal to 1.8×10^{-16} at 25°C

Since the degree of dissociation of water is so small, we may assume that the concentration of undissociated water is equal to the total concentration of water and is therefore constant, so we can rearrange Equation 3 and include the $[H_2O]$ term in a new constant on the right side of the equation.

$$[H^+][OH'] = K_a[H_2O] = K_w$$

where K_w is called the *ionic product* of water because it is equal to the product of two ion concentrations

At 25°C, it is equal to 1×10^{-14}. In pure water $[H^+] = [OH']$, therefore $[H^+] = 1 \times 10^{-7}$ g ions/liter and $[OH'] = 1 \times 10^{-7}$ g ions/liter. Since the product $[H^+]$ $[OH']$ is always constant at a fixed temperature and equal to 1×10^{-14} at 25°C, it follows that any change in concentration of one of the ions must be accompanied by a corresponding change in the other. For instance, if $[H^+]$ changes to 1×10^{-2} g ion/liter, then $[OH']$ must automatically become 1×10^{-12} g ion/liter.

The activity of enzymes and the physical state and biological activity of proteins in general are strongly influenced by hydrogen ion concentration, and so the measurement of $[H^+]$ is an important matter in molecular biology, biochemistry, and cell biology.

The method of expressing $[H^+]$ in exponential form involving negative powers of 10 is awkward in performing calculations. Sörensen suggested that another more convenient way would be to express $[H^+]$ in terms of a function that he called pH (hydrogen ion potential), which is defined as the negative logarithm of the hydrogen ion concentration.

$$pH = -\log [H^+] \tag{4}$$

This gives simple, positive numbers, which are much easier to handle. The hydrogen ion concentration in pure water is 1×10^{-7} g ion/liter, and its pH is therefore equal to 7.0 because, by definition, $pH = -\log (1 \times 10^{-7}) = -(-7) = 7.0$.

Ordinarily, pH is measured with a pH meter, and the pH value determined is converted into hydrogen ion concentration by calculation.

Example.   The pH of a solution is 4.52. What is its hydrogen ion concentration?

The value 4.52 for pH is substituted in Equation 4:

$$pH = 4.52 = -\log [H^+]$$

Multiplying through by -1 and rearranging gives:

$$\log [H^+] = -4.52$$

By definition, the whole-number part of a logarithm (the *mantissa*) must always be positive or zero, but it can never be negative, whereas the decimal part (the *characteristic*) can be positive, negative, or zero. Therefore, we must convert the value -4.52 into its numerically equal logarithm. This is $\bar{5}.48$ (i.e., minus 5 plus 0.48), which is equal to -4.52. We now turn to a table of logarithms to find the number whose logarithm is $\bar{5}.48$. This is 3.02×10^{-4} (i.e., the hydrogen ion concentration of the solution whose pH is 4.52 is 3.02×10^{-4} g ion/liter). Notice that pH values vary inversely as $[H^+]$, and that since pH is a logarithmic function, a change of 1 pH unit is equal to a tenfold change in hydrogen ion concentration.

All of this reasoning could, of course, equally apply to $[OH']$, and we could talk of the hydroxyl ion concentration and pOH. However, it is customary to use pH. One automatically determines the other; to every pH value, there is a corresponding pOH, and vice versa.

BUFFER ACTION

The macromolecular constituents of cells, especially enzymes and proteins in general, are very sensitive to changes in pH, and this is reflected in changes in their biological activity. The pH of the blood of a mammal is maintained close to 7.35, and a change of as little as 0.2 of a pH unit on either side of this may lead to serious malfunction or even death of the organism. In connection with this, it is instructive to calculate the change in pH produced when a known amount of acid or alkali is added to a known amount of water. First, we know that acids and bases such as HCl and NaOH are virtually completely dissociated into their ions in dilute aqueous solution and their dissociation constants approach infinity. For this reason HCl is called a strong acid and NaOH is a strong base.

Suppose that a very small amount of HCl, [e.g., 10^{-6} mole (about 4×10^{-5} g) is added to 1 liter of pure water (pH 7.0)]. Since the HCl is completely dissociated, the concentration of hydrogen ion will be 1×10^{-6} g ion/liter and its pH will be $-\log(1 \times 10^{-6}) = -(-6) = 6.0$. This very small amount of HCl has therefore sufficed to change the pH by one whole pH unit.

Living cells are able to resist large changes in pH; they can do so because they possess substances called buffers; Buffers are substances or mixtures of substances that undergo relatively little change of pH when acid or alkali is added and hence "buffer" the cell against pH changes. Evidently, pure water has practically no buffering ability.

Now consider a solution of a weakly dissociated acid HA. In aqueous solution there will be the following equilibrium.

$$HA \rightleftharpoons H^+ + A'$$

The pH of this solution will, of course, depend on $[H^+]$, and this, in turn, will deend on $[HA]$ and the extent of the dissociation. If we add H^+ ions to this solution, Le Chatelier's principle predicts that a change will take place that will tend to remove the added H^+ ions; the A' ions will combine with the added H^+ ions, thus giving the undissociated HA (i.e., much of the added H^+ ions have been removed from solution). If OH' ions were added the H^+ ions would combine with them to give the largely undissociated water molecule so that OH' ions are removed from solution. Hence, a solution of a weak acid or of a weak base tends to resist changes of pH when H^+ or OH' ions are added. Such solutions are said to have buffering capacity, and their resistance to pH changes is another example of Le Chatelier's principle.

An acid may be defined in general terms as a hydrogen ion donor; a base may be defined as a hydrogen ion acceptor. Weak acids and bases are incompletely dissociated into ions in aqueous solutions; when a weak acid dissociates reversibly into H^+ ions and anions, the anion functions as a hydrogen acceptor in the reverse reaction and hence is called the conjugate base of the acid.

$$HA \rightleftharpoons H^+ + A'$$

Undissociated acid   Conjugate base

For example, the acetate ion is the conjugate base of acetic acid. Likewise, a base and its protonated derivative give a "conjugate pair."

$$B + H^+ \rightleftharpoons BH^+$$

Conjugate acid of the base B

For example, the ammonium ion is the conjugate acid of the base ammonia.

The terms "strong" and "concentrated" when applied to acids and bases should not be confused. *Concentration* refers to the quantity of acid or base in solution *per unit volume*; *strength* refers to the extent to which an acid or base exhibits its properties relative to other substances, and this is determined by the extent of dissociation in aqueous solution. Strong acids and bases are completely or nearly completely dissociated, whereas weak acids and bases are slightly dissociated.

BUFFER SOLUTIONS

In practice, buffer solutions as used in biochemistry and molecular biology are mixtures of a weak acid and its conjugate base or a weak base and its conjugate acid.

A weak acid such as acetic acid is only slightly dissociated in aqueous solution.

$$CH_3COOH \rightleftharpoons CH_3COO' + H^+$$

Applying the law of mass action:

$$\frac{[CH_3COO'][H^+]}{[CH_3COOH]} = K_a \tag{5}$$

where K_a = the acid dissociation constant of acetic acid = 1.82×10^{-5}.

For the same reasons that pH values are preferred to hydrogen ion concentrations in calculations, pK_a values are used instead of dissociation constants, which are awkward to handle. The definition of pK_a is analogous to the definition of pH.

$$pK_a = -\log K_a \qquad (6)$$

The pK_a of acetic acid is therefore equal to $-\log(1.82 \times 10^{-5}) = 4.74$. Notice that the stronger the acid, the smaller is the pK_a value, which is analogous to the inverse relationship of pH to $[H^+]$. Rearranging Equation 5, we have:

$$K_a[CH_3COOH] = [H^+][CH_3COO']$$

so that

$$[H^+] = K_a \frac{[CH_3COOH]}{[CH_3COO']}$$

Taking logarithms of both sides and multiplying by -1 gives:

$$-\log[H^+] = -\log K_a - \log \frac{[CH_3COOH]}{[CH_3COO']}$$

that is,

$$pH = pK_a - \log \frac{[CH_3COOH]}{[CH_3COO']}$$

or

$$pH = pK_a + \log \frac{[CH_3COO']}{CH_3COOH]} \qquad (7)$$

Equation 7 is an important relationship and is known as the Henderson–Hasselbalch equation. Notice that the pK_a value becomes numerically equal to the pH at which the acetic acid—or any weak electrolyte—is half dissociated. (Here, $[CH_3COOH] = [CH_3OO']$; therefore, $[CH_3COO']/[CH_3COOH] = 1$ and $\log 1 = 0$, so that $pH = pK_a$.) This is an easy to remember and useful relationship that follows from the Henderson–Hasselbalch equation.

Suppose that a relatively large amount of sodium acetate is now added to a solution of acetic acid. Salts are completely dissociated in aqueous solution and, because the concentration of acetate contributed by dissociation of the acid is small (because the amount of acetic acid dissociated into H^+ and CH_3COO' ions is so small), the concentration of acetate in Equation 7 can therefore be set equal to the concentration of sodium acetate added without appreciable error. Equation 7 may therefore be rewritten:

$$pH = pK_a + \log \frac{[salt]}{[acid]} \qquad (8)$$

which is another form of the Henderson–Hasselbalch equation. Since pK_a is constant, Equation 8 tells us that the pH of the mixture of acid and salt depends on the ratio of the concentrations of acid and salt. The greater the absolute concentrations of acid and salt, the more efficient the buffer mixture will be (i.e., the greater will be its resistance to pH changes). If some strong acid is added to the mixture the excess H^+ ions will be "mopped up" by the acetate ion to give the largely undissociated acetic acid. If a strong base is added the hydroxyl ions will be neutralized by the acetic acid.

$$OH' + CH_3COOH \rightleftharpoons CH_3COO' + H_2O$$

There is no appreciable change in the ratio: $[CH_3COO']/[CH_3COOH]$ when neutralization of added acid or base occurs, so that pH is practically unaffected.

In mixtures in which the [salt]/[acid] ratio is equal to unity, $pH = pK_a$, and the buffering capacity against both strong acids and strong bases is optimal.

In biological experimentation tissue extracts, enzymes and other pH-sensitive materials are usually prepared in buffer solutions; commonly used buffer mixtures are $H_3PO_4 + NaH_2PO_4$, $NH_3 + NH_4Cl$, $NaHCO_3 + Na_2CO_3$, and $Na_2HPO_4 + NaH_2PO_4$.

The most important buffer substances in living cells are amino acids and proteins, which carry many weakly dissociated acidic and basic groups in their "side chains" (Chapter Three).

HYDROGEN BONDING

In the water molecule, the electron pairs of the two covalent bonds uniting the hydrogen atoms to the water molecule are not equally shared between the oxygen atom and the hydrogens. This is because of the strongly electron-attracting nature (electronegativity) of the oxygen atom. Because of this, the electrons tend to cluster around the oxygen atom, so that one end of the water molecule acquires a partial negative electric charge and the other end acquires a

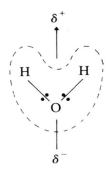

Figure 2-1   The water molecule.

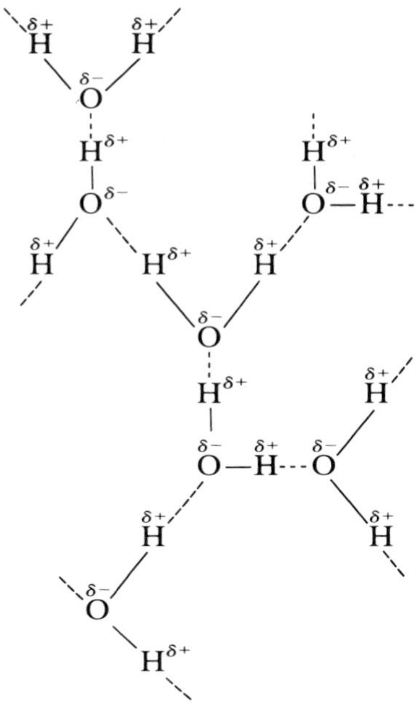

Figure 2-2   Hydrogen bonding of water molecules.

partial positive charge (Fig. 2-1). The entire water molecule is therefore electrically polarized. As a result, neighboring water molecules tend to attract one another by their unlike charges. The resulting rather weak electrostatic bonds are called hydrogen bonds (Fig. 2-2). The whole mass of molecules in a bulk of water are hydrogen-bonded in this way, and many of the properties of water are attributable to hydrogen bonding. One of these properties is that water is a liquid at room temperature, whereas its sulfur analog, hydrogen sulfide, is a gas. Sulfur is not as electronegative as oxygen, and so H_2S is hardly hydrogen-bonded. The high latent heat and high surface tension of water are also due to hydrogen bonding.

Hydrogen and oxygen atoms attached to nitrogen atoms and oxygen doubly bonded to carbon form polarized covalent bonds, and hydrogen bonding between such groups is therefore possible. Some of the more important types of hydrogen bonding are shown in Fig. 2-3. Of these, the hydrogen bond between a carbonyl oxygen and an amino hydrogen is of great importance in determining the structure of protein molecules and nucleic acids, as will be described in Chapters Four, Five, Six, and Fifteen. Note that a hydrogen bond always originates by electrons being drawn away from a hydrogen atom by virtue of its being bound to an electronegative atom. The resultant exposed positive charge on the hydrogen atom then forms a bond with a negative charge on an adjacent oxygen or nitrogen atom carrying a negative charge.

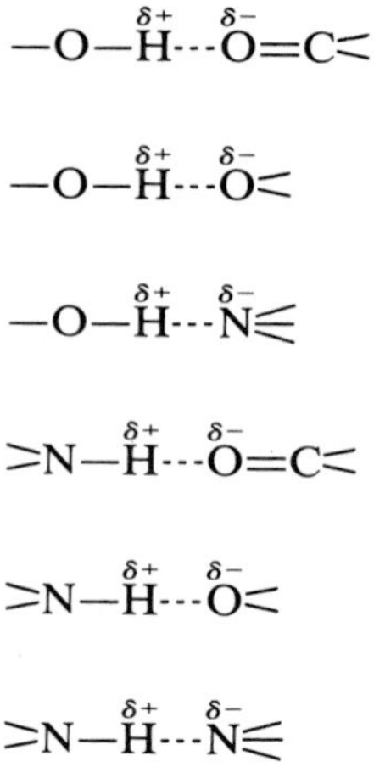

$$-O-\overset{\delta+}{H}\cdots\overset{\delta-}{O}{=}C{\Huge{<}}$$

$$-O-\overset{\delta+}{H}\cdots\overset{\delta-}{O}{\Huge{<}}$$

$$-O-\overset{\delta+}{H}\cdots\overset{\delta-}{N}{\Huge{\leqslant}}$$

$${>}N-\overset{\delta+}{H}\cdots\overset{\delta-}{O}{=}C{\Huge{<}}$$

$${>}N-\overset{\delta+}{H}\cdots\overset{\delta-}{O}{\Huge{<}}$$

$${>}N-\overset{\delta+}{H}\cdots\overset{\delta-}{N}{\Huge{\leqslant}}$$

Figure 2-3   Types of hydrogen bonds.

In contrast with —OH and —NH groups, the —CH group does not form a hydrogen bond, because the C and H atoms have about the same electronegativity and electrons are therefore not drawn away from the H atom as in —OH and —NH groups. Because of this, —CH groups are nonpolar and are said to be hydrophobic (water repelling). On the other hand, the —OH and —NH groups are polar; they are therefore water-attracting and described as hydrophilic. These different physical properties of —OH, —NH, and —CH groups have important consequences, that will be evident in the chapters on nucleic acid and protein structure.

Although the hydrogen bond is usually represented as an electrostatic attraction, there is thought to be some "overlapping" of electron "clouds" between the atoms involved, so that there is a small covalent component in the hydrogen bond.

THE LAW OF INCREASING DISORDER

It is common knowledge that if a highly organized structure such as a skyscraper, an automobile, or a cultivated garden is to be "kept up," effort (i.e., input of energy) is needed. We all know what happens if a cultivated garden is left to nature, and a skyscraper or an automobile will steadily deteriorate if left to time and the elements. No effort on our part is needed to bring about this deterioration, but the initial assemblage, building, or planting and cultivation involve energy input, as does the subsequent maintenance. The production of highly ordered structures from more random states and the maintenance of this order always involve energy input; a pile of bricks does not, of its own accord, spontaneously form itself into a house, nor does a packet of seeds plant itself into a garden. In these cases we are going from a disorderly or random state to a more orderly or organized situation, and such processes always require energy. They do not occur spontaneously. The reverse, in which orderliness gives way to disorderliness—such as a building gradually becoming a ruin, a disorderly pile of bricks, because of a lack of maintenance—is spontaneous and does not require expenditure of energy. This is one of the fundamental principles of thermodynamics, which is the branch of physical science that deals with energy and its transformations. Energy changes in systems are therefore accompanied by alteration of the relative orderliness or disorderliness of their constituent parts; this change in organization is measured by the thermodynamic function called entropy. The principle that spontaneous processes are accompanied by increased disorder or randomness (i.e., increase of entropy) is sometimes called the "law of increasing disorder." Living cells are much more highly organized than the randomly disposed atoms and molecules that were absorbed from the environment and incorporated into their structure. Hence, cells constantly need energy in order to maintain this complex organization. When the cell dies and ceases to extract energy from the environment, the law of increasing disorder takes over, and the cell disintegrates.

DIFFUSION

Many of the materials involved in the life of cells pass through the cell membrane by the process called diffusion. Diffusion results from the constant rapid random motion of molecules of a liquid or gas. If a crystal of a water-soluble substance (preferably colored so that we can see what is happening) is placed in a beaker of pure water (Fig. 2-4) and the temperature is kept constant to avoid convection currents, the crystal will first dissolve and form a concentrated solution. Gradually the color will spread into the bulk of the solution until, finally, it will be evenly spread throughout the water. This is an example of diffusion, and it results from the random motion of the solute (crystal) particles.

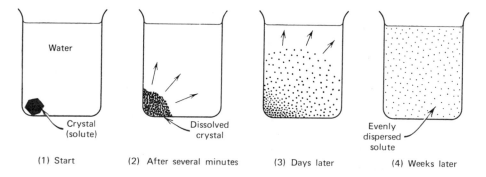

| (1) Start | (2) After several minutes | (3) Days later | (4) Weeks later |

Figure 2-4   Demonstration of diffusion.

It is again an example of the law of increasing disorder. First, the crystal, a highly ordered arrangement of atoms and molecules, dissolves and the molecules become more randomly arranged. Their concentration in one place in the beaker represents a more organized arrangement than if they were more randomly and equally distributed throughout the bulk of the water, so that they spontaneously diffuse outward until, finally, all are evenly distributed. During diffusion, the motion of individual particles is random, but there is a net movement of the entire bulk away from the center of high concentration to the area of low concentration until there is no net bulk movement when the concentration is equalized throughout the solution. Diffusion is a spontaneous process, and a substance always diffuses from an area of high concentration to one of low concentration; this is a result of random molecular motion. The reverse process never occurs spontaneously; the solute molecules show no spontaneous tendency to gather in one corner of the beaker and form a crystal again. To do so requires energy input. We can supply energy in the form of heat, which makes the water evaporate and the solution becomes more concentrated. When sufficiently concentrated, the solution will deposit crystals if it is allowed to cool. Another example of the spontaneous nature of diffusion is the spreading of an odor from one part of a room to another when a bottle of perfume is opened. The perfume molecules show no spontaneous tendency to go back into the bottle and concentrate themselves there. It is just as well that the random state that all diffusion tends toward is not spontaneously reversible. If it was, the air molecules in a room might all gather together in one corner, and the occupants of the room would suffocate.

OSMOSIS

The phenomenon of osmosis is a form of diffusion that is of special importance to living cells. Cells are usually surrounded by an aqueous environment in which various materials are dissolved in various concentrations and part of the cell interior itself is an aqueous solution of many different molecules. The cell membrane is permeable to water and, generally, it is permeable to small molecules but impermeable to large molecules. It is therefore an example of a differentially permeable membrane, sometimes called a "semipermeable" membrane.

Consider a hypothetical situation in which a membrane containing tiny pores separates two aqueous phases, either pure water and an aqueous solution or two aqueous solutions of the same solute, but of two different concentrations. Assume that the membrane pores are large enough for water molecules to pass through, but not large enough for the solute particles (Fig. 2-5). The membrane is therefore differentially permeable and allows solvent (water) molecules, but not solute particles, to penetrate it. In their random movement, the water

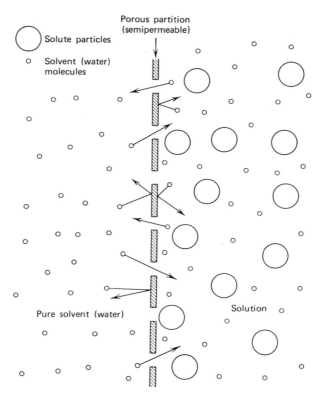

Figure 2-5   Osmosis.

molecules can pass through the pores (diffuse) in both directions. However, the number of water molecules per unit volume (i.e., their concentration) is not the same on both sides of the membrane. The solute particles on the solution side occupy space, and so there are fewer water molecules per unit volume on the solution side than on the pure solvent side. Therefore, following the law of diffusion, there will be a net movement of water molecules through the membrane from the pure water side to the solution side. This movement of water through a membrane permeable to solvent (water) molecules but impermeable to solute particles, from the pure water side into the solution (or from the weaker solution into the stronger solution), is called *osmosis*. It is a special case of diffusion in which solvent (water) molecules pass through a semipermeable membrane in the direction of higher to lower water concentration (or, to put it differently, the water diffuses toward the higher concentration of solute).

Osmosis can be demonstrated in the apparatus shown in Fig. 2-6. The height of the liquid column increases as water diffuses into the solution. The downward

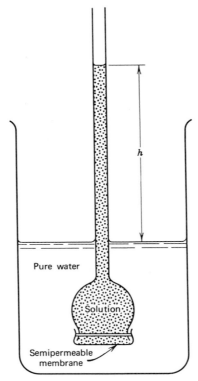

Figure 2-6 Apparatus for demonstrating osmosis and osmotic pressure. When the water in the tube stops rising, its height (h) is a measure of the osmotic pressure of the solution.

force (hydrostatic pressure) exerted by the column of water will force water molecules back through the membrane. Ultimately, this pressure will increase to a value that is just sufficient to force water molecules back through the membrane as fast as they are diffusing in the opposite direction through the membrane. The water level in the column will then remain stationary. The downward force exerted by the water column at the equilibrium point shows that osmosis is associated with a force, the measure of which is the height of the liquid column. This force is called *osmotic pressure*. The greater the difference in solute concentrations, the greater the osmotic pressure that is developed. Osmosis is a biologically important phenomenon and governs the flow of water into and out of cells. An *amoeba* in a freshwater environment absorbs water osmotically because of the higher concentration of solutes in its cytoplasm. It has to excrete the water with its contractile vacuole, a process called osmoregulation. If it did not do this, the cell would quickly expand and finally burst. Aquatic freshwater plant cells also absorb water, but there are no contractile vacuoles, and the rigid cellulose cell wall prevents the cell from bursting. A pressure called turgor pressure is therefore developed within the plant cell, and this helps to maintain the form and rigidity of the plant.

The reverse is true of marine organisms. Seawater has a salt concentration of about 3%, which is higher than the solute concentration of most cells. Therefore, water tends to be lost from the cells of marine organisms; although they are surrounded by water they are in danger of being dehydrated. To conserve water, energy has to be continually expended, because the direction of the osmotic flow has to be reversed; this is a nonspontaneous process. Such organisms have developed special energy-requiring processes for extracting the water they need from the seawater while excluding the unwanted salt.

DIALYSIS

Artificial selectively permeable membranes are useful in molecular biology and biochemistry because, with their aid, large molecules can be separated from small molecules [e.g., proteins (or other large molecules)] and can be purified

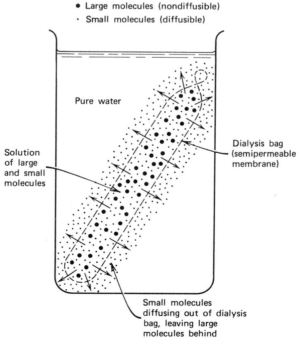

Figure 2-7   Dialysis.

and rid of contaminating salts and other small molecules. The solution of, say, protein with contaminating salts is placed in a leak-proof semipermeable membrane bag; this is immersed in pure water, which is periodically or continually changed (Fig. 2-7). The small molecules diffuse out; the large molecules do not, and they are retained in the bag. Ultimately, the large molecular species is left uncontaminated by the smaller molecules. This process, in which large molecules are separated from smaller molecules by diffusion through a semipermeable membrane, is called *dialysis*, and the membrane bag is called a dialysis bag. The water outside the membrane bag has to be changed; if it was not, the small molecules would merely diffuse into it until their concentrations outside and inside were the same, and there would then be no further net change. The water outside the dialysis membrane is called the dialysate, and it contains the small, diffusible molecules that pass through the dialysis membrane.

PART TWO

PROTEINS AND PROTEIN ASSEMBLIES: THE RELATION OF STRUCTURE TO FUNCTION

CHAPTER THREE
The primary structure of protein molecules

BIOLOGICAL MACROMOLECULES

Of all the chemical materials found in living cells, macromolecular substances are perhaps the most characteristic. The word "macromolecule" simply means a large molecule and implies molecular weights ranging from a few thousand to several millions. There are four types of macromolecule found in living systems: proteins, polysaccharides, nucleic acids, and lipids. The first three are polymers (i.e., they consist of very long chains built up from smaller units). They may be likened to a necklace in which the individual beads represent small molecular units that are strung together to form the polymeric macromolecule represented by the entire necklace. In the case of certain polysaccharides, the chain is branched. The lipids differ from the others in that they are not polymers. This chapter concerns proteins; the structure of other macromolecules will be described later.

PROTEINS

The Dutch chemist G. J. Mulder first used the word "protein," which was suggested by Berzelius in 1838. The word is derived from the Greek *proteios*, which means "first" in the sense of first order of importance, and Mulder recognized that the substance that he named "protein" was of primary importance among the many different kinds of substances found in living things. Proteins are the most obvious and possibly most characteristic macromolecular constituent of living matter. It may be argued that nucleic acids are more fundamental, since protein synthesis and structure ultimately depend on these. This is undeniably true but, nevertheless, it is through proteins that living systems express their characteristics most obviously.

AMINO ACIDS

Before describing the general properties of proteins, we must start at the beginning and consider the molecular units from which the giant protein molecule is built. These smallish molecular units are the α-amino acids, which have the general formula shown in Fig. 3-1.

About 20 different amino acids have been identified as constituents of proteins, and they differ from one another only in the nature of the "side chain" attached to the α-carbon atom and represented by R in Fig. 3-1. With the exception of glycine, in which R is a hydrogen atom, all are asymmetric molecules and are therefore optically active.

Amino acids may be classified according to the nature of their side chains. A purely structural classification is shown in Fig. 3-2. A structural classification is not entirely satisfactory from the functional point of view. For instance, the heterocyclic imidazole group of histidine is strongly basic in character; therefore, functionally, histidine belongs more logically with lysine and arginine, among the basic amino acids. A classification that is functional and includes imino acids, is given in Fig. 3-3.

The side chains of amino acids play important roles in protein structure, as we will see later. Because of this, their function, such as whether or not they contain *hydrophilic* (water-attracting) or *hydrophobic* (water-repelling) groups or ionizable or nonionizable groups, is more important to bear in mind than their purely structural features.

In a class by themselves are the compounds proline and hydroxyproline, which contain the imino (>NH) group instead of the amino group. They are therefore imino acids, but they are generally included in lists of amino acids, since they are commonly found as constituents of certain proteins, sometimes in large amounts.

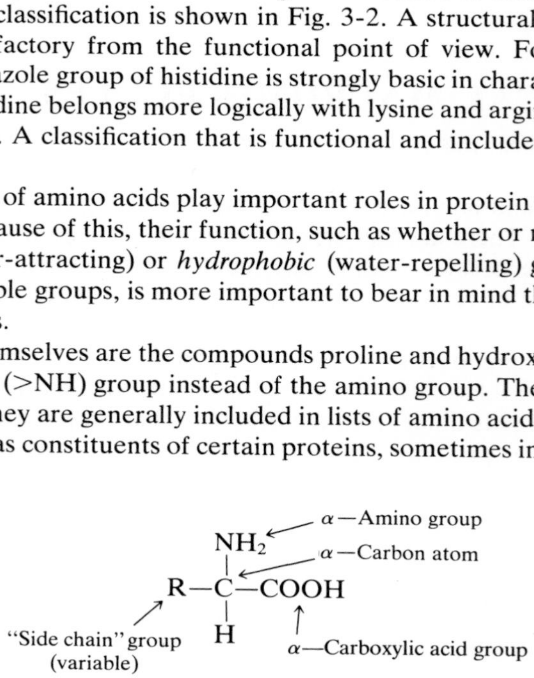

Figure 3-1   Structural formula of an α-amino acid.

General amino acid structure

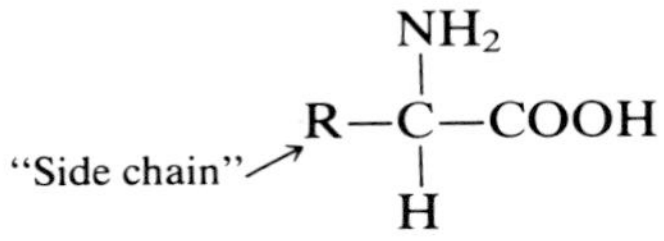

$$\underset{H}{\overset{NH_2}{R-\overset{|}{\underset{|}{C}}-COOH}}$$

"Side chain" →

Amino acid	Abbreviation	"Side chain" Groups (R)	Type
Glycine	Gly	$-H$	Hydrogen
Alanine	Ala	$-CH_3$	Hydrocarbon (aliphatic)
Valine	Val	$-CH\begin{smallmatrix}CH_3\\\\CH_3\end{smallmatrix}$	
Leucine	Leu	$-CH_2-CH\begin{smallmatrix}CH_3\\\\CH_3\end{smallmatrix}$	
Isoleucine	Ile	$-CH-CH_2-CH_3$ with CH_3 below	
Serine	Ser	$-CH_2OH$	Hydroxylic
Threonine	Thr	$-CH\begin{smallmatrix}OH\\\\CH_3\end{smallmatrix}$	
Aspartic acid	Asp	$-CH_2-COOH$	Acidic (carboxylic)
Glutamic acid	Glu	$-CH_2-CH_2-COOH$	

Figure 3-2 Classification of amino acids according to side chain structure.

Amino acid	Abbreviation	"Side chain" Groups (R)	Type
Arginine	Arg	$-CH_2-CH_2-CH_2-NH-\overset{\displaystyle NH_2}{\underset{\displaystyle \parallel NH}{C}}$	Basic
Lysine	Lys	$-CH_2-CH_2-CH_2-CH_2-NH_2$	
Glutamine	Gln Glu$-NH_2$	$-CH_2-CH_2-CO.NH_2$	Amido
Asparagine	Asn Asp$-NH_2$	$-CH_2-CO.NH_2$	
Cysteine	Cys	$-CH_2-SH$	Sulfur-containing
Cystine	(Cys)$_2$	$-CH_2-S-S-CH_2-\overset{\displaystyle NH_2}{\underset{\displaystyle H}{C}}-COOH$	
Methionine	Met	$-CH_2-CH_2-S-CH_3$	
Phenyl-alanine	Phe	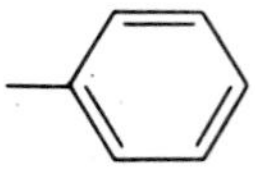	Aromatic
Tyrosine	Tyr	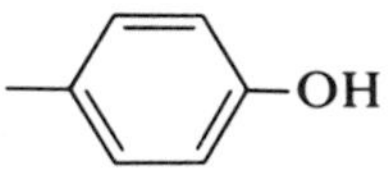	
Histidine	His	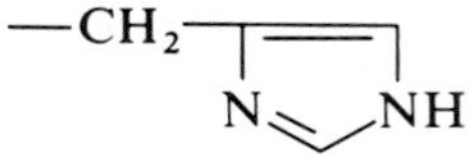	Hetero-cyclic
Tryptophane	Try	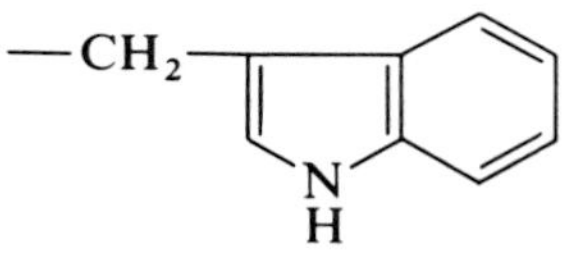	Aromatic +hetero-cyclic

Figure 3-2—*continued*

Imino acids	Abbreviation	"Side chain" groups
Proline	Pro	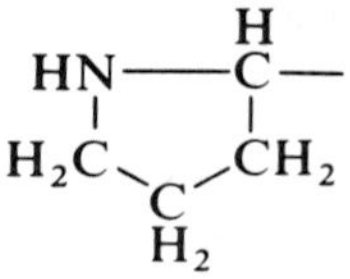
Hydroxyproline	Hypro	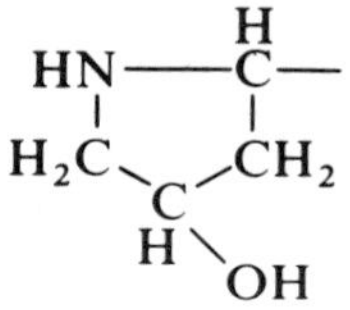

Figure 3-2—*continued*

Hydrophobic (water-repelling)

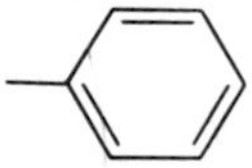 Phenylalanine	$-CH_2-CH-CH_3$ $\qquad\quad\ \ CH_3$ Leucine	$-CH-CH_2-CH_3$ $\ \ \ CH_3$ Isoleucine
$-CH_2-CH_2-S-CH_3$ Methionine	$-CH-CH_3$ $\ \ \ CH_3$ Valine	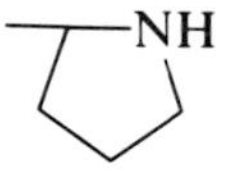 NH Proline

Figure 3-3 Functional classification of amino and imino acid side chain structure.

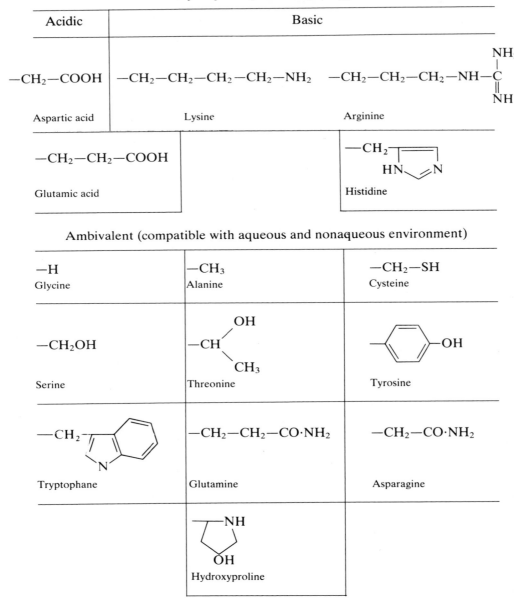

Figure 3-3—*continued*

FORMATION OF PEPTIDE BONDS

The process by which amino acids combine to produce the long-chain polymers that we call proteins may be regarded as a reaction between amino acids in which water molecules are eliminated. When two amino acids link up in this way, one of the α-amino hydrogen atoms on one of the amino acids combines with the hydroxyl group of the carboxyl group of the other amino acid. The remaining portions (or residues, as they are called) of the two amino acids thus become linked by a chemical grouping called a peptide bond (see Fig. 3-4). A glance at the formula of the resulting compound (dipeptide) formed from the two amino acid residues will show that there is a free amino group near one end and a free carboxyl group at the other end. Therefore, this process can occur repeatedly so that other amino acid residues can attach themselves at either end of the growing chain of residues. This process is sometimes called "dehydration synthesis," because a water molecule is eliminated every time a peptide bond is formed. When many amino acids have been linked together in this way, the resulting amino acid polymer is called a "polypeptide"; a protein consists of one or more very large polypeptides. Note that however many amino acid residues are linked together in a polypeptide, there is always a free amino group at one end and a free carboxyl group at the other. Proteins and polypeptides have a "backbone" in which there is the repeating —CO—NH—CH— group and, at intervals, attached to the —CH— groups are the R groups or side chains of the original amino acids (Fig. 3-5). (For what it is worth, the following rather amusing mnemonic for remembering the sequence of groups in the polypeptide backbone is given: COck—HeN—CHick.) The chemical properties of a protein or polypeptide will be largely determined by the nature and relative proportions of

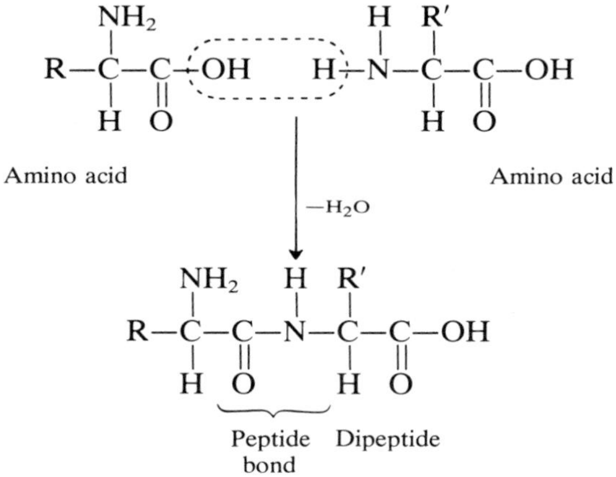

Figure 3-4   Formation of a peptide bond.

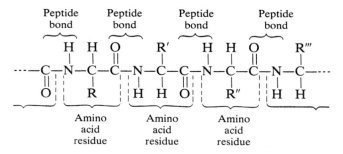

Figure 3-5 Part of a polypeptide chain.

these side chains. Note that when proline and hydroxyproline are incorporated into a polypeptide, the nitrogen atom in the peptide bond formed from the imino group will not carry a hydrogen atom

A phenomenal number of different proteins can be "built" from the 20 or so naturally occurring amino acids, according to the number in a given protein, the order in which they are strung together, and the relative proportions of each type. Not necessarily all of these amino acids will be present in every protein; some proteins have a restricted amino acid composition; this is reflected in the special properties of these proteins.

This may seem like a mere chemical detail, but the significance of this in proteins containing large proportions of the imino acids will become apparent in the chapter that deals with secondary structure of proteins.

PRIMARY STRUCTURE OF PROTEINS

The number of separate polypeptide chains that makes up a complete protein molecule and the precise sequence of amino acids in the chains are important aspects of the primary structure of a protein. In some proteins, the polypeptide chains may be covalently held together by disulfide bonds, which arise from interaction of the —SH groups of cysteine side chains on the separate polypeptide chains. The —SH group is easily oxidized; in the presence of oxygen the —SH groups of two cysteine molecules interact in an oxidative reaction to form a disulfide link or "bridge" and the amino acid cystine is formed.

$$\text{Cys-SH} \quad \boxed{\text{O}} \quad \text{HS-Cys} \longrightarrow \text{Cys-S-S-Cys} + H_2O$$

Cysteine	Cysteine	Cystine

The —SH groups of two cysteine residues on separate polypeptide chains may similarly interact to form a disulfide bond joining the two polypeptide chains. In this case it is called an interchain disulfide bond. Intrachain disulfide bonds are formed when two cysteine —SH groups on the same polypeptide chain interact. The presence of cystine in a protein hydrolysate is therefore indicative of the presence of disulfide bonds in the original protein and possibly, but not necessarily, of more than one polypeptide chain. More rarely, polypeptide chains may be held together by the phosphate diester linkages that occur in certain phosphoproteins. The primary structure of a protein is therefore a description of its covalent structure. This is illustrated in Fig. 3-6, which shows the primary structure of a hypothetical protein molecule. Now that the primary structure of a protein has been defined, two questions naturally suggest themselves. First, how

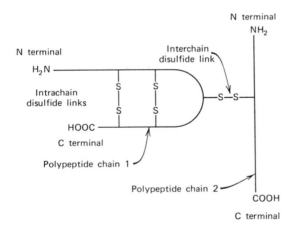

Figure 3-6 Primary structure of a hypothetical protein molecule. A complete description of the *primary structure* of a protein molecule includes the following: (1) *number* of polypeptide chains (two in this case); (2) *number* and *types* of amino acids in the separate polypeptide chains; (3) *amino acid sequence* in each chain; (4) *number, position,* and *nature* of *interchain* linkages (in this case there is one *disulfide* interchain group); and (5) *number, position,* and *nature* of *intrachain* linkages (in this case there are two *disulfide* intrachain linkages in polypeptide chain 1). The primary structure of a protein molecule is therefore its *covalent structure.*

is primary structure elucidated and, second, what is the significance of primary structure?

A necessary preliminary to the determination of primary structure is the discovery of what amino acids are present in the given protein and then the determination of how much of each is present. This may be achieved by breaking down the protein into its constituent amino acids by a hydrolysis reaction such as boiling with acid or alkali or by subjecting it to enzyme action. The separate amino acids must then be separated and identified. Until a few years ago this was a laborious and inefficient process; the introduction of the simple but powerful technique of paper chromatography revolutionized analysis of complex chemical mixtures. So important is this technique that some time will be spent here in describing its essential features.

A typical paper chromatographic separation of an amino acid mixture is performed as follows. A small drop of the amino acid mixture is placed near one end of a strip of absorbent filter paper; the type known as "Whatman No. 1" is a favorite with biochemists. The spot is allowed to dry, and the paper strip is suspended in an airtight jar with its lower end dipping into a trough that contains some kind of solvent mixture. A frequently used mixture consists of butanol, acetic acid, and water. Alternatively, the trough may be placed near the top of the chamber. The upper end of the strip dips into the solvent, and the free end is bent over the edge of the trough so that it hangs vertically downward. Whichever arrangement is used, the solvent will travel through the paper strip by capillarity and, at the same time, each amino acid will move with the solvent along the strip at a characteristic rate, depending largely on the chemical properties of the amino acid, the composition of the solvent, and the absorptive power of the filter paper. When the solvent reaches the end of the filter paper, the paper is removed and dried and then sprayed with a solution of some chemical, such as "ninhydrin," which gives a color reaction with amino acids. When this is done, a series of colored spots will be seen on the paper at various distances from the starting point. Each spot corresponds to a particular amino acid, and the distance traveled from the starting point or "origin" is characteristic of each amino acid in a given solvent system and helps in its identification. The amino acids present in the original mixture may be identified by comparing the pattern of colored spots on the paper strip (or chromatogram, as it is called) with a similar one run simultaneously in which a spot of a reference mixture of known amino acids was used. The concentration of each amino acid may also be determined if a drop of the mixture of a known volume was used at the start. All that is necessary is to cut out each amino acid spot from the chromatogram, dissolve the colored spot in a suitable solvent, and determine its concentration spectrophotometrically.

Amino acids may also be separated by electrophoresis. In this technique, the amino acid mixture is applied to a moist, buffered paper or gel strip, and a high-voltage electric current is passed through. Individual amino acids will have different net electric charges at the particular pH used and will therefore migrate in the electric field, the direction and speed of the migration will depend on the nature and magnitude of the charge on the amino acid. A separation will thus be achieved. Excellent analyses of amino acids may be achieved by a combination of chromatography and electrophoresis.

It may be justifiably said that many of the major biochemical discoveries of recent years have resulted from the application of chromatography and its various modifications in one form or another to biochemical problems. Recently, G. Leaf has drawn attention to a new chromatographic system devised by P. B. Hamilton that has very high resolving power (i.e., it has the ability to separate into discrete spots mixtures of substances that "run" at very nearly the same speed on a chromatogram). It reveals as many as 150 "ninhydrin positive" substances in normal urine, for example. The commonly used chromatographic systems are of much lower resolving power. The danger of interpretation errors with the conventional systems has been ignored and, to use Leaf's words, "in studies involving the analysis of biological fluids, chromatographic separations are interpreted in terms of a relatively small number of common amino acids and related compounds."

There is, of course, much more to primary structure than merely determining the relative amounts of each amino acid present. Five main steps are involved; these are determination of:

1. Amino acid composition of the protein.
2. Molecular weight.
3. Number and length of polypeptide chains.
4. Amino acid sequence of the constituent polypeptide chains.
5. Location of cross-linking groups.

The method will be illustrated by describing the determination of the primary structure of the protein hormone insulin.

DETERMINATION OF THE PRIMARY STRUCTURE OF INSULIN

The elucidation of the amino acid sequence of insulin was achieved in 1955 by F. Sanger and his colleagues. This was the first protein for which the amino acid sequence was determined. It was chosen, first, because of its great biological importance as one of the many hormones regulating the sugar metabolism of the body and, second, because a good deal about its chemistry was already known. Its ready availability in purified form and its relatively simple primary structure were further factors contributing to success in determining its primary sequence. The successful solution of a biological problem is greatly dependent on the choice of a suitable material or test organism. The choice of insulin was therefore a good one.

The molecular weight of the insulin molecule is 6000; this indicates the presence of about 50 amino acid residues. The task of determining the precise sequence, if any, of the amino acid residues in insulin was thus a formidable one. In order to understand how Sanger approached the problem of amino acid sequence determination, remember that all proteins and polypeptides possess a free amino group at one end of the chain and a free carboxyl group at the other end. Hence a polypeptide chain is said to have an N-terminal and a C-terminal end, respectively, and the two terminal amino acids are referred to as the N- and C-terminal amino acids. One of Sanger's contributions was the introduction of an organic reagent that had the property of reacting specifically with free amino groups. This substance was 2,4-dinitrofluorobenzene, or DNFB. The free amino group of the N-terminal amino acid of a peptide reacts with DNFB, as shown in Fig. 3-7, so that the N-terminal amino acid is "tagged." The bond formed between the dinitrobenzene group and the amino acid is relatively resistant to the hydrolysis reaction to which the derivative is then subjected to break it down into its constituent amino acids. The dinitrophenyl derivative of the N-terminal amino acid is yellow and therefore is easily visible on a paper chromatogram when the protein hydrolysate is subjected to paper chromatography. The N-terminal amino acid can be identified by comparing its chromatographic behavior with the dinitrophenyl derivatives of known amino acids. This technique is called end group analysis. A disadvantage of using DNFB in end group analysis is that the dinitrophenyl derivatives of certain amino acids such as glycine and proline are rapidly destroyed during the hydrolysis of the DNFB derivative of the polypeptide. Quantitative determination of these amino acids is therefore not feasible. The reagent 2-fluoronitro pyridine has been introduced and found to be preferable to DNFB in that the corresponding amino acid

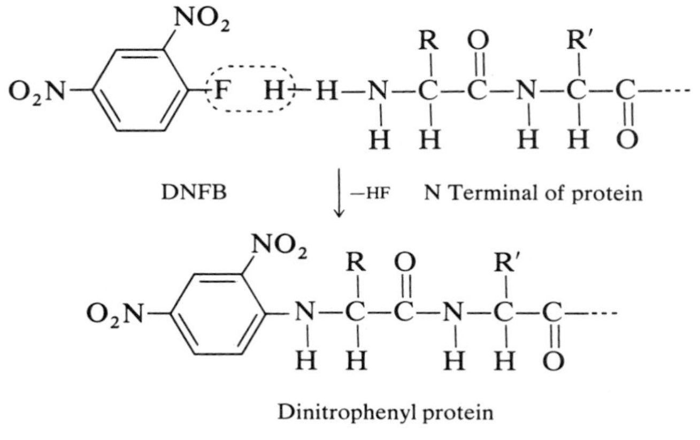

Figure 3-7   "Labeling" of the N-terminal amino acid of a protein with dinitrofluorobenzene (DNFB).

derivatives are generally stable at 100°C. Only the derivatives of serine, threonine, and glutamic acid suffer some decomposition, and even this is markedly reduced at lower temperatures.

Sanger began his approach to the problem of determining the primary structure of insulin by first finding whether there was one or more separate polypeptide chains in the molecule. He knew that among the components of insulin were three molecules of the amino acid cystine. This suggested that perhaps more than one polypeptide chain was present and that these chains were held together by disulfide bridges. Sanger later was able to show that there were, in fact, two polypeptide chains in the insulin molecule. He was able to separate these intact by reductively breaking the disulfide bonds. The next task was the determination of the amino acid sequence of the two separated chains. When faced with this, perhaps Sanger thought of the old proverb, "The journey of a thousand miles begins with a single step." The first step in this long journey was the identification of the N-terminal amino acids of both polypeptide chains by the DNFB end group analysis technique. The two polypeptide chains were named the A chain and the B chain. Sanger found that the N-terminal residue of the A chain was glycine, and that of the B chain was phenylalanine. The C-terminal residues were identified by taking advantage of the fact that the enzyme carboxypeptidase specifically catalyzes the hydrolysis of a peptide bond adjacent to a free carboxyl group. The liberated C-terminal amino acid can then be identified chromatographically. The C-terminal amino acids of the A and B chains were shown to be asparagine and alanine, respectively.

The A chain was shown to contain four cysteine residues and the B chain was found to contain two.

From all of these data the following preliminary structure for insulin was suggested:

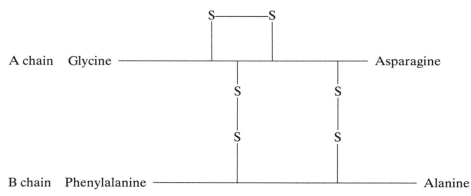

The determination of the intermediate sequences of the A and B chains was achieved by partial hydrolysis using acids, alkalis, or enzymes, followed by separation of the resulting peptide fragments. Sanger subjected the peptide fragments resulting from hydrolysis of the A and B chains to N-terminal and

C-terminal amino acid analysis. They were then further degraded until the amino acid sequences of each of these smaller peptides revealed overlapping sequences of residues that enabled reconstruction of the original polypeptide to be made. The complete primary structure of bovine insulin is shown in Fig. 3-8.

Apart from the accumulation of valuable information regarding techniques, Sanger's great achievement, by demonstrating that the seemingly impossible could be done, provided investigators with a psychological stimulus that encouraged attempts at the elucidation of the primary structure of other proteins. The second protein—and the first enzyme molecule—to have its primary

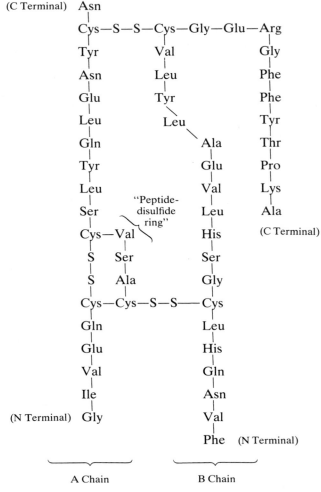

Figure 3-8 Primary structure of the bovine insulin molecule.

structure determined was ribonuclease, which consists of a single polypeptide chain of 124 amino acid residues. Other proteins of which the primary structures are now known are the tobacco mosaic virus protein (157 residues), egg white lysozyme (129), myoglobin (151), the α, β, γ, and δ chains of human hemoglobin (141, 146, 146, and 146 residues, respectively), the α and β chains of horse hemoglobin (141 and 146, respectively), the α and β chains of many other hemoglobins, cytochrome c from over 40 different organisms (103 to 112), and chymotrypsinogen (246).

Enzymes that have been used in the degradation of proteins prior to amino acid sequence determination are trypsin, chymotrypsin, and pepsin. The specificities of these proteases are such that the resultant digests of protein contain large peptides, and such mixtures may be difficult to purify and analyze. Some of the large peptide fragments may even be water-insoluble, especially if they contain a large proportion of hydrophobic residues. A protease has now been introduced that overcomes these difficulties. This enzyme is thermolysin, which is extracted from the bacterium *Bacillus thermoproteolyticus*; its specificity is unlike that of any of the other proteases. It catalyzes hydrolysis of peptide bonds on the N-terminal side of hydrophobic residues and hence cleaves many more bonds than any of the other proteases. Its great usefulness lies in the secondary degradation of the large peptides resulting from the tryptic or chymotryptic digestion.

VARIATIONS IN PRIMARY STRUCTURE AND MOLECULAR EVOLUTION

Biologists have always been interested in detecting structural homologies and analogies among the various organs and systems of different groups of organisms. By so doing, evolutionary relationships between organisms may be indicated. Today, attempts are being made to detect evolutionary relationships between the enormous variety of molecules and biochemical processes that are found in living cells.

Because their primary structure has been worked out in so much detail, the cytochromes c molecules of different species may be compared and possible evolutionary significance perceived. Cytochromes c are found in the tissues of practically every organism, and they all have identical biochemical characteristics. The primary structures of more than 40 cytochromes c from a wide variety of organisms are known. Horse heart cytochrome c was the first to have its primary structure worked out, and the molecule was found to be a single polypeptide chain of 104 amino acid residues. The cytochromes c of all vertebrates subsequently investigated (except the tuna fish) were found to contain 104 amino acid residues, whereas the cytochromes c of all invertebrates so far studied contain more than 104 amino acid residues.

In comparing the primary structures of vertebrate and invertebrate cytochromes by aligning them side by side, the question of where to place the extra residues of invertebrate cytochromes in relation to the vertebrate cytochromes arose. These extra residues might equally be considered as being in the middle of the chain or at one or other end. The biological function of cytochromes c is closely connected with its heme group, and this was adopted as a reference point in aligning the different cytochromes c molecules when comparing them. When they were aligned with the heme groups coincident, it was obvious that all cytochromes c were closely related and that the extra residues of the invertebrate cytochromes appeared at the N-terminal end of the molecule. The results of a comparative structural study of the different cytochromes revealed the following.

1. Some regions of the chain were identical over a wide range of organisms, for example, the important region involved in the binding of the heme group. In all of the cytochromes studied, 35 amino acid residues were found not to vary.

2. Among the variable amino acids, many different classes were noted. In many cases what may be called "conservative" substitutions may occur, for example, a basic residue like lysine may be replaced by another basic residue, such as arginine. Likewise, hydrophobic residues were replaced by other hydrophobic residues. "Radical" substitutions also occur in which an amino acid is replaced by another with completely different physicochemical characteristics. Yet, again, there were places where both "conservative" and "radical" substitutions occurred.

3. Generally, it was found that the nearness of the donor organisms from the evolutionary viewpoint was roughly reflected in the similarity of their cytochromes c molecules.

An explanation of all this is possible if it is assumed that mutations have occurred randomly during evolutionary time that have determined the differences in primary structure. Presumably, natural selection will eliminate mutations that produce cytochromes that do not function properly. From this it follows that the residues that exhibit radical substitutions from organism to organism are probably not vital for normal structure and function, whereas those that are invariant or where conservative substitutions occur are probably of critical structural and functional importance.

PRIMARY STRUCTURE AND BIOLOGICAL FUNCTION

What relation is there between the primary structure of a protein and its biological function? In the case of insulin there is, as yet, not much that can be said. The primary structure of insulin from a given species does not vary from molecule to molecule. However, small variations have been observed among the primary structures of insulins from different mammalian species, although the insulins are indistinguishable in their biological effects. Many of these differences occur in the part of the molecule known as the "peptide-disulfide ring"; these are summarized in Fig. 3-9. Since variation in the amino acid sequence of the peptide-disulfide ring does not appear to be reflected in appreciable differences of activity, it appears that the amino acid sequence in this part of the molecule may not be critical for biological activity. The disulfide "bridges" of the insulin molecule are somehow involved in its biological activity. Insulin is peculiar among proteins in that its disulfide bridges are unusually reactive in the native state. The biological activity of the insulin molecule parallels the reactivity of these disulfide bonds; this, in turn, depends on the structural integrity of the molecule. This reactivity of the disulfide groups is consistent with the idea that they are involved in biological activity in another way; a metabolic regulator is unlikely to be effective unless it is both rapidly available and can be rapidly destroyed. Some evidence indicates that the first step in the degradation of the insulin molecule is reductive cleavage of the disulfide bonds. It has been suggested, therefore, that the mode of action of insulin might involve attachment of the molecule to cell membranes by thiol-disulfide interactions.

Much more definite and fascinating examples of structure-function relationships are afforded by the polypeptide hormones. Some of these are quite small

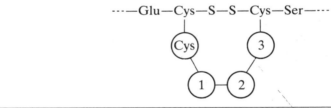

Source	1	2	3
Beef	Alanine	Serine	Valine
Horse	Threonine	Glycine	Isoleucine
Sheep	Alanine	Glycine	Valine
Pig	Threonine	Serine	Isoleucine

Figure 3-9 Variations in the peptide-disulfide ring of insulin from different mammalian species.

molecules such as oxytocin, which consists of only 9 amino acid residues. Others are larger, for example, the molecule of adrenocorticotrophic hormone (ACTH), which consists of 39 amino acid residues.

A hormone called gastrin initiates secretion of gastric juice by the walls of the stomach. Of the 17 amino acid residues that constitute the molecule of gastrin, only the four C-terminal residues are necessary for biological activity. In fact, this tetrapeptide alone possesses all of the properties of the natural hormone.

There are interesting structure-function relationships among the peptide hormones of the posterior pituitary gland—oxytocin and vasopressin. Both of these exhibit three properties: first, the oxytocic function, which means stimulation of contraction of unstriated muscle; second, antidiuretic, which means stimulation of resorption of water by the kidney; and third, vasopressor, the increasing of blood pressure. Figure 3-10 shows the primary structure of oxytocin and vasopressin from different vertebrate sources and their relative activities with respect to these three functions. Clearly, substitution of one amino acid by another at positions 2, 6, and 7 has marked effects on the physiological properties of the hormones. The overall shape of these molecules also appears to be important from the viewpoint of function. R. S. Snart and N. N. Sanyal studied the interaction of polypeptide hormones with lipid monolayers in Langmuir surface trough experiments. Assuming that lipid monolayers are suitable models of cell membranes, Snart and Sanyal suggested, on the basis of their results, that the structures formed by interaction of hormones and monolayers have the effect of providing "pores" that help the movement of water and other small molecules. Polypeptide hormones have been shown to have biological effects that are in agreement with what would be expected if "pore" size had been changed. In one of these experiments the existence of a 6:1 lipid/polypeptide ratio in one of these films was found. This led to the suggestion that six lipid molecules were adsorbed at each peptide link associated with the peptide ring of the hormone. It is interesting to note that the peptide ring is an essential feature for phsyiological activity in these hormones.

The hormone molecules would need to displace cellular proteins from the polar groups of the lipids within the membrane, and this could readily occur by a competitive effect. Such displacement reactions could be helped by the disulfide "bridge" of the molecule attaching itself to cellular protein. In fact, evidence suggests that the disulfide bonds play an important part in the attachment of hormone to protein thiol groups in the membrane. The disulfide bonds in the insulin molecule also may possibly be involved in attaching the molecule to cellular membranes.

Perhaps the most interesting and complex structure-function relationships are found among the three hormones produced by the anterior lobe of the pituitary gland. These are ACTH and the two melanophore stimulating hormones α-MSH and β-MSH. Apart from its main function of controlling steroid hormone

Compound	Primary structure	Activity (units per gram)		
		Smooth muscle	Blood pressure	Antidiuretic
Oxytocin (beef)	$\overset{\displaystyle S\text{———}S}{\text{Gly—Leu—Pro—Cys—Asn—Gln—Ile—Tyr—Cys}}$	450	5	2
Vasopressin (beef)	—Arg— —Phe—	200	400	400
Vasopressin (pig)	—Lys— —Phe—	5	300	200
Vasotocin (frog)	—Arg— —Ile—	100	100	200
Isotocin (fish)	—Ile— —Ser—Ile—	150	0	0

Figure 3-10 Primary structures of oxytocin and vasopressin from different animal species. [From H. K. King, *Sci. Progr.*, 54 (421), 1966.]

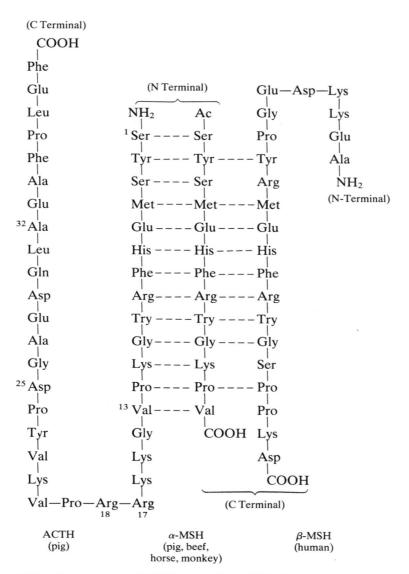

Figure 3-11 Structures of ACTH, α- and β-MSH. The molecules are aligned so as to emphasize the similarities in primary sequences (indicated by dotted lines).

production by the adrenal cortex, ACTH also has a slight melanophore-stimulating effect. It is very interesting to note that the primary structure of the molecules of these three hormones are remarkably similar (Fig. 3-11). It can be seen that the amino acid sequence of α-MSH exactly matches the first 13 residues of ACTH. There is just one point of difference—the first residue of a-MSH is acetylated. If the N-terminal serine of ACTH is acetylated, the resulting compound is devoid of ACTH activity. That the N-terminal residue of α-MSH is acetylated is suggestive. If 19 amino acid residues are removed from the C-terminal end of the ACTH molecule, the remaining 20-residue fragment possesses practically the same biological activity as the intact molecule. One cannot help but wonder what, if any, is the function of this 19-residue fragment?

Further chain shortening results in profound changes of biological activity. A thousandfold drop in ACTH activity occurs if the arginine residues at positions 17 and 18 are removed. Further reduction in chain length to 13 residues results in a molecule with hardly any ACTH activity. However, it *does* possess marked· MSH activity because this fragment is, in fact, α-MSH.

The smallest peptide exhibiting any effect on melanophores consists of 5 amino acid residues and has the following structure.

$$His—Phe—Arg—Try—Gly$$

This sequence occurs in both α- and β-MSH.

Hormonelike substances are formed locally in blood or tissues in response to injury. Since they are not secreted by special glands at a distance from the target tissue, as are the other hormones, and since they are produced exactly where they are needed, these agents have been called "local hormones." Among these are the physiologically active peptides called kinins. The first to be prepared in a state of purity was a nonapeptide called Kallidin I, or Bradykinin, which was made by the action of trypsin on globulin. These peptides cause pain, raise weals on the skin, increase permeability of capillaries, cause dilation of blood vessels, and act on various smooth muscles, generating either contraction or relaxation. In short, they have the properties of histamine, but are much more powerful. Interestingly, aspirin and certain other antiinflammatory drugs strongly inhibit the action of Bradykinin on smooth muscle, but do not inhibit the same reaction when produced by histamine. Studies of this phenomenon might give insights into the mechanism of action of aspirin, which has always been obscure. Relatively minor alterations in the structure of Bradykinin greatly affect its activity. At first it was thought by Elliott and his co-workers that Bradykinin was an octapeptide, with the structure shown in Fig. 3-12*b*, but when this was synthesized it was found to be inactive and was called "Boguskinin." A compound having the structure in Fig. 3-12*c* was synthesized, and this had weak kinin activity. The only difference between this and Boguskinin is that one of the two adjacent proline residues in Boguskinin has had its position changed. The nonapeptide shown in Fig. 3-12*a* was synthesized and found to have full activity,

N Terminal   C Terminal

$$Arg-Pro-Pro-Gly-Phe-Ser-Pro-Phe-Arg \quad (a)$$

Bradykinin

$$Arg-Pro-Pro-Gly-Phe-Ser-Phe-Arg \quad (b)$$

Boguskinin

$$Arg-Pro-Gly-Phe-Ser-Pro-Phe-Arg \quad (c)$$

$$Lys-Arg-Pro-Pro-Gly-Phe-Ser-Pro-Phe-Arg \quad (d)$$

Kallidin II

Figure 3-12 Primary structures of kinins.

so this structure was assigned to Bradykinin. The decapeptide made by attaching a lysine residue to the arginine in position 1 of Bradykinin also has high kinin activity and is known as Kallidin II (Fig. 3-12d).

A kinin similar in its action to Bradykinin has been isolated from wasp venom; its structure is shown in Fig. 3.13. The "active region" of the molecule has an amino acid sequence identical with that of Bradykinin, so it may be considered a higher homologue of it.

A polypeptide called Mellitin, consisting of 26 amino acid residues, has been isolated from bee venom (Fig. 3-14); its structure is reflected in its physiological properties. Note that one end of the molecule (residues 21 to 26) is basically hydrophilic, whereas the remainder is predominantly hydrophobic. It is therefore like a soap and should orient itself at an air–water or oil–water interface. In fact, it lowers the surface tension at an air–salt solution interface to an extent comparable with the hemolytic substances lysolecithin and digitonin. It increases permeability of erythrocyte membranes and is hemolytic, but there is no strict correlation between surface activity and hemolytic activity, because the basicity of the molecule, among other things, is involved in this property.

C Terminal

$$Arg-Pro-Pro-Gly-Phe-Ser-Pro-Phe-Arg$$
$$|$$
$$Gly$$
$$|$$
$$Arg-Leu-Lys-Lys-Lys-Asn-Thr-Tyr$$

N Terminal

Figure 3-13 Structure of a wasp kinin.

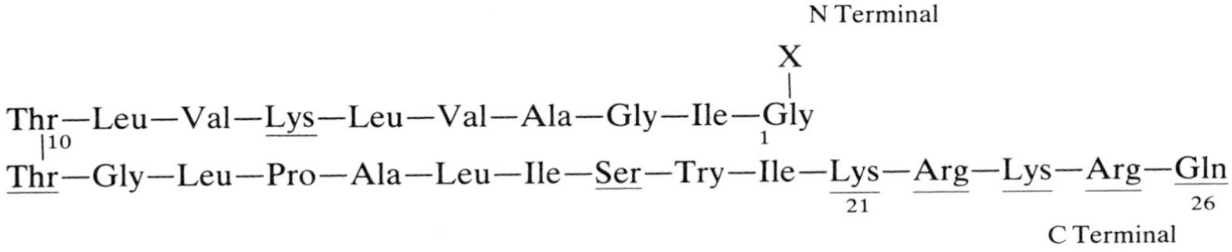

N Terminal

$$X$$
$$|$$

Thr—Leu—Val—Lys—Leu—Val—Ala—Gly—Ile—Gly
|10   1
Thr—Gly—Leu—Pro—Ala—Leu—Ile—Ser—Try—Ile—Lys—Arg—Lys—Arg—Gln
                                              21   26

C Terminal

Figure 3-14 Structure of mellitin. The side of the molecule on the left is mostly hydrophobic. Amino acids with hydrophilic side chains are underlined. The right side (residues 21–26) is the hydrophilic strongly basic part. (X may be H or a formyl group.)

Before leaving the subject of structure-function relationships in physiologically active polypeptides, the existence and significance of a substance that has been considered the pathological counterpart of a hormone is worthy of mention. Nakahara and Fukuoka isolated a toxic principle from malignant tumors that they showed to be unique to cancer cells and named it "toxohormone." Its most obvious biological effect is the depression of the enzyme catalase in the livers of animals bearing malignant tumors; the same effect is produced in healthy animals when injected with toxohormone. It has been isolated from a wide variety of tumor types and qualifies as a hormone insofar as it is produced at a site (tumor) remote from the target tissue (liver) and is carried to the target tissue in the bloodstream.

Toxohormone is characterized by other biological effects, all of which appear to be related to a deep-seated disturbance of iron metabolism in the affected animal. Chemical studies strongly indicate that it is a polypeptide consisting of at least 32 amino acid residues. Apparently, the molecule contains no more than 12 different amino acids that include arginine, phenylalanine, leucine, lysine, and glutamic acid. Tyrosine, histidine, tryptophane, and the two sulfur amino acids methionine and cystine appear to be absent.

The medical and biological significance of toxohormone would appear to lie in a "newer pathology of cancer," to use Nakahara and Fukuoka's expression. Traditionally, the approach to the medical problem of cancer has been the customary histological preoccupation with the cancerous growth itself. This is understandable, since the tumor is the immediate threat to health and life. The suggested new approach is directed toward a study of the widespread disturbance of body function caused by toxohormone and an anticancer rationale aimed at combating its effects.

Cancer has been regarded as a "molecular disease," because certain biochemical changes seem to be associated with it such as gene and enzyme "deletions." Complete chemical characterization of toxohormone and further biochemical research into its systemic effects may provide the basis for a new molecular approach to the study of the clinical and biological problem of cancer and perhaps an entirely new rationale of chemotherapy directed toward its eradication.

Apart from the structure-function relationships outlined above, the significance of primary structure also lies in its directive influence in determining the three-dimensional secondary and tertiary folding of protein molecules. This will be considered in the following chapters.

CHAPTER FOUR
The secondary structure of protein molecules

In the previous chapter the primary structure of proteins was described without anything being said about the actual three-dimensional shape or configuration of the molecules. The polypeptide chain configurations shown in the figures were quite arbitrary; the kinks and bends were arranged so that the structures fitted the page conveniently. Knowledge of the amino acid constitution of proteins and of the amino acid sequences by themselves is hopelessly insufficient to account entirely for the observed biological and physical properties of protein molecules. As an example, consider the fact that most proteins fall into one of two major classes, which are distinguished by their gross physical characteristics. First, there are the fibrous proteins, which are characteristically water-insoluble. They form the structural components of the animal body. Examples are keratin (the protein of hair, hoof, and horn), myosin (striated muscle), collagen (connective tissues and cartilage), and fibrin (the blood clotting protein). Then there are the globular proteins, so-called because studies of their physical properties indicate that their molecules are more or less globular in shape. They are water-soluble and include proteins such as the albumins and globulins of the blood plasma, hemoglobin, and the enzymes. Clearly, properties such as these and many others cannot be adequately explained solely in terms of amino acid sequence. We must envisage a higher three-dimensional organization of the protein molecule.

The polypeptide "backbone" is not rigid. As represented on paper, the covalent bonds between the carbon and nitrogen atoms in the backbone are

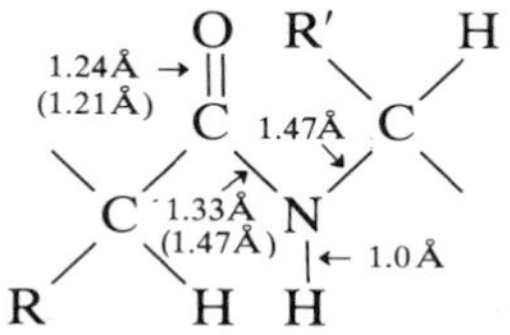

Figure 4-1 Dimensions of the peptide bond. The figures in parentheses indicate what the bond lengths would be if resonance was not occurring.

single, and therefore free rotation about all of them should be possible. However, physical measurements show that the covalent bond joining the nitrogen atom to the carbonyl carbon has a length of 1.33 Å, which is shorter than one would expect for a single covalent C—N bond. Also, the double bond of the carbonyl group is 1.24 Å, which is longer than expected. These measurements are shown in Fig. 4-1. The deviation of these bond lengths from the expected values indicates that resonance is occurring and that the C—N bond therefore is not a typical single bond. Free rotation around this bond is highly restricted and, in fact, the two atoms of the peptide bond and the four atoms attached to these all lie in the same plane. The covalent bond joining the peptide nitrogen atom to the α-carbon atom (i.e., the one to which the characteristic side group of the amino acid residue is attached) and the covalent bond joining the carbonyl carbon atom to an α-carbon atom are normal single covalent bonds and free rotation about them can occur. The polypeptide chain may therefore assume many random configurations because of the flexibility conferred by free rotation about these covalent bonds, but not as many as would be possible if free rotation were possible about all of the single covalent bonds in the polypeptide backbone. At room or body temperature, molecules are in constant rapid motion, and a polypeptide chain may be expected to exhibit constantly changing random configurations, similar to a long piece of thread in swirling water. The physical properties of many polypeptides in solution, as deduced from studies of diffusion rates and sedimentation rates in the ultracentrifuge, are consistent with this "random coil" model of a polypeptide chain. Physical measurements indicate that in many soluble proteins the molecule is not in the form of a randomly coiled thread, but is compact and more or less spherical in shape. Obviously, then, the polypeptide chain must be intricately folded in such molecules, and forces of some kind are presumably involved in stabilizing it. The characteristic biological properties of globular proteins are usually completely altered or destroyed by treatments that would be expected to disrupt the intricately folded molecule, such as heating and exposure to certain chemicals. These agents convert the globular molecule to the random configuration, a process called denaturation. It is usually irreversible. The compact folding of the globular protein molecule is therefore the almost exclusive determinant of its biological properties.

The structure of protein molecules can be studied at three levels, as suggested by Linderstrom–Lang in 1952. First, there is the amino acid sequence, or primary structure. Then two levels of folding of the molecule follow. The polypeptide chain is frequently found to be coiled into one of several helical configurations, exhibiting different types of symmetry; this is called secondary structure. Finally, the secondarily coiled chain may be further bent and twisted into a more or less spherical, three-dimensional structure that is characteristic for a particular native protein molecule. This is called tertiary structure. These three levels of structure are also characterized by the types of chemical bonds and other forces that stabilize them. In the primary structure there are the strong covalent bonds uniting the chains of atoms that constitute the polypeptide backbone. The secondary coiling is stabilized by the much weaker hydrogen bonding forces (as will be described later), and the tertiary structure is held together by electrostatic, hydrophobic, and Van der Waals forces.

A WORD ABOUT SCREW SYMMETRY

The polypeptide chains of fibrous proteins are frequently secondarily coiled into various configurations that are characterized by possession of different types of *screw symmetry*. Screw symmetry simultaneously implies translation along, and rotation around, a common axis. Reference to a simple example (Fig. 4-2)

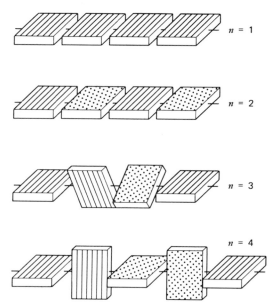

$n = 1$

$n = 2$

$n = 3$

$n = 4$

Figure 4-2 Screw symmetry (See text.) (Redrawn and slightly altered from C. U. M. Smith.)

should make this clear. A screw axis possessing n repeating units (e.g., a chain of amino acid residues) in each complete turn is said to have n-fold screw symmetry. The first diagram, where $n = 1$, illustrates the special case where there is translation but no rotation. This represents simple translational symmetry.

The other examples in Fig. 4-2 are all cases where n is an integer (2, 3, 4,) and they are said to exhibit twofold, threefold, and fourfold screw symmetry, respectively. Sometimes the number of repeating units in one complete turn of a screw is not an integer. Such a screw is said to be nonintegral. Examples of twofold, threefold, and nonintegral screw symmetry are provided by the three main types of protein secondary structure about to be described.

THE α–HELIX

The physical technique known as X-ray diffraction has been a powerful tool in elucidating the structure of fibrous and globular protein molecules. Although a detailed description of the method is beyond our scope, a brief outline of the principles involved in its use to study protein structure is given in Chapter Six. By 1933 it had been established that one of the most prominent features of the X-ray diffraction patterns produced by many proteins indicated interatomic spacings of from 5 to 5.5 Å. Clearly, as a glance at Fig. 4-3 will show, such a result could not be produced by extended polypeptide chains. As was first noted by Astbury, this must mean that the polypeptide chain is coiled or folded in some sort of regular pattern. One might ask how such folding could be stabilized. It will be remembered that the carbonyl oxygen atoms and the imino hydrogen atoms in the polypeptide backbone are capable of forming hydrogen bonds with each other (Chapter Two). Infrared spectroscopy had shown that virtually all of the hydrogen bonding capabilities of the polypeptide chain in fibrous proteins were satisfied. In addition, the hydrogen bonds were shown to be running parallel to the protein fibres. Armed with these and other pertinent facts, Linus Pauling and his co-workers attempted to solve the problem of how polypeptide chains are secondarily coiled by making use of accurate molecular models.

The correct model would have to satisfy three criteria. First, it must be a conformation of minimal potential energy. Second, it must permit maximum hydrogen bonding. Third, it must result in minimal distortion of valency bonds and angles. Pauling and his group finally discovered a pattern that satisfied these requirements and that was in accord with all the known bond angles, bond lengths, and other physical data.

In order to understand the approach to the problem, we must consider the hydrogen bonding capabilities of the polypeptide chain (Fig. 4-4). Theoretically, one might anticipate that any given carbonyl group of the peptide backbone is potentially capable of forming a hydrogen bond with any of the imino groups. This can be understood by remembering that the flexibility of the chain could

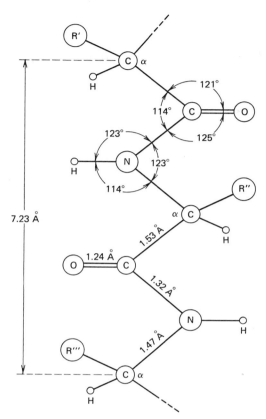

Figure 4-3   Dimensions of a fully extended polypeptide chain.

bring any imino group in close enough contact with the carbonyl group for hydrogen bonding to occur. Of the various possibilities, Pauling and his co-workers found, on the basis of their studies with molecular models, that a helical configuration of the polypeptide chain with a repetitive hydrogen bonding pattern involving a ring of 13 atoms (Fig. 4-5) was the most satisfactory. It was the only one that satisfied all of the criteria. The carbonyl group of 1 amino acid residue would therefore form a hydrogen bond with the imino group of the peptide bond 4 residues further along. This helical configuration of the poly-peptide chain was designated the α-helix, and its form and dimensions are illustrated in Fig. 4-6. The coils of the "spring" are held together by hydrogen bonds. The type of helix shown in Fig. 4-6 is right-handed, and this is the type usually found in proteins. When a polypeptide chain is coiled into an α-helix, the characteristic "R" side groups of the amino acid residues project outward from the helix (i.e., they do not play a direct role in its structure). It follows that, at

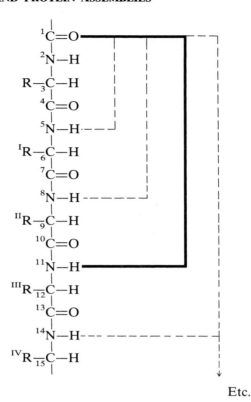

Etc.

Figure 4-4   Possibilities for intramolecular hydrogen bonding in a polypeptide chain. The hydrogen bond indicated by the heavy line is the type found in the α-helix. (See text, also Fig. 4-5.)

least to a first approximation, there is no restriction as to the types of amino acid that can be incorporated into an α-helix (but see later). Those with bulky side groups would be expected to be accommodated as easily as those with small side groups. The pattern of hydrogen bonding in the α-helix prevents free rotation about the covalent bonds in the polypeptide backbone. Consequently, a polypeptide molecule completely coiled into an α-helix is structurally a long, thin, rigid rod.

An unexpected feature of the α-helix that was not in accord with previously known helical structures was the fact that its symmetry is of the nonintegral screw type. In each turn of the helix, there are 3.6 amino acid residues. Because of this, the α-helix is sometimes referred to as the 3.6 residue helix.

As shown in the last chapter, the proline and hydroxyproline amino acid residues in a polypeptide chain will not possess a hydrogen atom on the nitrogen

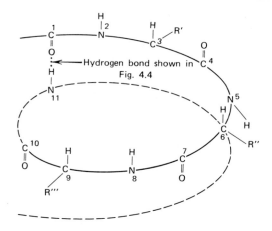

Figure 4-5   Hydrogen bonding pattern of the α-helix.

atom, which they contribute to the peptide link, and therefore cannot form a hydrogen bond with a carbonyl group. The presence of proline and hydroxy-proline will therefore inhibit the pattern of hydrogen bonding, which stabilizes the α-helix and will lead to the formation of other types of secondary structure in polypeptide chains. When present in larger proportions, the imino acids can completely prevent α-helix formation, as will be shown later.

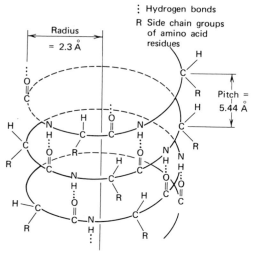

Figure 4-6   The α-helix (schematic). 3.67 amino acid residues per turn of the helix.

THE RELATION BETWEEN AMINO ACID SEQUENCE AND α-HELIX CONTENT

Previously, it was stated that there is virtually no restriction on the types of α-amino acids that can be incorporated into an α-helix. It appears, however, that certain amino acids do not favor α-helix formation. D. R. Davies examined a number of proteins and determined α-helix content from optical rotatory dispersion data, and the amino acid content was determined. There seemed to be no correlation between the mole percentage of any particular amino acid and the α-helix content.

A different approach was then tried. Previous work by E. R. Blout et al. had shown that certain polyamino acids did not form α-helical structures. These amino acids were therefore classified as "nonhelix formers." They are valine, iso-leucine, serine, threonine, cysteine, and proline. Davies then examined certain proteins and looked for a correlation between their α-helical content and the mole percentage of certain of these amino acids, not taken singly but in groups. This time, there seemed to be definite correlations. The best (as judged from inspection of the graphs) were for (ser + thr + val + ile + cys) and for (ser + thr + val + ile + cys + pro). Figure 4-7 shows the plot of the (ser + thr + val + ile + cys) group against helix content. Such a result should perhaps be interpreted with caution, but the correlation seems to be too strong to be due merely to

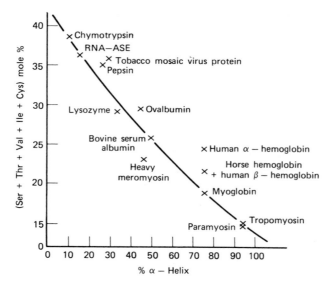

Figure 4-7   Correlation between mole % of (Ser + Thr + Val + Ile + Cys) and α-helix content in various proteins. [Redrawn from D. R. Davies, *J. Mol. Biol.*, *9* (505), 1964.]

chance. Interesting data emerge from studies of proteins in which the exact amino acid sequence and three-dimensional structure are known. Among these are the proteins myoglobin and α- and β-chains of human hemoglobin and lysozyme. D. A. Cook placed the distribution of the total of 569 amino acid residues of these four proteins into four categories. These were the random region, the 3 residues at the N-terminal end of each helical segment, the 3 residues at the C-terminal end of each helical segment, and the remaining residues, which are all in the helical segments. The observed and expected (calculated) frequencies of certain amino acids were compared in different regions. As an example, of the 569 residues, 167 are in the random regions. If the distribution of alanine were uniform, it would be expected that of the 64 alanine residues in this sample, the number in the random region would be:

$$\frac{64 \times 167}{569} = 18.8$$

The observed and expected distributions of amino acid residues were studied in these four regions of the proteins. By comparing the observed and expected frequencies of the different amino acid residues, it was possible to form an estimate of the "helix-breaking" or "helix-favoring" properties of each residue. Alanine, valine, and leucine were all found to occur significantly more frequently than expected in helical regions and hence are "helix-favoring." Asparagine, aspartic acid, and glycine occurred significantly more frequently than expected in random regions and hence are "helix breakers." For some of the remaining amino acid residues, there was no significant difference between observed and expected frequencies in helical and random regions.

Naturally, if some individual amino acids were found in unusually large numbers in any one region, this would lead to a paucity of these residues in the other three regions. In random regions, alanine and valine were found less frequently than expected but, at the ends of helical segments, they occurred with expected frequency.

In random regions, leucine was also found with somewhat less than the expected frequency but, in three N-terminal regions where the expected frequency is 10 times, it occurs only 3 times, a most significant change. Asparagine is not often found in helical segments; neither are aspartic acid and phenylalanine. The actual frequencies of the remaining amino acid residues in helical and random regions do not differ significantly from the expected frequencies.

A comparison of the expected and observed distributions of various amino acids in the different regions enabled Cook to draw the following conclusions.

1. Aspartic acid, glutamic acid, and threonine occur preferentially at the N-terminal ends of helical segments.
2. Lysine, arginine, and histidine occur preferentially at the C-terminal ends of α-helical segments.

3. Alanine, valine, and leucine seem to favor helix formation
4. Asparagine, aspartic acid, and phenylalanine seem to be "he x-breakers."
5. Participation of hydrophobic amino acids in an α-helix appears t be partly a function of side chain size.

THE STABILITY OF THE α-HELIX
(i) The strength of hydrogen bonding

The hydrogen bond is quite weak (1 to 7 kcal/mole) when compared with covalent bonds (15 to 150 kcal/mole). Although disruptive thermal agi ion forces are energetically comparable in strength with the hydrogen bon (approximately 1 kcal/mole at 20 to 37°C), the presence of numerous hyc gen bonds in a biological macromolecule may result in a stable structure the physiological temperature at 37°C. Quite small changes in the environme nay, however, lead to instability. Biological molecules are not to be though of as inflexible and unchangeable; they are able to undergo subtle change even within the narrow physiological range of conditions.

It should not be forgotten that when a randomly coiled polypeptide in in aqueous conditions assumes an α-helical configuration, the formation f the intrachain hydrogen bonds involves the breaking of hydrogen bonds betw n the carbonyl and imino groups of the polypeptide and the surrounding wat mole-cules. It follows, therefore, that the α-helical configuration is like to be assumed only if the intrachain hydrogen bonds are stronger than the I lrogen bonds formed with the molecules of solute or solvent in the envi ment. Generally, synthetic peptides in aqueous solution assume an helical configuration possessing only marginal stability. If urea is added to ar queous solution of a polypeptide or protein, the α-helix tends to uncoil as ie urea concentration is increased. This is because the urea molecules forn tronger hydrogen bonds than water molecules. Hence, strong urea solutions have as protein denaturants in bringing about unfolding of the α-helix.

(ii) The influence of ionizable side groups

As previously pointed out, the side groups of the amino acid resid s do not participate directly in the α-helix structure. Their nature and relative mbers of the different types in a given polypeptide chain may, howeve narkedly influence its stability. Consider, for example, poly-L-glutamic acid, synthetic polypeptide consisting entirely of L-glutamic acid residues. In aquec solutions at pH values below 4.0, where the side group carboxyl groups are t ionized, the molecule has the physical characteristics of the α-helix. At pH ues above 6.0, it has the properties of the random coil. If physical prope s such as

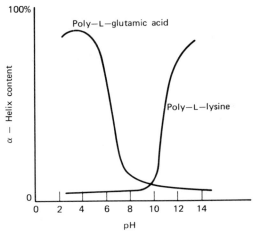

Figure 4-8   Effect of pH on α-helix content of poly-L-glutamic acid and poly-L-lysine.

viscosity and optical rotation are continuously measured between pH values of 4.0 and 6.0, a major change in both of these properties of the solution occurs in the region of pH 6.0, which is indicative of a sharp transition between helical and random coil configurations. What happens is that the side chain carboxyl groups ionize simultaneously at the critical pH value and thereby acquire a negative charge. The simultaneous mutual repulsion of the similar charges on the ionized carboxyl groups causes disruption of the α-helix and its reversion to the irregular open chain form of the polypeptide. A similar phenomenon is observed in the case of poly-L-lysine, except that the effect of changing pH is, of course, opposite from that observed in the case of poly-L-glutamic acid (Fig. 4-8). The helix-random chain transition may also be brought about by varying the temperature.

(iii) Thermodynamic considerations

As explained in Chapter Two the changes in a "system" that occur spontaneously are those involving an increase of entropy, which is known as the "law of increasing disorder." A simple analogy is the case of a glass jar filled with alternate layers of light and dark sand. If the jar is shaken, this orderly arrangement will be destroyed and is hardly likely to be restored by continued shaking.

The α-helix is a more highly ordered structure than the random coil configuration of a protein molecule. Yet, as we have seen, under appropriate conditions the α-helical configuration is assumed spontaneously in aqueous solution. This is the reverse of what might be anticipated. The explanation is that in this example, there is a compensating factor in the form of a decrease in total

energy when the carbonyl-imino hydrogen bonds are formed; t'
spontaneous formation of the more highly ordered structure.

The structure, properties, and features of the α-helix have be
some detail because this form of secondary protein structur
among proteins. It occurs in diverse protein types such as the tou
insoluble structural proteins (e.g., keratin), the contractile pro
and the water-soluble globular proteins.

OTHER INTERNALLY HYDROGEN-B
POLYPEPTIDE HELICES

As mentioned earlier, free rotation cannot occur about t'
because of resonance, but it can occur about the covalent
α-carbon atom and the imino nitrogen atom, and about t'
linking the α-carbon atom and the carbonyl carbon atom in p
degree of rotation at every α-carbon atom is similar, the poly
assume the form of a helix.

Because free rotation is restricted to the α-C—NH and α-
places restrictions on the number of possible helical configu
peptide chain. With the aid of computers and molecul
Ramachandran and his co-workers investigated the possibl
tions of polypeptide chains. The different orientations of the
of the peptide group—the "amide plane"—relative to the of
possible helices are more or less favorable for hydrogen b
owes its stability to the fact that its structure of 3.6 amino a
of the helix brings the peptide carbonyl oxygen atoms int
they are close to and point almost directly to the peptid
which they form hydrogen bonds; in the α-helix these
strained.

Other helices are the γ-helix, with 5.13 residues per t
residues per turn) and the 3_{10} helix (3.0 residues per
bonding patterns in these differ from that of the α-helix

In addition to the α-helix, short stretches of 3_{10} helic
found in the oxygen-binding proteins hemoglobin and
enzyme lysozyme and other globular proteins. What may
π-helix has been found in myoglobin.

THE TRIPLE HELICAL STRUCTU
COLLAGEN

The fibrous protein called collagen has a peculiar an
composition in which three amino acids predominate

hydroxyproline. The analytical figures of collagens from different sources vary somewhat, but about one-third of the collagen molecule always consists of glycine residues, and proline and hydroxyproline together account for about one-quarter of the total amino acid residues present. It is interesting to note that collagen is the only protein so far studied that contains hydroxyproline and hydroxylysine. As we will see, the structural characteristics of collagen are related to this unusual and highly restricted amino acid composition. In chapter three it was pointed out that both proline and hydroxyproline do not possess α-amino groups. Instead, they have the imino group. When these imino acids are incorporated into a polypeptide chain, the nitrogen atom of the imino group will not carry a hydrogen atom and, therefore, will be unable to form a hydrogen bond with the carbonyl group of another peptide bond. Hence, a polypeptide chain containing a high proportion of imino acids would not be expected to form an α-helical type of secondary structure. The formation of the hydrogen bonding pattern to which the α-helix owes its stabilization will be impossible. Furthermore, the bulky pyrrolidine side groups of the imino acid residues prevent close approach of the carbonyl group to the backbone nitrogen, which is a feature of the α-helix. Not surprisingly, it is found that collagen molecules possess a secondary structure that is non α-helical. However, this does not preclude the formation of other types of helical structure. The restricted amino acid composition of collagen does, in fact, determine a type of helical secondary structure, which is unique to itself.

X-ray diffraction studies of collagen show that, among other things, it has the very small repeat distance of 2.86 Å. Although quite good X-ray diffraction pictures of collagen had been available for many years, the working out of the secondary structures led to many false trails. Early attempts to solve the problem were based on two objectives. First, an explanation had to be found for the strong X-ray diffraction reflection, which indicates the 2.86 Å repeat distance and, second, a way had to be found to accommodate the large number of imino acid residues in the structure. In addition, most investigators felt that the structure should be able to explain the fact that collagen fibrils were virtually inextensible. All of the early models proved to be erroneous and did not take into account the fact that the features of the X-ray diffraction pattern for collagen indicated a helical structure. The latter was not realized prior to the publication in 1952 of the calculations of Cochran et al., which permitted the direct examination of the X-ray diffraction pattern for helical structure.

Since collagen contains a high proportion of glycine and proline, one might anticipate that a profitable approach to the elucidation of the secondary structure of the collagen molecule would be to study the configurations of polymeric molecules consisting of glycine residues only (polglycine) and of L-proline residues only (poly-L-proline). Polyglycine can be precipitated out of solution in two forms called polyglycine I and II, respectively. X-ray diffraction studies indicate that polyglycine I is in the form of fully extended parallel polypeptide chains, the

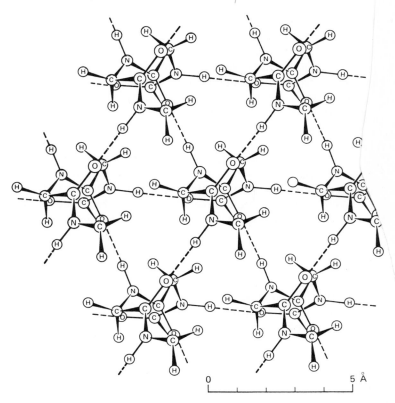

Figure 4-9   Structure of polyglycine II. This is an "end-on" view. The das... lines represent hydrogen bonds. Each polypeptide chain is hydrogen-bonded... six neighboring parallel chains. [From A. Rich and F. H. C. Crick, *Nature, 176* (780), 1955.]

so-called β-configuration with twofold screw symmetry. Polyglycine II is quite different. It is helical with threefold screw symmetry, and the chains run parallel to one another. Each is hydrogen-bonded to its six neighbors (Fig. 4-9). Left- and right-handed helices occur with equal probability, since glycine does not possess an asymmetric carbon atom. X-ray studies show a repeat distance of 3.1 Å Similar structures have been proposed and largely accepted for artificially produced fibers of poly-L-proline. Unlike the polyglycine II helix, the poly-L-proline II helix is necessarily left-handed, since a right-handed helix of the required dimensions cannot be constructed from L-proline. One might reasonably infer that because collagen consists largely of glycine and proline residues, it may have a secondary structure similar to the helical forms of polyglycine II and poly-L-proline II. However, the 3.1 Å repeat distance of the two amino acid polymers does not agree with the 2.86 Å repeat distance of collagen. Any model of the collagen secondary structure must give an explanation of the 2.86 Å repeat distance.

The following generally accepted model for the secondary structure of the collagen molecule was proposed almost simultaneously in 1955 by Rich and Crick and by Cowan et al. To understand it, one must imagine two polyglycine II chains lying parallel to one another (Fig. 4.10) and held together by interchain hydrogen bonds in the manner shown. These two left-handed helices are then twisted around each other in a right-handed sense to give a superhelix. An important feature of this model is that the twist can be done in a way that ensures the residues following each other every 2.86 Å in accordance with the X-ray diffraction data instead of 3.1 Å. If, now, a third amino acid chain is added either behind or in front of the other two and twisted in the same sense as the first two, a model results that is in even better accord with the physical data. The two alternative structures generated, depending on whether the third chain is placed behind or in front of the other two, are referred to as collagen I and collagen II, respectively. The amino acid chains of collagen do not, however, consist of

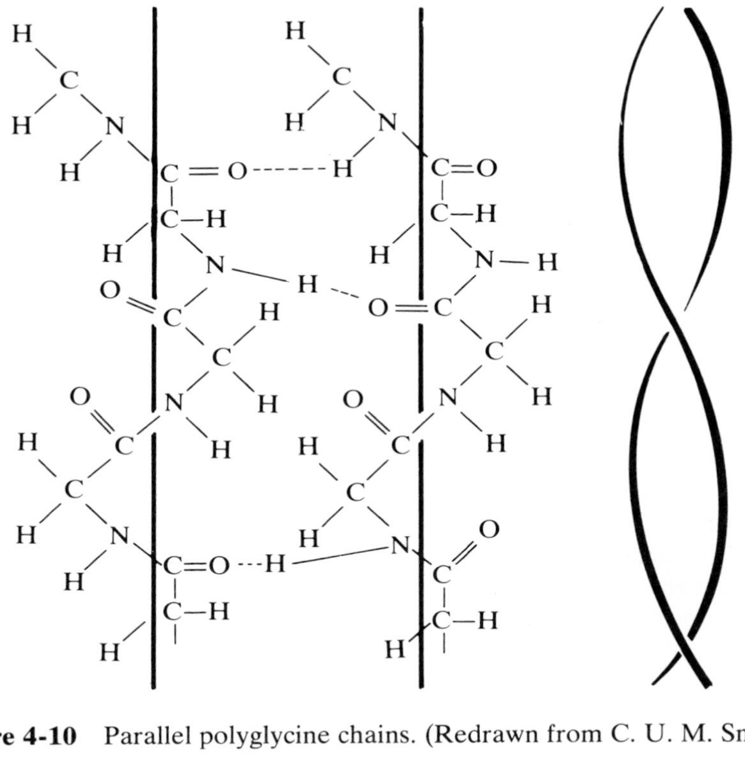

Figure 4-10   Parallel polyglycine chains. (Redrawn from C. U. M. Smith.)

glycine or L-proline residues only. Studies of controlled acid hydroly
collagen show that peptides having the sequence -gly-pro-hypro-gly- a
pro-ala- occur frequently in collagen. If sequences such as -gly-pro-hy
are incorporated into the collagen II structure in place of the polygl
polyproline, a structure results that is even more satisfactory than co
This collagen II is a compact and tightly bound structure. The three cl
held to one another by interchain hydrogen bonds and lie very close t
Their axes are separated by a distance of 4.5 Å, and each is 2.6 Å f
common central axis. Since the packing of the chains is so close, ev
amino acid residue must be a glycine. This is so because glycine has no s
and is the only residue that can be accommodated. Thus, the empirica
that all collagens contain 33% glycine is explained. Further confirmation is
provided by analyses of the amino acid sequences of collagens that show that
there is at least one glycine residue in every tripeptide. The opposite senses of
the separate amino acid helices and the superhelix, coupled with the interchain
hydrogen bonding, gives a structure that is like an inextensible three-stranded
rope and so is consistent with the well-known and unique properties of collagen
fibers.

THE β-CONFIGURATION: STRUCTURE OF
SILK FIBROIN

Fibroin (silk protein), like collagen, has a restricted amino acid composition that
is reflected in its peculiar secondary structure. There is a great preponderance of
amino acids, such as glycine, alanine, and serine, with small side chains. Almost
half of the amino acid residues of silk are glycine, and the remainder consist
mostly of serine and alanine. Analysis of peptides derived from partial digestion
of silk reveals a predominant sequence pattern in which glycine residues alternate with alanine or serine. There is a basic 6-residue unit that repeats over long
stretches of the chain: (gly-ser-gly-ala-gly-ala)$_n$. X-ray studies of *Bombyx mori*
silk fibers have shown that the molecular structure is highly ordered and that it
appears to repeat itself every 7.0 Å. This indicates that the polypeptide chains
are fully extended, the so-called β-configuration characterized by twofold screw
symmetry. The 7.0 Å repeat distance is, however, shorter than that predicted by
Pauling and Corey for a fully extended amino acid chain (7.2 Å).

In 1955, Marsh proposed the now generally accepted structure for silk, which
is known as the antiparallel pleated sheet (Fig. 4-11). In this model, the sheets
consist of polypeptide chains lying in physically parallel array but antiparallel in
the sense that the sequence of groups of the adjacent polypeptide backbones
runs in opposite directions. Viewed from the edge, as in Fig. 4-12, the chains
have a zigzag or pleated structure, so that alternate α-carbon atoms are not
coplanar. The chain is therefore not fully extended, and this probably accounts

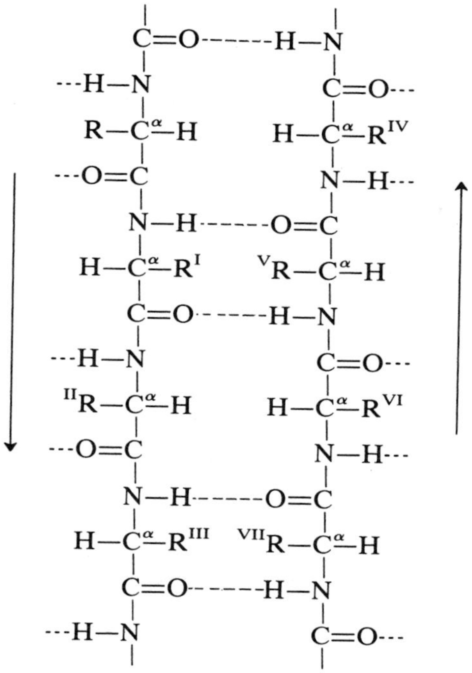

Figure 4-11 Antiparallel pleated sheet structure of silk. The polypeptide chains are not fully extended and have a zigzag appearance when viewed in the plane of the paper. The side chain groups (R) attached to the α-carbon atoms project in a direction perpendicular to the plane of the paper in alternate "up" and "down" directions.

for the repeat distance of 7.0 Å being somewhat shorter than that expected for a fully extended chain. In this configuration, the carbonyl and imino groups of the polypeptide backbone project from the chain in a direction perpendicular to its length, with the result that the extended zigzag chains form pleated sheets by becoming hydrogen-bonded together. This results from interaction of the respective carbonyl and imino groups on the adjacent chains (Fig. 4-11). The side groups attached to the α-carbon atoms of the amino acid residues project upward and downward from the polypeptide chains in a direction perpendicular to the plane of the pleated sheets. Since every second amino acid residue is glycine, and because the α-carbon atoms of the amino acid chains are not coplanar, it follows that all of the glycine side groups will be on one side of a given polypeptide chain and all of the alanine side groups will be on the opposite side. Consequently, successive layers of pleated sheets will be packed with ananine to alanine and glycine to glycine; the result is that the sheets will be

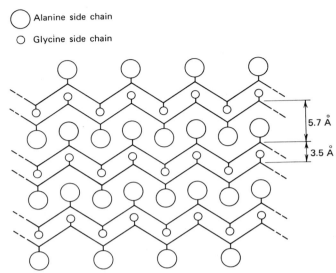

Figure 4-12   Edge view of several stacked, antiparallel pleated sheets of silk. The polypeptide chains are in the almost fully extended condition (the β-configuration) and adjacent antiparallel chains are hydrogen-bonded to each other through their CO and NH groups (not shown here because they are perpendicular to the plane of the paper). The resulting β-pleated sheets are stacked and held together by weak Van der Waals forces. The small alanine and glycine side chains permit close packing. The diagram is therefore a side or edge view of a few stacked, β-pleated sheets. (From Marsh, Corey, and Pauling, *Biochim. et Biophys. Acta, 16*:1 1955.)

separated by spacings that are alternately 5.7 Å and 3.5 Å, as shown in Fig. 4-12. The significance of the high content of amino acids with small side chains will now be apparent, since only these will permit the close packing of the pleated sheets. Sixty percent of the structure of *Bombyx mori* silk conforms to this pattern, which is found in the so-called "crystalline regions" of the fiber. These ordered (crystalline) regions alternate with disordered (amorphous) regions that contain a greater variety of amino acids, including those with large side chains as well as the three primary ones.

In summary, the three-dimensional molecular organization of silk is characterized by a directional bonding pattern, covalent in one direction and hydrogen-bonded in the second direction, with weak Van der Waals forces constituting the bonding in the third direction. The highly ordered regions in which this pattern of bonding occurs alternate with amorphous regions.

The physical properties of silk can be nicely correlated with its molecular organization and the forces holding it together. The almost fully extended polypeptide chains lie parallel to the fiber length and so account for the almost completely inextensible nature of silk fibers. The fiber is very strong, since the resistance to stretching is borne by the covalent bonding of the polypeptide chain. The sheets are held together by the much weaker Van der Waals forces; this gives silk its flexibility. What little extensibility silk possesses appears to be conferred by the amorphous regions, because silks with the greatest proportion of amorphous regions exhibit the greatest extensibility.

SUMMARY

It is apparent that there are three basic configurations of the polypeptide chain in fibrous proteins, each of which is characterized by a particular form of screw symmetry. Each of the three fiber types possesses distinctive physical properties that are associated with the configurational symmetry adopted by the polypeptide chains. Flexibility and suppleness (silk) are associated with twofold screw symmetry, the single nonintegral α-helix is found where strength with flexibility is required, and great tensile strength is achieved by twisting together three threefold screw axes to produce an unstretchable "rope," as in collagen. The amino acid composition and sequence determines which one of these three configurations the polypeptide chain will assume. We have seen that proline and hydroxyproline inhibit formation of the α-helix. Polyglycine, which lacks side groups, may be crystallized in either the extended β-configuration with twofold screw symmetry or in the helical form possessing threefold screw symmetry. The two classes of fibers possessing a high glycine content are those that possess these two types of screw symmetry. The sequence (-gly-ser-gly-ser-gly-ala–)$_n$ induces the β-pleated sheet structure, and the sequence (-gly-X-pro-)$_n$ induces the collagen triple helix.

The α-helix with its nonintegral screw symmetry possesses considerable intrinsic stability and lack of steric strain. Therefore, one considers what factors *prevent* its formation instead of those that generate it because, in the absence of specific factors such as a large proportion of imino acids in the primary structure, the α-helix seems to be the most stable configuration.

CHAPTER FIVE
Fibrous protein assemblies. Structural proteins

Fibrous proteins form the structural foundation of the animal body. They have a tough, fiberlike organization and, as previously mentioned, they are typically water-insoluble. Some of them are very widely distributed throughout the animal kingdom. A good example is collagen, which is found in all animal phyla from the coelenterates onwards. Also very widely distributed are the remarkable contractile proteins of vertebrate striated muscle which, acting in conjunction with the skeletal structures, permits movements of the animal body. Not so widespread are the keratins, the proteins of vertebrate hair, horn, hooves, beaks feathers, and scales. The keratins, along with one of the proteins of muscle (myosin), epidermin, and fibrin (the blood-clotting protein) are characterized by giving a similar X-ray diffraction picture; therefore they were classed together as the "k.m.e.f." proteins by Astbury (these being the first letters of keratin, myosin, epidermin and fibrin). They all possess helical secondary structure of the nonintegral screw type. Peculiar rubberlike proteins are found in elastic fibers and in the elastic structures of certain arthropods such as the elastic hyaline structures of the thoracic skeleton of grasshoppers.

COLLAGEN

Because of its almost universal occurrence in the animal kingdom, the fibrous protein collagen merits discussion first. The characteristic properties of collagen fibers are their inextensibility and great tensile strength, which make them ideal material for tendons. Tendons consist of parallel bundles of collagen fibers. They join muscles to bones in vertebrates and transmit the mechanical pulls of muscular contraction to the bony levers of the skeleton; this makes animal movement possible. Collagen occupies the spaces between muscle fibers and the cells of several organs. The thick mat of collagen fibers in the skin is called "corium"; corium makes the skin tough and forms the leather of animal skins after the process of tanning.

The foundation of the vertebrate skeleton is itself made from cartilage, a form of collagen and, as will be shown later, a unique feature of the supramolecular organization of collagen is responsible for the precipitation of calcium salts in the process of bone formation (ossification).

The transparent cornea of the eye is composed of collagen in which the fibers are arranged in parallel stacked sheets, each sheet running in a direction at right angles to its immediate neighbor. It is presumably this peculiar arrangement of fibers that gives the cornea its glasslike transparency.

In the chapter on protein secondary structure, collagen was shown to be an example of a protein with a restricted amino acid composition in which a large proportion of the total amino acid residues consist of glycine, proline, and hydroxyproline, with relatively much smaller amounts of the other amino acids. This type of amino acid composition determines the characteristic secondary structure of collagen of three intertwined open helices hydrogen-bonded together to give the unstretchable "three-stranded rope." Here we see reflected

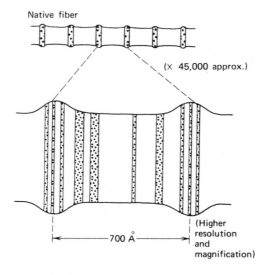

Figure 5-1   Banding pattern of collagen fibers.

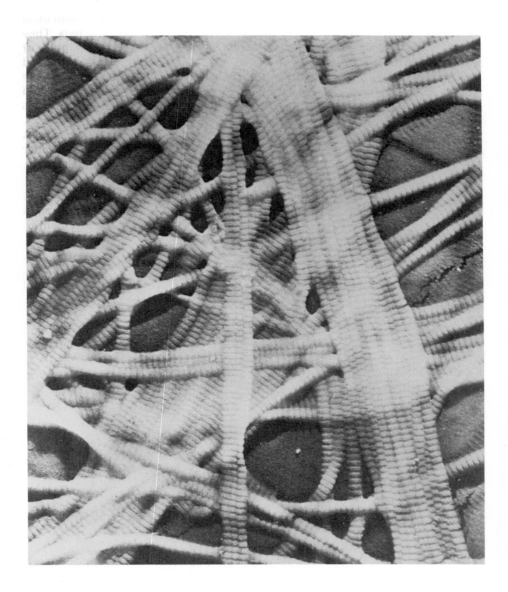

Figure 5-2a Electron micrograph of native collagen fibers. (Courtesy of Dr J. Gross.)

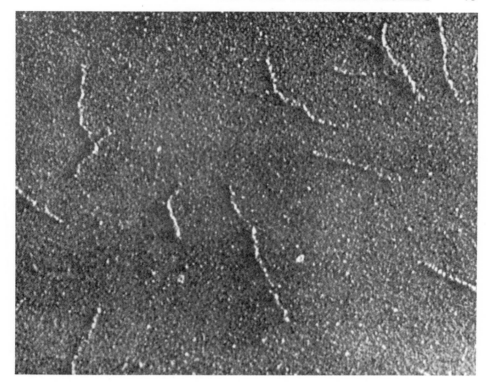

Figure 5-2b   Tropocollagen molecules. (Electron micrograph by Dr C. E. Hall.)

at the molecular level the gross physical properties of collagen fibers, great tensile strength and inextensibility.

When isolated collagen fibers are metal-shadowed and examined with the electron microscope, they are seen to possess a characteristic and repetitive pattern of transverse bands spaced at intervals of about 700 Å. Higher resolution reveals the presence of much finer bands between the major bands in asymmetrical arrangement (Figs. 5-1 and 5-2a). It is natural to wonder what relation this banding pattern bears to the underlying molecular organization of the collagen fibers.

Fresh collagen fibers dissolve in dilute acetic acid and, when properly prepared, this solubilized form of collagen is seen to consist of tiny rodlike particles under the electron microscope. They are about 2800 Å in length, about 14 Å wide, and have a distinct beaded appearance that is asymmetrical. These particles are the molecular units of which collagen fibers are composed and are known as tropocollagen (Fig. 5-2b). The tropocollagen molecule is polarized because of its asymmetrical beaded structure and so can be described as having a "front" and "back" end. Tropocollagen can be induced to reaggregate spontaneously, and a variety of fibrous structures can be produced, the characteristic appearances of which depend on the conditions under which the reaggregation occurs. Neutralization of the acetic acid induces formation of insoluble fibrous aggregates closely similar to the native fibers with major bands at intervals of about 700 Å. If glycoprotein is added to the acetic acid solution of tropocollagen, a fibrous form is produced with major bands at intervals of 2800 Å, the so-called "fibrous long spacing" (FLS) form (Fig. 5-3). There is an interband pattern of finer bands that has a symmetrical design. A third type of banded aggregate called the "segment long spacing" (SLS) form is induced by adding ATP to the acid solution of tropocollagen. This results in the formation of short lengths of reaggregated material that look something like single segments of the FLS form. The major bands are spaced at intervals of 2800 Å with a fine band structure in between with an asymmetrical pattern (Fig. 5-3). See also Fig. 5-4.

How are the tropocollagen molecules arranged in these different fibrous assemblies? J. Gross postulated that in native collagen fibrils, the tropocollagen molecules line up in straight "head to tail" arrangements, all facing in the same direction with individual linear aggregates overlapping their immediate neighbors by one-quarter of the length of a tropocollagen molecule (Fig. 5-5a). Thus,

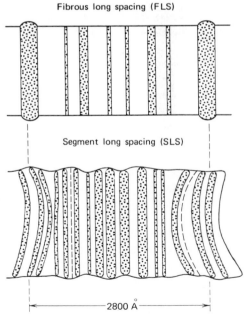

Figure 5-3   Reconstituted forms of collagen.

the 700 Å major bands and the asymmetric minor bands of the fibrils are generated by the collective effect of all the single linear aggregates in staggered parallel array. In the FLS form there are similar parallel arrays, but they lie strictly in register without the quarter length overlap, thus accounting for the 2800 Å major banding pattern. The rows of tropocollagen molecules are not all pointing in the same direction, but are randomly orientated, and so the symmetrical banding pattern is formed (Fig. 5-5*b*). In the SLS form, the tropocollagen molecules lie parallel and in register, all pointing in the same direction. The result is the 2800 Å banding pattern and the asymmetrical pattern of minor bands (Fig. 5-5*c*).

The FLS and SLS forms of reconstituted collagen are laboratory curiosities and are not found in nature. This raises the question about whether there is something unique about the native form of collagen that is related to its biological function. If native collagen fibers are placed in a saturated solution of calcium phosphate and then examined under the electron microscope after suitable preparation, minute crystals of hydroxyapatite (a crystalline form of calcium phosphate) are found within the fibers. The crystals have a regular arrangement at intervals along the fibers. The same regular arrangement of hydroxyapatite crystals is seen at certain specific sites in the collagen fibers of calcifying cartilage when examined with the electron microscope. It seems, therefore, that it is a specific surface configuration of the native collagen fibers that is responsible for the deposition of calcium phosphate from a saturated solution. As is well known, cartilage forms the protein framework of the skeleton in young vertebrates; later it is replaced by calcium phosphate to form bone in the process called ossification. The significance of the native type of collagen fiber would therefore appear to be its role in the precipitation of calcium salts in the vertebrate skeleton, because the FLS and SLS forms of collagen do not precipitate hydroxyapatite crystals from calcium phosphate solutions

BIOSYNTHESIS OF COLLAGEN FIBERS

Collagen fibers are formed by the fibroblasts of the connective tissues. The precursor tropocollagen molecules are synthesized within the fibroblasts and are then extruded into the extracellular space. The tropocollagen molecules spontaneously align themselves in single head-to-tail aggregates called protofibrils, and the protofibrils then aggregate in staggered parallel bundles with quarter length overlap to form the native-type collagen fibers with the characteristic 700 Å major band pattern (Fig. 5-6).

"BIOLOGICAL AGE" AND COLLAGEN

Some individuals of a given animal species seem to age more quickly than others; it appears that "biological age" is not strictly correlated with "calendar

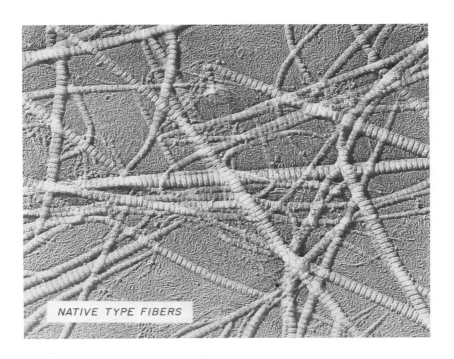

NATIVE TYPE FIBERS

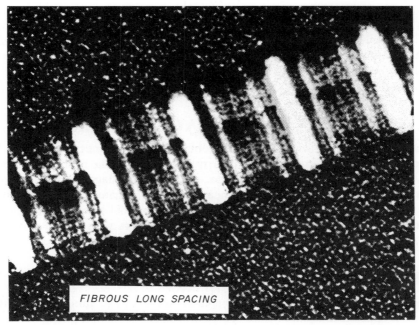

FIBROUS LONG SPACING

SEGMENT LONG SPACING

Figure 5-4 Electron micrograph of reconstituted forms of collagen. (Courtesy of Dr J. Gross.)

age." It is well known, for instance, that among humans, there are those who seem "older than their years" while others are relatively "young" at advanced ages.

Interest has recently focused on collagen as an indicator of the biological age of an animal because this substance, once it is formed in the body, is not renewed; it undergoes slow progressive changes in its molecular organization that affect function as time goes on. Measurement of the extent of these changes by appropriate tests enables an objective estimate of the biological age of the donor animal to be made. One of these tests is measurement of the ability of collagen fibers to develop mechanical tension under certain conditions. F. Verzar experimented with collagen fibers from the tails of young rats. The fibers were suspended in neutral salt solutions with weights up to 3 g suspended from them. When the temperature was raised to 60°C, the fibers contracted strongly if the weights were not too heavy. The minimum weight needed under these isotonic conditions to prevent contraction at a given temperature was measured, and the experiment was repeated with rats of various ages. It was found that the older the rat, the heavier was the weight needed to prevent contraction of the collagen fiber. When the experiments were repeated under isometric conditions, the fiber started to relax when it attained maximum tension and eventually broke

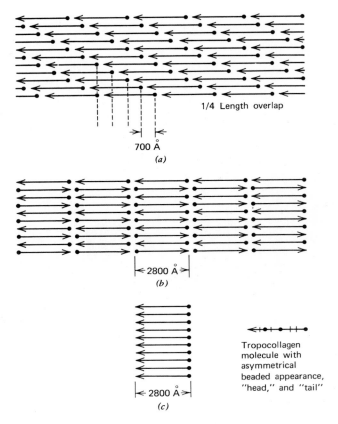

1/4 Length overlap

700 Å

(a)

2800 Å

(b)

Tropocollagen
molecule with
asymmetrical
beaded appearance,
"head," and "tail"

2800 Å

(c)

Figure 5-5 Arrangement of tropocollagen molecules in native collagen fibers and in the FLS and SLS reconstituted forms of collagen. (*a*) Native fibers, (*b*) FLS form. (*c*) SLS form.

if the temperature was maintained constant. Slight lowering of the temperature caused the fiber to "relax" completely, and the original zero value of the tension was recovered. Young fibers relaxed more quickly than older fibers, and the temperature at which maximum tension developed appeared to be correlated with the age of the fiber. Certain chemicals were also capable of inducing similar contraction phenomena in collagen fibers.

The cause of the primary contractions appeared to be the sudden "melting" at 58°C of the hydrogen bonds that link together the hydroxyproline and the hydroxylysine residues in the triple helix structure. This makes the fibers lose their highly ordered arrangement, and they become rubbery and elastic. The increased tension developed by older fibers seems to be due to an increase in the

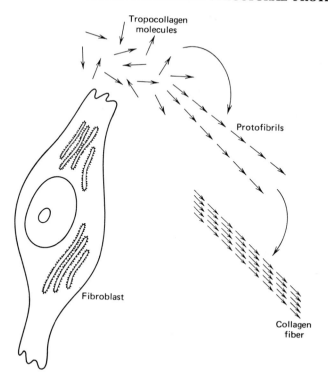

Figure 5-6   Biosynthesis of collagen fibrils.

number of cross links among the tropocollagen molecules. When allowed to stand *in vitro*, denatured collagen becomes more and more insoluble because of the increase in the number of cross-linking hydrogen bonds. Because collagen fibers are not renewed, the number of hydrogen bond cross links may similarly be expected to increase *in vivo* during the life time of an animal. In a contracted fiber at 58°C, the hydrogen bonds are broken, but there are stronger covalent bonds that remain. These are probably ester bonds. They would also be expected to increase with age, and their greater strength is probably responsible for the increased tension developed in older collagen fibers in these experiments.

It seems likely that these slow progressive changes in the molecular organization of collagen are responsible for the degenerative changes of old age and also for certain diseases. The damaging aftereffects of rheumatic fever on the heart valves, the bony changes associated with rheumatoid arthritis, and the clouding of the cornea of the eye (cataract) all appear to be due to abnormal deposition of collagen or to changes in its molecular organization. For this reason these conditions are sometimes called "collagen diseases."

ELASTIN

Elastin is the protein constituent of elastic fibers, which are found in the intercellular spaces of vertebrate connective tissues. It is relatively abundant in the walls of large arteries, the trachea, bronchi, and the ligamentum nuchae, but it is sparse in the skin and tendons. Elastic fibers are cylindrical in shape; they are peculiar in that they do not have a banded appearance and the molecules do not appear to be orientated with any regularity. Therefore elastin belongs neither to the k.m.e.f. nor to the collagen class of proteins.

Elastic fibers appear to consist of two components, an amorphous matrix and a fibrous component consisting of microfibrils about 110 Å in diameter. The amorphous material consists of the peculiar rubberlike protein elastin. The molecules of elastin appear to lack a hydrogen-bonded secondary structure, but there is a tertiary structure consisting of a three-dimensional network of randomly coiled polypeptide chains joined together by covalent crosslinking groups. This molecular organization is clearly reflected in the properties of elastic fibers. They are rubberlike and stretch easily and quickly and rapidly regain their former length and shape when the stretching force is removed. They are therefore well adapted to their function as the major components of elastic tissues and ligaments.

(i) Amino acid composition

Like collagen, one-third of the amino acid residues are glycyl, but there is no evidence to indicate that they are regularly distributed. Almost 95% of the total amino acid residues of elastin are nonpolar (e.g., glycine, alanine, proline, isoleucine, leucine, phenylalanine). There is approximately 1 to 2% of hydroxyproline. The basic, dicarboxylic and hydroxy amino acids and the sulfur amino acids and tyrosine are present in small amounts. Tryptophane seems to be absent. Sequence studies have been hindered because, among other things, the low proportions of lysine and arginine and of tyrosine and phenylalanine have limited the use of trypsin and chymotrypsin, respectively, for breaking the polypeptide chains into smaller fragments.

(ii) Nature of the cross-linking material

Protein chains are commonly cross-linked by disulfide bonds formed by the oxidation of cysteine residues, but elastin does not contain appreciable amounts of cystine. The mechanism of cross-linking in the protein chains of elastin was therefore something of a mystery at one time.

Elastic fibers have a characteristic yellow color and show a blue-white fluorescence in ultraviolet light. Many investigators have worked on the problem of the identity and biological role of the yellow pigment, among them Partridge and Davis, who suggested that the pigment might be the cross-linking material that determines the tertiary structure of elastin. Partridge, Elsden, and Thomas performed chromatographic analysis of peptides obtained by enzymatic degradation of elastin. They isolated a bright yellow fluorescent material, analysis of which led them to the conclusion that the yellow chromophore was situated at the centre of a cross-linked structure carrying at least three or four peptide chains. Chromatographic fractionation of acid hydrolysates of the peptides resulted in isolation of two ninhydrin-positive substances that were different from all known amino acids. Subsequent analytical work enabled these investigators, in 1963, to publish almost complete structures for the two compounds, which were named desmosine and isodesmosine because they were found to be the crosslinking material responsible for the tertiary structure of elastin (the words are derived from a Greek word meaning "bond"). The structures are shown in Fig. 5-7. The desmosines are formed from four lysine molecules; structurally, they are pyridinium derivatives that contain amino acid side chains capable of crosslinking peptide chains. The two compounds are

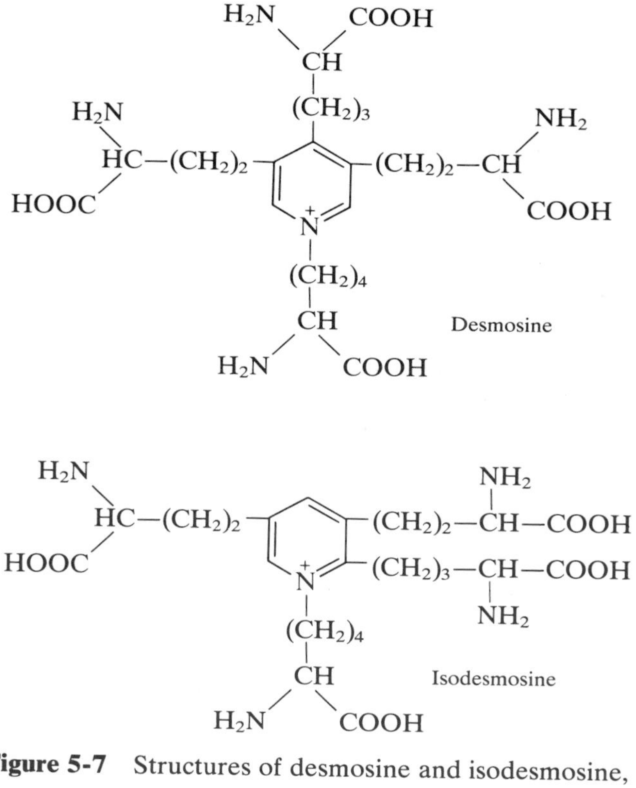

Figure 5-7   Structures of desmosine and isodesmosine,

considered to be isomeric. The lengths of the individual carbon chains attached to the pyridinium rings are still uncertain. Until the discovery of the desmosines, the only known mechanism of covalently linking polypeptide chains together was the disulfide bond.

(iii) Formation of elastic fibers

Electron microscope studies show that during the earliest stages of development of ligaments and tendons, only the fibrous component is present, and there is no trace of the amorphous component. Like collagen, the fibers are formed outside of the cells in which the protein precursor molecules are synthesized but, in this case, there are indications that the assembly of the first-appearing fibrous component may be under cellular control. Later, the amorphous component appears until, in the adult animal, it is the major constituent of elastic fibers.

Regarding the precursor protein molecules of elastic fibers, it seems certain that the protein from which the microfibrils are formed is not an elastin precursor. The two proteins themselves differ greatly, and the amino acid compositions are quite different. In severely copper-deficient pigs, the cross-linking of the protein precursor of elastin appears to be inhibited. The elastic fibers in the aorta of such animals are abnormal, and a water-soluble protein can be extracted from them; its amino acid composition is closely similar to that of elastin except that it has a much larger proportion of lysine than elastin and contains no desmosine. This is what would be expected in an elastin precursor on the assumption that elastin is formed by cross-linking of precursor protein molecules at lysine sites, which are the precursors of the desmosine and isodesmosine cross-linking groups. The precursor protein is therefore called tropoelastin; it appears that copper deficiency prevents formation of the cross links, presumably because cupric ions are required by the enzyme (lysyl oxidase) involved in forming the cross-linking compounds (Fig. 5-8).

Elastin by itself would not seem to have any natural propensity to form fibers, and it seems likely that the microfibrils play a part in shaping the elastic protein into a fibrous form. The elastic fibers of ligaments and tendons originate, like collagen fibers, from their precursor proteins, synthesized by fibroblasts, whereas the smooth muscle cells in the middle layer of the arteries synthesize the precursors of elastin, the microfibrils, and collagen.

RESILIN

A previously undescribed structural protein was recognized by Weis-Fogh in 1960. Thirteen years earlier, LaGreca had observed elastic hyaline structures in the thoracic skeleton of grasshoppers, and Weis-Fogh suggested that these played a part in insect flight. He considered that they stored elastic energy during

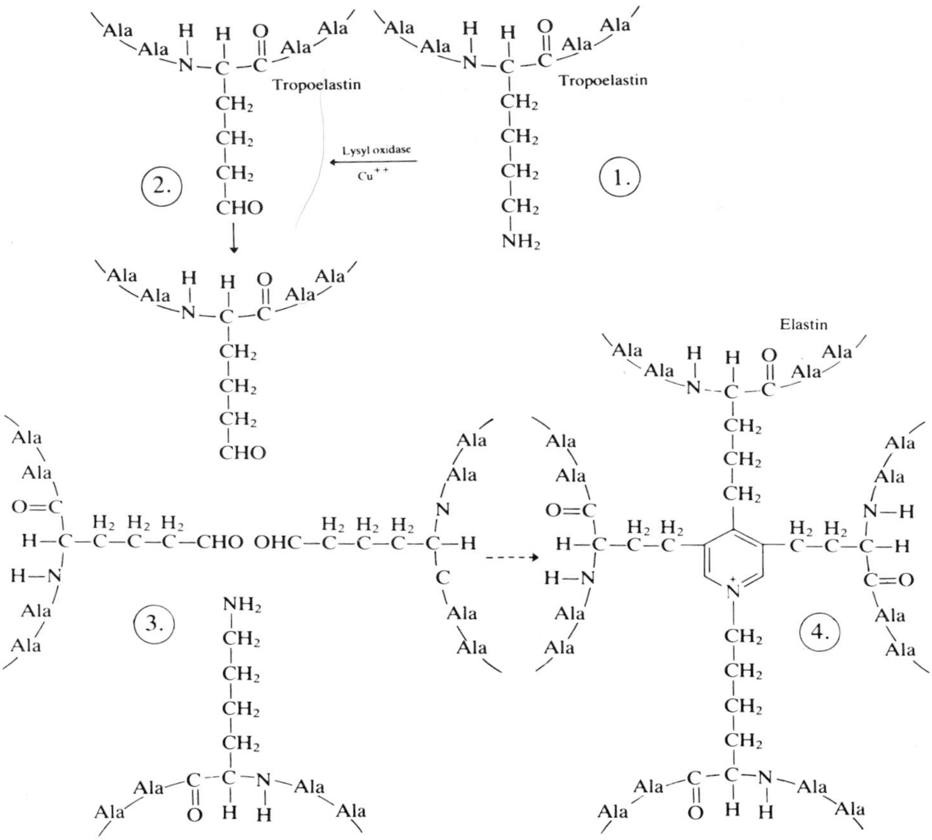

Figure 5-8 Synthesis of elastin. (1) This is a short segment of a tropoelastin chain with a lysine residue surrounded by alanine residues (Ala). (2) Under the influence of lysyl oxidase and cupric ions, the lysine side chain is converted to the aldehyde derivative. (3) The close proximity of three aldehyde groups and a lysine side chain involves them in a series of reactions that lead to formation of desmosine and, then, to: (4) Elastin.

the downstroke of the insect wing, which was then relaxed during the upstroke. Weis-Fogh and his colleagues showed that an important component of the elastic structures was an elastic protein that was named resilin. Resilin is of interest for various reasons. It is an almost ideal protein rubber. Its composition differs in important respects from elastin and, as a result, ideas regarding requirements for elasticity at the molecular level have had to be reexamined. Then, as a result of its three-dimensional network of cross-linked, randomly coiled polypeptide chains with a large proportion of certain polar groups, it has properties that may make it a model for cell membrane proteins. Its polar groups not only confer elasticity, but also provide a gel filtration system. Both attributes seem to be necessary for cell membrane structure.

As well as being found in all the winged insects examined by Weis-Fogh, resilin has also been found in crayfish. Three types of arthropod structure have been employed as sources of resilin for study; the ligaments of the prealar arm and of the main wing hinge of the desert locust (*Schistocerca gregaria*, Forskål), and the elastic tendon of the dragonfly (*Aeshna cyanea*).

(i) Amino acid composition

Resilins from the three sources just mentioned have closely similar amino acid compositions. Methionine, cystine, and hydroxyproline are absent, and there is less than 0.3% of tryptophane. In its amino acid makeup, resilin resembles collagen more closely than it resembles elastin; in all three of these more than one-third of the amino acid residues are glycyl, and between 89 and 156 residues per 1000 are prolyl.

(ii) Physical structure of resilin

On the basis of studies of the mechanical and optical properties of resilin over a large range of strain and swelling and at different pH values, Weis-Fogh concluded that resilin is a three-dimensional network of randomly coiled polypeptide chains. These are cross-linked; there are about 60 amino acid residues between two junction points in dragonfly resilin and about 40 residues in locust resilin. Secondary structure is practically nonexistent, but chemical cross-linkages provide the three-dimensional tertiary structure.

(iii) Nature of the cross-linking material

The nature of the cross-linking material was the subject of a careful study undertaken by Andersen, who had previously noticed that the rubberlike structures of insects fluoresce a blue-white color in ultraviolet light. He concluded that the cross-linking material may be responsible. Complete acid hydrolysis of resilin from the prealar arm and wing hinge ligaments of the desert locust, followed by paper chromatography of the hydrolysate, revealed two distinct fluorescent spots that were named substances I and II. Physical and chemical investigations of these compounds led Andersen to propose structural formulas for substances I and II, shown in Fig. 5-9. He was able to establish that these compounds were involved in the cross-linking of polypeptide chains. The cross-linking material has been found in other arthropods such as the crayfish and

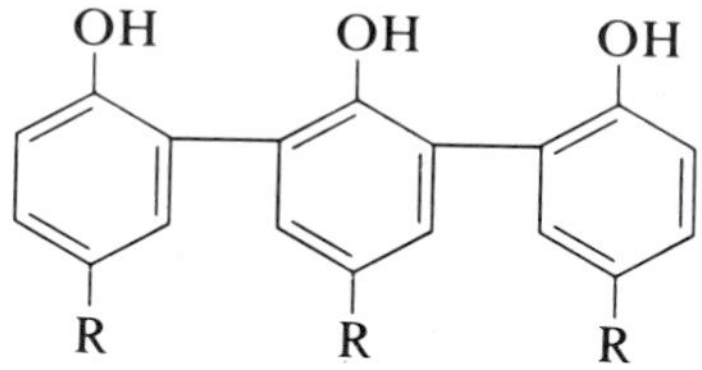

Substance I (trityrosine)

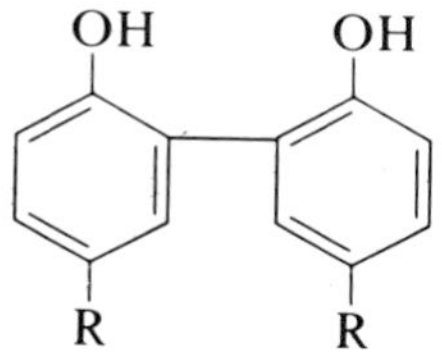

Substance II (dityrosine)

Figure 5-9   Structures of substances I and II in resilin. (R is the alanyl side chain of a tyrosine residue.)

always in association with resilin and no other protein. Substances I and II thus specifically indicate the presence of resilin.

Andersen proposed a mechanism by which the cross-links are formed. The cells that synthesize resilin are thought to manufacture a precursor called pro-resilin which is devoid of secondary and tertiary structure. Simultaneously, a peroxidaselike enzyme is secreted that has the effect of producing reactive tyrosyl free radicals from tyrosyl residues. The free radicals on one chain react with those on another, forming dimeric cross-links that cause the protein to assume its tertiary structure. However, not all observations are consistent with this model, and further work will be needed before the cross-linking mechanism can be considered completely known.

KERATIN

Keratin is the tough, insoluble protein in many vertebrate exoskeletal structures such as hair, scales, feathers, hoofs, horns, beaks, and claws. It is characterized by a relatively high sulfur content (up to 5% dry weight), the precise amount depending on the source. This is due to a large proportion of the combined amino sulfur-containing amino acids cystine and methionine. Keratins are classified as "soft" if they are pliable, as in human skin epidermis; in soft keratins the sulfur content is low (1% dry weight). "Hard" keratins such as those of hoofs and horns contain a high proportion of sulfur (up to 5% dry weight).

Because of its economic importance, wool keratin has been studied most of all, and the keratin of hair has also received detailed study.

(i) Structure of the α-Keratin of hair and wool

W. T. Astbury's X-ray diffraction pictures of the 1930s indicated that wool and hair keratin protein has an α-helical structure. Unstretched wool fibers gave a pattern indicating a repeat distance of 5.1 Å, and stretched fibers gave a pattern corresponding to a repeat distance of 3.3 Å. These were called the α- and β-patterns, respectively; the first indicates the presence of the helically coiled polypeptide chains, and the second indicates that the chains are fully extended. Generally, the α-pattern is given by unstretched mammalian keratins but, in those of reptiles and especially bird feathers, the β-pattern as well as the α-pattern is found in the unstretched state. Accordingly, these are known as α- and β-keratins, respectively.

The insolubility of keratins and their resistance to enzymatic attack are due to intermolecular disulfide linkages provided by cystine residues (Fig. 5-10). The amino acid compositions suggests that disulfide bonds may occur as frequently as one for every four turns of the α-helix. These linkages are responsible for the return of a wool fiber to its natural length after being stretched; the fiber can be stretched to as much as twice its normal length. The intramolecular hydrogen bonds of the α-helices are broken during stretching, but these would not be capable of restoring the fiber to its former length when the stretching force

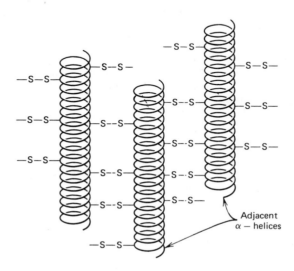

Figure 5-10   Intermolecular disulfide bonds in keratin.

ceased. The intermolecular disulfide bonds provide resistance to stretching to begin with. Disulfide bonds can be broken and remade by reversible reduction and oxidation; this is the chemical basis of the "permanent waving" of human hair. After breaking of the disulfide bonds, the hair is set in the desired pattern; the disulfide bonds are remade and retain the shape of the hair.

Breaking of the disulfide linkages by reduction or oxidation yields water-soluble materials. P. Alexander and his co-workers showed that wool keratin could thus be separated into a large, water-soluble, sulfur-rich fraction consisting of globular molecules and a smaller water-soluble fraction low in sulfur consisting of fibrous molecules from which fibers could be regenerated. The fibrous low-sulfur fraction was further fractionated by W. G. Crewther and B. S. Harrap with the enzyme pronase, which splits it into two components, one containing "cores" made up of helical protein and the other consisting of nonhelical fibrous "tails" that contained most of the sulfur. Useful though these data are they are about as informative regarding hair and wool structure as the statement that the Empire State Building is made of bricks, concrete, and steel girders is about the architectural style of that edifice.

The best approach to understanding wool and hair is to proceed from the macroscopic to the microscopic. As an example of an α-keratin, the structure of hair and wool fibers will be described.

Light microscopy reveals that the shaft of a mammalian hair or wool fiber is covered with a layer of flattened scale cells; within these are longitudinal bundles of elongated dead cortical cells, all of which are longitudinally packed with parallel bundles of macrofibrils about 2000 Å in diameter (Fig. 5-11). The electron microscope is required to show finer detail, and it reveals that the macrofibrils are bundles of microfibrils that are 75 Å in diameter and that they are embedded in an amorphous matrix. These electron microscope studies seem to offer visual evidence for Alexander's belief, based on the chemical fractionation of keratin, that the keratin fibers are made up of a fibrous component (corresponding to the microfibrils) embedded in a cementing material that holds them together (consisting presumably of the high-sulphur globular fraction). Within the microfibrils is yet finer detail; B. K. Filshie and G. E. Rogers introduced a staining technic that "brought out" internal structures within the microfibrils, as seen in transverse sections of fibers with the electron microscope. They interpreted the appearance as indicating that within the microfibril were 11 "protofibrils" (diameter 20 Å) that were arranged apparently in a pattern of nine peripheral and two central protofibrils (Fig. 5-11). Interestingly, this is the same pattern of arrangement of protein fibers within the cilia and flagella of eucaryotic cells (see Chapter Twenty).

The existence of protofibrils has been questioned by D. J. Johnson and J. Sikorski, principally on the basis of the variable number of electron-dense areas within the microfibrils—supposedly protofibrils in transverse section—as seen when the electron microscope is slightly defocused. The technique of image

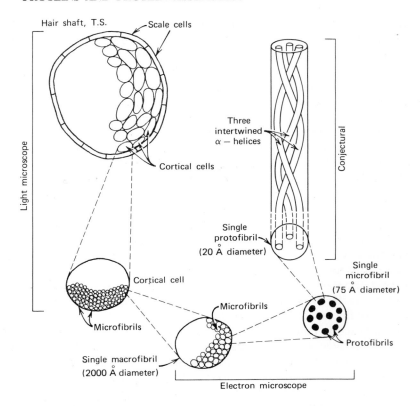

Figure 5-11   Structure of hair and wool fibers.

averaging was applied to the electron micrographs of microfibrils and has led to abandonment of the "9+2" structure, which has now been replaced by one in which there is an electron-dense outer ring and an electron-dense axial core, separated by a less dense circular region. This model appears to have been accepted since its proposal in 1970 by R. D. B. Fraser and others. Protofibrils are not necessarily "thrown out" in this model; the problem is how they are arranged within the "ring and core" structure.

This is about as much as can be resolved with the electron microscope. Protofibrils are much too large to be α-helices, so there has been some discussion as to the number and arrangement of α-helices in the protofibrils. F. H. C. Crick's protofibril model of two or three intertwined, slightly distorted α-helices appears to be most widely accepted (Fig. 5-11). On the whole, the X-ray evidence indicates that there are two intertwined α-helices. The slight distortion or tilt of the helices reduces the usual 5.5 Å repeat distance indicated by X-ray diffraction to the observed 5.1 Å. A speculative view suggests that the two helices may be formed by end-to-end fusion in series or parallel of the "cores" of the low sulfur proteins and that the "tails" may regulate the packing to give the 9 + 2 structure within the microfibril and probably also fill the space between the protofibrils.

Microfibrillar structure in α-keratins from various mammals exhibits considerable uniformity. The observed differences between different keratins is a result of the way in which microfibrils are packed and variations in the proportions and composition of the sulfur-rich material binding them together.

If we take the average annual growth rate of human hair to be about 6 in., calculation shows that nine and one half turns of α-helix are being produced each second.

(ii) Structure of feather β-keratin

This has not been studied as closely as α-keratin. The x-ray diffraction pattern of feather keratin resembles that of β-keratin in some ways, but it is much more complicated. Two different models have been advanced by several groups of investigators to account for this pattern. One of these assumes that ellipsoidal units of structure are disposed in a two-dimensional net. In the second model, the existence of rodlike fibrils with helical properties is proposed. The actual finding of rodlike fibrils about 35 Å in diameter in feather keratin with the electron microscope adds credence to the latter model. Apparently, there are no high- and low-sulfur protein components in feather keratin, so that β-keratin seems to have a more uniform chemical composition than α-keratin. The basic fibrous structural units are the unusually fine microfibrils only about 35 Å in diameter, and they appear to be embedded in a matrix, as seen in cross sections with the electron microscope; but, as noted, there appears to be no protein fraction to account for its existence, as is the case with α-keratin.

The X-ray evidence indicates the presence of α-helical structure, and S. Krimm et al. suggested that the β-chains in feather keratin are distorted to give a helical configuration, 10 of which associate to give a hollow cylindrical structure. Although consistent with much of the X-ray evidence, this model is not in agreement with some of it, and it has been judged to be inconsistent with the known 35 Å diameter of the microfibrils seen in electron micrographs.

E. Suzuki, working in the laboratory of R. D. B. Fraser, showed by infrared spectroscopy that about half of feather keratin exists in the β form, and the remainder is a random chain structure, the latter probably forming the matrix in which the microfibrils are embedded. On the basis of this finding and X-ray diffraction data, D. A. D. Parry, T. P. MacRae, and R. D. B. Fraser showed that in feather keratin, the microfibrils are constructed from two intertwined helical protofibril strands made up of β-crystallites (short segments of polypeptide in the β-configuration joined end to end), with the irregular strands laterally

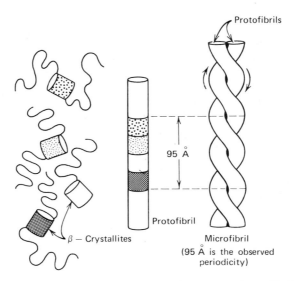

Figure 5-12   Structure of feather keratin. (Redrawn from R. D. B. Fraser.)

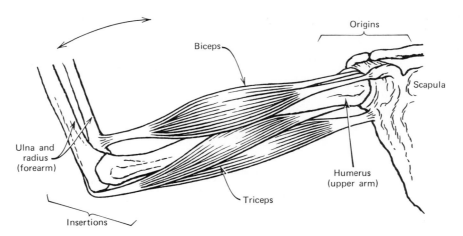

Figure 5-13   Origin and insertion of skeletal muscles (human biceps and triceps).

disposed (Fig. 5-12). The two protofibril strands are in an antiparallel arrangement (i.e., one strand goes "up" and the other goes "down").

VERTEBRATE STRIATED MUSCLE

Much of the bulk of the body of a vertebrate consists of striated muscle, so-called because the fibers of this type of muscle are seen to possess alternate light and dark transverse striations when examined with the microscope. To the lay person, it is familiar as the "red meat" of an animal's carcass, which is used as food. Striated muscles are closely associated with the endoskeleton. Typically, an individual muscle is somewhat spindle- or cigar-shaped; at both ends, it is connected by tendons to bones. One end of the muscle will usually be attached by tendons to a relatively stationary bone, the origin of the muscle, while the other end is attached by tendons to a movable element, the insertion.

The characteristic contractility of striated muscle actuates the bones that function as a system of levers, and this enables animals to move about. Muscles usually occur in antagonistic pairs, one opposing the effect of the other, as with the biceps and triceps muscles of the human forearm (Fig. 5-13).

(i) Fibrous organization and microscopic structure of striated muscle

The fibrous nature of striated muscle is easily demonstrated by stripping off a piece of, say, the gastrocnemius or sartorius muscle of a dead frog and shredding it longitudinally with dissecting needles. The muscle will be seen to consist of fine unbranched fibers easily visible to the naked eye and in parallel bundles. Each fiber is an elongated cylinder about 0.05 mm in diameter with blunt, rounded ends. When examined under medium power of the light microscope, the fibers are seen to exhibit a regular repetitious pattern of alternating light and darker bands or stripes (Fig. 5-14).

Structurally, a striated muscle fiber is a coenocyte, which means that it is formed by end-to-end fusion of many cells into a continuous tube. The intervening cell membranes disappear during development. Careful microscopic examination of individual adult fibers reveals the presence of many nuclei that once belonged to the individual cells. In higher vertebrates these lie just beneath the membrane (sarcolemma), enveloping each muscle fiber.

Muscle fibers themselves are made up of still finer threads called myofibrils that are about 1 to $2\,\mu$ in diameter. They also exhibit alternating light and darker segments; the combined effect of numerous myofibrils in parallel array with their light and dark zones in perfect register give the larger muscle fibers their characteristic striated appearance.

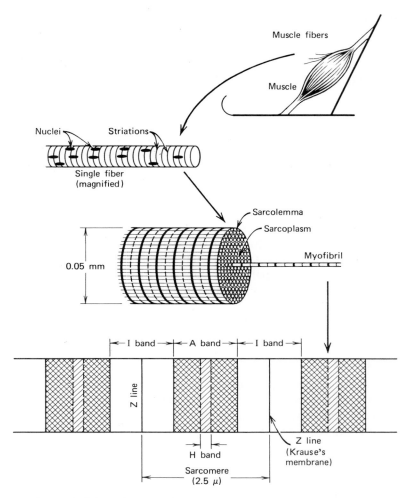

Figure 5-14   Microscopic structure of striated muscle fibers.

Critical higher-power light microscope examination of a relaxed myofibril reveals the presence of a delicate line in the middle of each light area orientated at right angles to the length of the myofibril. This is known as Krause's membrane, and it has been established that the structure between two consecutive Krause's membranes (i.e., a dark band and half of a light band on either side of it) is the contractile unit of the myofibril. It is called a sarcomere; its appearance and the names of its parts are shown in Fig. 5-14. The dark bands stain well with various dyes, but the light bands do not. When myofibrils are examined microscopically between crossed polarizing filters (i.e., the "dark field" position), the stainable bands shine brightly. This indicates a high degree of directional molecular orientation, and these bands are therefore designated "A" bands, because they are anisotropic. Substances showing this effect (e.g., many chemical crystals and various biological materials) do so because they "rotate the plane of polarized light" and therefore appear bright when viewed between crossed polarizing filters. The nonstaining bands look dark under these conditions and therefore cannot possess such a high degree of directional molecular orientation. They are therefore designated "I" bands, because they are isotropic. Presumably the differing stain affinities and appearances in the polarizing microscope of the A and I bands reflect underlying physical and probably chemical differences, which is, in fact, the case as we will see later.

When a fully extended myofibril and a contracted myofibril are examined and compared under the microscope, differences in their appearances are seen. In the contracted sarcomere:

1. The distances between the Krause's membranes (Z lines, as they are also called) are shorter (i.e., the sarcomere as a whole is shorter).
2. The I bands are narrower.
3. The H zones are narrower or have disappeared (depending on the degree of contraction) (see Fig. 5-15).

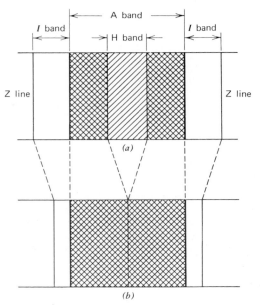

Figure 5-15 Light microscope appearance of (*a*) relaxed and (*b*) contracted sarcomeres. (1) I Bands shorten, (2) H bands diminish and disappear, and (3) A band practically unchanged, when sarcomere contracts.

In both extended and contracted myofibrils, there is no appreciable difference in the width of the A bands.

(ii) Ultrastructure of the sarcomere

Examination of ultrathin sections of striated muscle in the electron microscope reveals a highly organized system of long, straight filaments within the sarcomere (Fig. 5-16). The A bands correspond to regions composed of relatively thick filaments, all arranged parallel to each other and to the long axis of the sarcomere. The I bands, likewise, are occupied by parallel longitudinal filaments that are thinner than those of the A band. At one end the I band filaments appear to be attached to the Z lines; the other (free) ends interdigitate with the thick filaments in the A band. This is shown in Fig. 5-17, which also shows why the H zone is somewhat less "optically dense" than the rest of the A band; it is because the free ends of the fine I band filaments do not extend right to the center of the A band in the extended sarcomere. Also shown in the same figure are the "bridges" revealed by high-resolution electron microscopy between the thick and thin filaments in those regions where they overlap within the A bands.

Transverse sections of the vertebrate sarcomere reveal an ordered paracrystalline arrangement of the filaments (Fig. 5-18). In the region of interdigitation

Figure 5-16 Electron micrograph of ultrathin section of sarcomeres. (Courtesy of Dr H. E. Huxley, J. Biophys. Biochem. Cytol. *3*:631–648, 1957.)

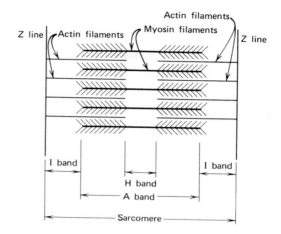

Figure 5-17 Ultrastructure of the sarcomere.

this takes the form of a repetitive pattern of six thin filaments hexagonally arranged around each thick filament; each thin filament is surrounded symmetrically by three thick filaments. There are twice as many thin filaments as thick ones.

Before attempting to describe the role of these structures in the contraction and relaxation of the sarcomere, we should look at the chemical composition of striated muscle.

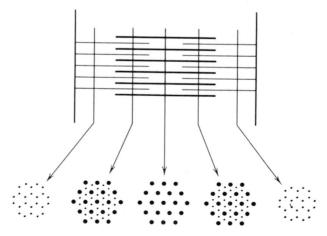

Figure 5-18 Ultrastructural organization of sarcomere fibers as seen in transverse sections.

(iii) Chemical composition of striated muscle

There are four main protein constituents of striated muscles. These are myosin, actin, tropomyosin, and troponin. The most abundant are myosin and actin. Myosin and actin differ in molecular structure and properties in several respects, and they will therefore be described separately.

When fresh muscle tissue is minced and extracted for 20 min with $0.5M$ potassium chloride buffered at about pH 7.0, myosin is dissolved contaminated with actin, but may be further purified by fractionation procedures with solutions of lower ionic strength. Myosin constitutes about 50 to 55% of the protein of mammalian myofibrils, and has an α-helical structure. Its molecular weight is about 470,000.

(iv) Myosin

Electron microscope studies of myosin preparations in the early 1960s showed that the molecular unit of myosin is an elongated asymmetrical object about 1600 Å long and shaped like a tadpole (Fig. 5-19). This confirmed the results of earlier chemical and physicochemical work. The swollen "head" end is connected to the thinner "tail" by a flexible "neck" region that is sensitive to the enzyme trypsin. In 1953, Szent-Györgi showed that treatment of myosin with this enzyme splits the myosin molecule into two dissimilar fragments. One, corresponding to the head end, is a globular structure called heavy meromyosin (HMM) and is unusual as a component of a fibrous protein in that it has enzymatic properties. It catalyzes the hydrolysis of ATP to give ADP and inorganic phosphate. This property naturally led to the idea that ATP is the immediate source of energy for muscular contraction, but the HMM also catalyzes the dephosphorylation of ITP, GTP, UTP, and CTP. By using a new electron microscopical technique, Slayter and Loewy showed in 1967 that the

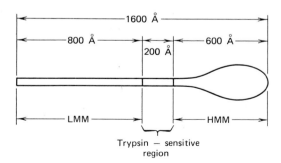

Figure 5-19   The myosin molecule (simplified).

globular head of the myosin molecule itself consists of two subunits. The thinner tail region (the other fragment released from myosin upon trypsin treatment) is called light meromyosin (LMM) and is believed to be made up of two intertwined α-helices forming a supercoil that is probably stabilized by side chain interactions. The most recent view is that the entire myosin molecule therefore apparently consists of two closely associated polypeptide subunits that are among the largest known. They can be separated by nonenzymatic means such as treatment with 12M urea. (These *subunits* should not be confused with the HMM and LMM *fragments* that result from trypsin treatment of the myosin molecule.)

Giese mentions that an interesting feature of the amino acid composition of myosin is the abundance of dicarboxylic acids so that at near pH 7.0, the molecule has a high charge density. To this property the molecule probably owes its ability to retain the fibrous form because of the repulsion of each part of single molecules and the mutual repulsion between separate molecules.

(v) Aggregation of myosin molecules

H. E. Huxley showed that at low ionic concentrations, myosin molecules form filamentous aggregates *in vitro* of variable length but of fairly constant diameter and with "knobbly" ends (Fig. 5-20). The length of the smooth central zone is constant (about 1500 Å), whereas the ends with the projections are of variable length. These appearances have been explained by assuming that myosin molecules come together in parallel fashion with the LMM portions lying end to end and the HMM components pointing in opposite directions. Addition of further molecules increases the overall length of the aggregate and of the "knobbly" ends, but does not affect the length of the centerpiece (Fig. 5-21). The whole structure is reminiscent of two stacks of golf clubs arranged with the ends of the handles coming together in the center. The fact that the projections at the ends of the myosin aggregates are the HMM "heads" is supported by the observation that LMM molecules aggregate to form smooth filaments lacking projections.

There is a striking resemblance between the "knobbly" filamentous myosin aggregates and the thick A band filaments and their cross bridges in the sarcomere. That the thick sarcomere filaments are, in fact, composed of myosin molecules is demonstrated by extracting myosin from muscle fibers with potassium chloride solution of appropriate strength and then examining the myofibrils with the electron microscope. The thick filaments show as "ghosts" after this treatment, and this indicates that they consist of myosin. Similarly, the thin filaments are missing from muscle fibers from which actin has been extracted. Immunochemical techniques in which muscle fibers are treated with fluorescent antiactin and antimyosin antibodies confirm that myosin and actin are localized in the thick and thin sarcomere filaments, respectively.

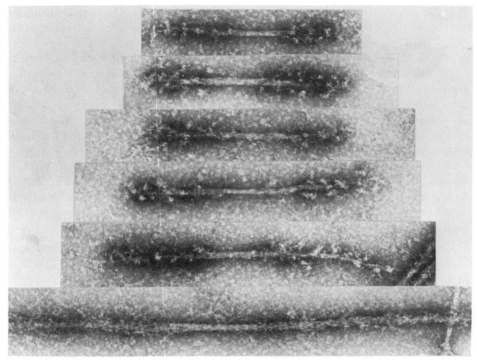

Figure 5-20 Filaments formed by *in vitro* polymerization of myosin molecules. (Courtesy of Dr H. E. Huxley, J. Mol. Biol. *7*:281–308, 1963.)

(vi) Actin

This protein can exist in two different forms that are (in *in vitro* experiments) interconvertible. One of these, fibrous actin, consists of long threads, each of which is formed by the helical intertwining of two finer threads (Fig. 5-22). These are long strings of globular protein subunits that can be considered as the

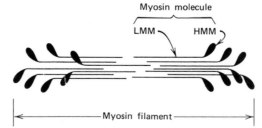

Figure 5-21 Aggregation of myosin molecules.

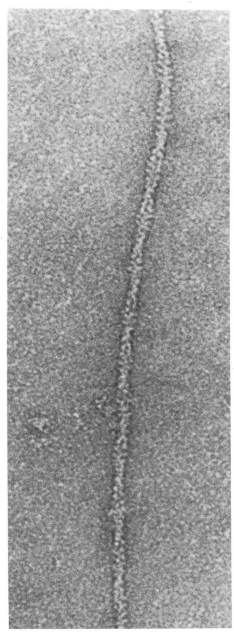

Figure 5-22 A double helical F-actin thread. (Courtesy of Dr H. E. Huxley, J. Mol. Biol. *6*:46–60, 1963.)

molecular units of actin. The fibrous form or F-actin can be easily and reversibly dissociated into these separate subunits, called globular actin or G-actin. Although G-actin molecules are spherical, they behave as if they were polarized (i.e., each has a head and a tail). The individual actin molecules are oriented head to tail in the F-actin threads and, therefore, these are also polarized, a property that appears to be necessary to muscular contraction.

(vii) Tropomyosin and troponin

The two other proteins, tropomyosin and troponin, are associated with the actin filaments (Fig. 5-23). The tropomyosin molecules are long and thin and consist of two α-helical polypeptide chains wound around each other. They line up end to end. According to Ebashi, two tropomyosin threads formed in this way are on the surface of the actin threads, each lying near the helical groove between the two actin strands. A single tropomyosin molecule stretches over seven actin units, and there is one troponin molecule for each tropomyosin. Troponin is a more compact roughly spherical molecule and straddles the tropomyosin molecule a little way from one end, probably. The actin threads and the associated tropomyosin and troponin correspond to the thin filaments, attached at one end to the Z line, that are seen in ultrathin sections of the sarcomere when examined with the electron microscope.

(viii) Interaction of actin and myosin

Mixing of solutions of actin and myosin results in a considerable increase of viscosity and also an increase in a physical property known as flow birefringence. Both phenomena indicate the formation of a complex between the actin and myosin. In this complex, the actin and myosin are combined in definite proportions. Examination of the complex—known as actomyosin—with the electron microscope reveals that it consists of a tangled mass of F-actin threads to which are attached the heavy meromyosin components at regular intervals in an

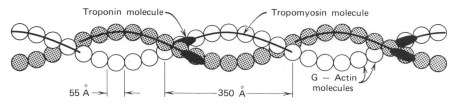

Troponin molecule · Tropomyosin molecule

G — Actin molecules

55 Å ⟶ ⟵ 350 Å ⟶

Figure 5-23 Structure of actin filaments. (Redrawn from Murray and Weber.) (For greater clarity, the two intertwining chains are shown slightly separated, and the G-actin units of one chain are shaded.)

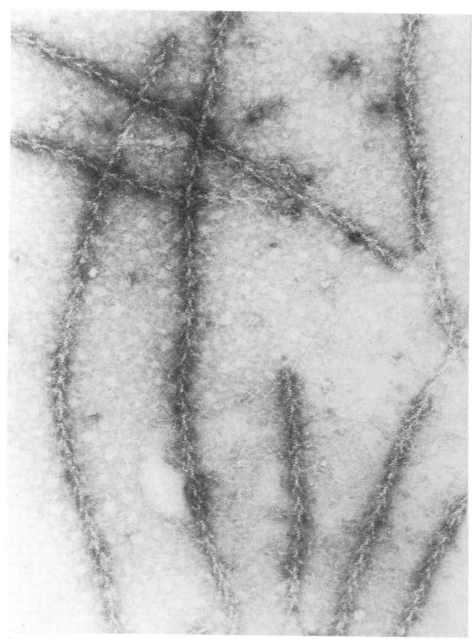

Figure 5-24 Featherlike aggregates of heavy meromyosin molecules and F-actin filaments. (Courtesy of Dr H. E. Huxley, J. Mol. Biol. 7:281–308, 1963.)

oblique orientation so that a featherlike or "arrowhead" appearance is created. A similar appearance is seen in complexes resulting when actin and heavy meromyosin are mixed (Fig. 5-24). When ATP is added to actomyosin, the viscosity suddenly decreases, but more slowly returns to its former high value. The decrease in viscosity indicates that the complex has been dissociated. Direct visual evidence of this was obtained by Huxley in 1963, when he showed that in actomyosin preparations treated with ATP, only actin threads were visible in the electron microscope. ATP has been shown to bind to the actomyosin. The ATP-ase activity of the myosin catalyzes the hydrolysis of ATP to ADP and inorganic phosphate and, therefore, the viscosity increases again as ATP disappears. This ATP-induced cycle of viscosity changes can be repeated several times.

In 1934, Weber showed that threads of actomyosin could be made by forcing a solution of the complex through a hypodermic needle into a solution of low ionic strength. In 1941, Szent Györgi made the exciting observation that actomyosin threads containing more actin than those prepared by Weber and in the presence of $0.05M$ potassium chloride and $10^{-4}M$ magnesium ion would contract when ATP was added. However, this was not a good analog of muscular contraction, because the threads did not only undergo lengthwise shortening, but shortened in every direction. Later, actomyosin threads were prepared in which the molecules were more highly oriented; these not only shortened in the presence of ATP, but could raise a small weight attached to one end if the other end was fixed (i.e., they could perform work). The earlier actomyosin threads did not do this; they merely stretched when weights were attached in the presence of ATP. Although these later threads were still not perfect analogs of muscular contraction, the general belief was that the mechanisms underlying the contraction phenomena observed in them qualitatively approximated those of actual muscle fibers. A model of the mechanism of muscular contraction, as well as accounting for the phenomena of contraction, must also give a satisfactory explanation for the observed differences in the banded appearance of extended and contracted myofibrils as seen in the light microscope. The generally accepted model is the sliding filament concept of Hanson and Huxley. Another, the "helical gear" model, will also be discussed.

(ix) The sliding filament model of sarcomere contraction

Even a casual glance at the ultramicroscopic organization of a sarcomere, as shown in Fig. 5-17, will suggest that shortening or contraction of the sarcomere could be brought about by the sliding of the thin (I band) filaments in between the thick (A band) filaments. That this does occur is borne out by comparison of electron micrographs of longitudinal sections of extended and contracted

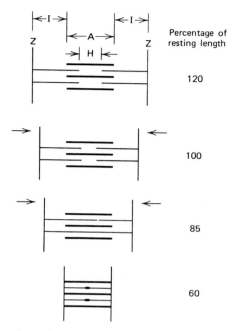

Figure 5-25   The sliding filament model of sarcomere contraction. (Redrawn from Finean, 1961.)

sarcomeres. The appearances are exactly in accord with this idea. In addition, this sliding filament mechanism gives a perfectly natural explanation for the narrowing of the I bands and H zones, the unaltered appearance of the A bands and the shortening of the sarcomere as a whole (Fig. 5-25). Presumably, the combined effect of shortening of most or all of the sarcomeres in the muscle fibers leads to shortening of the muscle fibers and contraction of the muscle as a whole.

What causes the thin filaments to slide between the thick filaments? The electron microscope studies of actomyosin complexes clearly indicate that actin filaments and the HMM component of myosin have an affinity for one another and interact reversibly in some way. To explain the mechanism of contraction, we must remember that the neck region between the LMM and HMM parts of the myosin molecule is flexible and that the "bridges" between the thick and thin filaments in the sarcomere (Fig. 5-26) are presumed to be the HMM components of the thick myosin filaments. These engage with regularly arranged depressions in the thin actin filaments and, by rapid flicking movements, rather like a finger, pull the actin filaments further into the bundles of thick filaments in a ratchetlike manner (Fig. 5-27). It is similar to the way in which an earthworm pulls itself along its burrow by engaging its bristles (setae) with the rough burrow walls.

in contraction that was no longer influenced by calcium ions, and uncontrolled ATP hydrolysis proceeded until all the ATP was exhausted.

The following account of the events of muscular contraction at the molecular level is that given by Murray and Weber.

The first step in contraction is the attachment of an ATP molecule to the globular head of the myosin molecule. The myosin-ATP combination passes over into a "charged" intermediate form that tends to bind strongly with actin. As soon as this happens, the charged ATP-myosin (which on its own is more or less stable) rapidly breaks up, and ATP is hydrolyzed with release of energy. This energy is what causes the wriggling motion of the myosin heads (cross-bridges) of the thick filaments which, engaging with the actin filaments, draws the latter into the bundles of myosin filaments. The release of the cross-bridges happens when another ATP molecule attaches to the actin-myosin combination. This ATP-actin-myosin complex then breaks down to give free actin plus ATP-myosin, and the cycle can start again (Fig. 5-28). Notice that two myosin-actin complexes take part in this cyclic scheme; the high-energy ("active") one that is formed when the charged myosin-ATP combination links up with an actin molecule; the low-energy one formed when the ATP splits with release of energy. This stays as it is until another ATP molecule binds with it. If little or no ATP is available, this low-energy actin-myosin complex is stable and is responsible for the postmortem rigor mortis stiffening of muscle. Murray and Weber call the low-energy, actin-myosin combinations "rigor complexes."

The process that is calcium-sensitive is the combination of the charged ATP-myosin with actin, which results in the "active complexes." Since calcium regulation depends on the presence of tropomyosin and troponin, it would seem that these proteins somehow prevent the active complex from forming when calcium is lacking. It is known that calcium ions bind to troponin; Huxley and Hazelgrove suggest that maybe lack of calcium could result in a configurational change or movement of the troponin molecule, causing the tropomyosin to be pushed away from the helical groove. This might result in a steric prevention of contact of the charged ATP-myosin with actin. There is a further observation that needs to be explained, however; rigor complex formation is *not* influenced by calcium ions. Murray and Weber therefore reasoned that if calcium is absent, only the rigor complexes should be formed. If, then, a mixture of myosin only and myosin with bound ATP was added to actin filaments, and if calcium was absent, only rigor complexes would form, and no hydrolysis of ATP should be observed. When this was done, they found to their surprise that ATP was, in fact, hydrolyzed, showing that the ATP-myosin was reacting with actin despite the absence of calcium. Subsequently, they discovered that regulatory control in the presence of high concentrations of ATP was normal, but was progressively lost as ATP concentration was reduced. Murray and Weber speculated that the presence of rigor complexes might somehow "turn on" the actin fibers with respect to their tendency to bind ATP-myosin. Since the result of the binding of

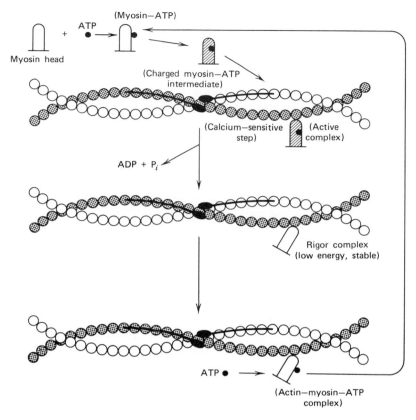

Figure 5-28   Molecular events in the contraction of muscle. (Redrawn after Murray and Weber.)

ATP to a rigor complex is the breakdown to ATP-myosin and actin, it follows that when ATP concentration is high, the number of rigor complexes on an actin filament at a given instant must be low, and vice versa. If, then, myosin and actin are mixed in varying ratios in the presence of a constant ATP concentration, the number of rigor complexes can be readily varied. As expected, Murray and Weber found that a high myosin/actin ratio resulted in a disappearance of calcium control, and a low myosin/actin ratio resulted in its reappearance. Moreover, they found in addition that rigor complexes not only turned on actin when calcium ions were absent, but also modulated the activity of actin complexes in the presence of calcium ions that were already turned on. This was indicated by hydrolysis of ATP proceeding more rapidly than normal. This is called a "potentiated" state of actin.

The actin filaments do not therefore appear to be associations of independent actin molecules giving mere inert "handholds" for the myosin cross-bridges. On the contrary, it seems that here is a highly cooperative system, since rigor complexes, as has been shown, can affect the activity of the thin filaments. There must be communication between the actin molecules because somehow, an actin molecule involved in a rigor complex must be transmitting this information to an actin molecule not so involved and therefore able to carry out ATP hydrolysis. Tropomyosin was shown to play an important role in this cooperativity and message sending. Calcium ions bind to troponin, and this information is conveyed by tropomyosin to the actin molecules associated with the tropomyosin. Only those actin molecules in association with both tropomyosin and troponin were found to be calcium-sensitive.

Tropomyosin is always found in contractile systems, perhaps because it converts the actin molecules that are independent of one another into a cooperative system.

The subject of cooperative action of proteins will be further discussed in the later chapters dealing with the globular proteins.

(xii) Nonstriated ("smooth") muscle

Certain types of muscle lack prominent cross-striations and hence are called "smooth" or nonstriated. Electron microscopy reveals various types of organization in smooth muscle, many of which, like striated muscle, possess thick and thin filaments, but the Z lines and periodicity are absent.

Invertebrates such as mollusks and annelids possess smooth muscles with faint oblique striations. Thick and thin myofilaments are joined by cross-bridges, and the myofilaments are contained within longitudunal fibrils that run obliquely in parallel helices in the smooth muscle fibers. In 1961, Hanson and Lowy found that the paramyosin muscle of the oyster had an organization like this and that each thick filament was surrounded by 12 thin filaments. Striated muscle generally contracts to about 80% of the resting length, whereas these smooth muscles can contract to about 30% of the resting length. An example of extreme contraction is the body wall muscle of the earthworm. In this, it is thought that, as in striated muscle, there is a system of thin fibers interdigitating with thick fibers. Not only do the thick fibers slide relative to thin fibers, but also relative to each other, so that extreme contraction is possible.

Vertebrate smooth muscle fibers show almost no recognizable organized structure in the light microscope. The electron microscope does not appear to show any form of two-filament type of organization with overlapping of fibers, but rabbit uterus muscle does contain myofibrils that are packed with seemingly uniform longitudinal filaments. Smooth muscle cells from the gut lack myofibrils, but possess fine longitudinal filaments. However, thick and thin myofilaments have been found in the smooth muscle of the fowl.

(xiii) Evolutionary aspects—a speculation

Between 1950 and 1960, Loewy and other investigators prepared extracts at high ionic strength from a plasmodial slime mold, *Physarum polycephalum*, the properties of which resembled actomyosin preparations made from vertebrate striated muscle. Later, in 1968, Hatano and Tazawa succeeded in making a purified actomyosin extract from *Physarum*; the results of their studies of this material indicated that actinlike and myosinlike components were, indeed, present. Nachmias and Huxley made similar preparations and examined them with the electron microscope. The characteristic filamentous featherlike or arrowhead structure was seen, and Nachmias, Huxley, and Kessler showed that in preparations treated with ATP, only thin filaments (actin) with a beaded appearance were visible. According to Nachmias and Huxley, the similarity of the slime mold actin and vertebrate actin suggests that part, at least, of the actin molecule has been conserved in the process of evolution. With this in mind and because many kinds of cytoplasmic streaming are known in various types of eucaryotic cells, they further propose that muscular contraction may have had its evolutionary origin in cytoplasmic streaming.

CHAPTER SIX
Tertiary structure. Globular Proteins

The water-insoluble fibrous proteins that form the structural basis of the animal body are found typically in the connective tissues that bind together the different bodily structures and in the skeleton. In addition to the structural proteins, there is a second class of proteins that performs catalytic and transport functions that are essential to life processes. "Life," as we know it, would be inconceivable without catalytic proteins (enzymes); the myriad chemical reactions associated with life phenomena usually proceed much too slowly under the conditions of temperature and pH usually found in living cells to be useful without being catalytically accelerated. Another necessity is that various materials such as oxygen, trace metals, and hormones must be transported to the areas in cells or tissues where they are needed.

The functions of catalysis and transport are performed by proteins that differ from the structural proteins in two important respects. First, the individual molecules are compact instead of fibrous and often approximate a sphere in shape—hence their name, globular proteins. Second, they are water-soluble, a property that comes as no surprise in protein molecules with catalytic and transport functions; it is difficult to imagine catalysis and transport being performed by water-insoluble fibrous protein molecules. The globular proteins can be divided into two major groups. Those that bind specifically with certain other molecules and exert catalytic effects are called enzymes; those that exhibit specific binding but no catalytic activity are designated emphores.

ARCHITECTURE OF GLOBULAR PROTEIN MOLECULES: TERTIARY STRUCTURE

Physicochemical studies indicate that the polypeptide chains comprising many globular proteins possess α-helical secondary structure to a significant extent and also short segments of γ-, π- and 3_{10}-helix. Evidently, the helical structure must somehow be bent and folded to produce a compact molecule. At first, this may seem strange, because it was previously pointed out that physicochemically, the α-helix is an elongated, relatively stiff rod. Globular proteins often exist largely but not entirely in the α-helical form; part of the polypeptide chain remains in the random coil form, usually owing to the presence of amino acid residues in these regions which, as noted earlier, inhibit α-helix formation. The flexible random segments of the polypeptide chain therefore alternate with the "stiffer" α-helical segments. The random segments permit bending of the entire polypeptide chain into a compact roughly globular molecule; the type of folding that is superimposed on that of the secondary (α-helical) coiling is called tertiary folding or tertiary structure. It constitutes a third level of structural organization within protein molecules and is the basis of the specific three-dimensional structure of globular proteins. The biological activity and properties of globular proteins are almost exclusively determined by this three-dimensional configuration, and it is characteristic for any given globular protein. If this folding pattern is altered or destroyed, biological activity decreases or disappears. The efficient stabilization of the pattern of tertiary folding or coiling is therefore a matter of great importance in the structure of globular protein molecules.

FORCES STABILIZING TERTIARY FOLDING

The side chain groups of amino acid residues on neighboring α-helical segments are brought sufficiently close together by tertiary coiling for the side chain groups to interact. Noncovalent bonding forces of different types result from these interactions. There are at least three types (see Fig. 6-1).

1. *Electrostatic bonds* (salt bridges, salt bonds) (e.g., between the oppositely charged groups of basic and acidic side chains).
2. *Hydrogen bonds* (e.g., between the hydroxyl group of tyrosine and a carboxyl group).
3. *Hydrophobic bonds*, the most important stabilizing forces of tertiary structure; they involve interaction between the hydrocarbon side chains of, for example, phenylalanine, leucine and iso-leucine.

The origin of the attractive forces in electrostatic bonds and hydrogen bonds is easily understood, but why should there be an attraction between the uncharged side chains? To understand hydrophobic interactions, consider an analogy.

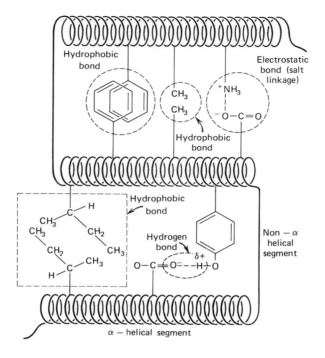

Figure 6-1 Forces stabilizing tertiary structure of protein molecules.

Everyone knows that "oil and water don't mix." If mineral oil (hydrophobic) is added to water, it floats on the water as an upper layer, and vigorous shaking (i.e., expenditure of energy) is needed to disperse the oil layer in the water as small globules. When the shaking ceases, the oil globules rapidly coalesce to form larger ones, and a separate, upper oil layer soon reappears. Evidently, it is energetically more favorable for hydrophobic droplets suspended in an aqueous medium to coalesce than to remain apart. The hydrophobic side chains of certain amino acid residues of globular proteins in the aqueous cellular environment are somewhat analogous to oil droplets in their behavior; if at all possible, they will coalesce with others. Effectively, the water is squeezing them together and a hydrophobic bond is formed between the two side chains. There will therefore be a strong tendency for hydrophobic side chains to orientate themselves toward the inside of the globular molecule away from the aqueous exterior, and a similar tendency for the hydrophilic groups to remain on the outside. The final pattern of tertiary folding will thus be partly determined by the balance between these forces. Most of the hydrophobic groups will be on the inside forming an oily core. Most of the charged and electrically neutral hydrophilic groups will be on the outer surface and will confer water-solubility on the globular protein molecule. Physicochemically, the globular protein molecule may be considered a solubilized oil drop.

DETERMINANTS OF THE PATTERN OF TERTIARY FOLDING

Upon reflection it would seem that a single polypeptide chain consisting of alternate α-helical and random segments could be folded into an almost unlimited variety of tertiary structures, each one having different biological properties. Actually, we find that a native globular protein of given primary structure always possesses one unique pattern of tertiary folding. We know that the two-dimensional linear template of the DNA molecule ultimately determines the amino acid sequences of polypeptide chains. We might also postulate another genetically determined template that is responsible for guiding the correct folding pattern of the polypeptide chain. Such a template would have to have three-dimensional characteristics, but this is inconsistent with modern knowledge of the linear organization of genetic material.

Another idea is that of all the possible tertiary structures, the one of least energy and therefore of greatest thermodynamic stability under physiological conditions will be the most favored. That this is so in one isolated case was demonstrated by C. Anfinsen. He showed that an enzyme, bovine pancreatic ribonuclease, was denatured by $8M$ urea, with a resulting loss of enzyme activity, but that the denaturation was spontaneously reversible and full enzymatic activity was recovered when physiological conditions were restored. (Enzyme activity, as will be shown later, is dependent on a critical configuration of certain chemical groupings in a restricted part of the surface of the enzyme molecule that is maintained by tertiary folding. Denaturation, or a pattern of folding other than that of the native protein, results in a disruption of the critical arrangement of these groups and hence in loss of catalytic activity.) The molecule of ribonuclease is a single polypeptide chain of amino acid residues with four intrachain disulfide links at specific positions (Fig. 6-2). A urea concentration of $8M$ reduces the disulfide bridges to sulfhydryls, and the characteristic pattern of folding of the ribonuclease molecule is disrupted; this results in loss of enzymic activity. If the $8M$ urea is removed and physiological conditions restored, the disulfide bonds and the native tertiary structure spontaneously re-form. Denaturation of protein molecules is, however, usually irreversible. Ribonuclease is an unusual case in which re-formation of the native secondary and tertiary structure after denaturation occurs spontaneously and virtually completely. The rather small molecule of ribonuclease is special in that the spontaneously assumed, thermodynamically most stable configuration also happens to be the catalytically active configuration. However, this is not universal among proteins; the biologically active tertiary configurations of higher

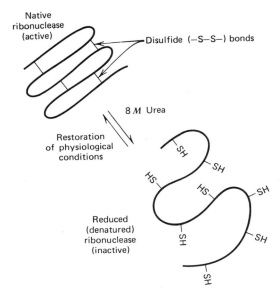

Figure 6-2 Anfinsen's experiment on the reversible denaturation of ribonuclease (schematic).

molecular weight proteins are not always necessarily the most thermodynamically stable.

Certain other globular proteins exhibit partial recovery of biological activity following reduction of their disulfide bonds and "unfolding." Examples are egg white lysozyme (4 disulfide bonds, 50 to 80% recovery), *E. coli* alkaline phosphatase (2 disulfide bonds, 80% recovery), and human serum albumin (17 disulfide bonds, 50% recovery).

Theoretically, any one of the eight sulfhydryl groups of the denatured ribonuclease can combine with any of the remaining seven and, mathematically, one can show that there are 105 different modes of recombination of the sulfhydryl groups. Only one of these will be the pattern characteristic of the enzymatically active ribonuclease molecule. Therefore, if the recombination was a strictly random process, the resulting mixture would be expected to have somewhat less than 1% of the original ribonuclease activity, because only about 1 in every 105 re-formed molecules would possess the pattern of folding of the native ribonuclease. In fact, almost complete enzyme activity is restored. Therefore, the renaturation process cannot be random, but is directed in some way, which results in formation of the specific pattern characteristic of ribonuclease. Apparently in cellular protein biosynthesis, the primary structure of a protein molecule automatically determines the secondary and tertiary structure by the sequence of amino acid residues of the primary structure. Secondary (α-helical) coiling spontaneously occurs under the right conditions in specific segments of the polypeptide chain where it is not inhibited by imino acids or helix-breaking amino acids. Tertiary coiling is determined by the mode of side chain interactions on adjacent α-helical segments, which produces a folding pattern of maximum stability. This, as we have seen, is one in which the bulk of hydrophobic residues are turned toward the center of the molecule and most of the hydrophilic groups are on the outside; that is, it is a pattern in which hydrophobic and hydrophilic interactions are maximized. Thus, the entire spontaneous process of formation of secondary and tertiary coiling is ultimately directed and can be thought of as being programmed in automatic stepwise manner by the amino acid residue sequence of the primary structure, the final pattern being sometimes but not always the one of greatest stability within the cellular environment.

X-RAY DIFFRACTION ANALYSIS

The technique known as X-ray diffraction has been a powerful and important tool in the elucidation of the three-dimensional structure of globular protein molecules. The principle can be explained by reference to an analogy with visible light because both light and X-rays are electromagnetic radiations. For most practical purposes, light can be considered to travel in straight lines; indeed, that part of optical science that deals with designing of lenses and optical instruments depends on the method of ray tracing and assumes straight line propagation of light. There are, however, certain phenomena that cannot be explained by straight line propagation or ray tracing methods. One of these phenomena is diffraction. When a beam of monochromatic light (i.e., light of nearly all one wavelength) passes through a slit or by a single straight edge, careful observation shows that contrary to ray tracing theory, light can bend around corners. The shadow of the edges of a slit on a screen are not perfectly sharp geometrical shadows, but are diffuse, and close examination reveals that there are several faint light and dark fringes parallel to the slit edge shadow. This is due to the phenomenon of diffraction, which is a result of the wave nature of light. If waves in water with straight fronts strike a wall or screen in which there is a slit, the waves do not proceed in a straight line through the slit, but spread out in a circular fashion (Fig. 6-3). In just the same way, light waves spread out when passing through a slit, but the slit must be very narrow for the effect to be noticeable.

If a narrow beam of white light (wavelength about 0.5 μ) is passed through a diffraction grating (i.e., a thin sheet of glass on one side of which are engraved very fine straight parallel lines at about 1000 per mm), an interesting effect is seen. A screen placed on the other side of the grating will show the expected white spot

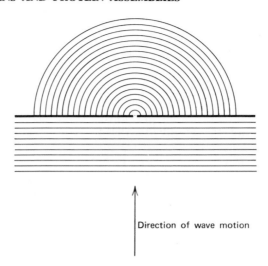

Direction of wave motion

Figure 6-3 Diffraction of water waves through a narrow slit.

of light made by the main beam passing in a straight line through the grating but, in addition, on either side of the white spot and clearly separated from it will be seen two identical spectra. These diffraction spectra, as they are called, and the white light spot all lie in a straight line that is at right angles to the direction of the lines ruled on the grating. If the grating is rotated, the spectra will rotate around the central white spot. The distance of the spectra from the central spot depends on the closeness of the lines ruled on the grating. The closer they are (i.e., the finer the grating), the further apart are the spectra. Hence, the orientation of the diffraction spectra is determined by the orientation of the parallel lines, and their distance apart is inversely related to the distance between the lines.

If two diffraction gratings are placed on top of one another so that the two sets of rulings are at right angles and a beam of light is passeed through them, the central spot on the screen will now be surrounded by two pairs of spectra disposed at right angles to each other. If the rulings on the two gratings differ with respect to the distance between the engraved lines, this will result in the two sets of spectra being at different distances from the central spot. Three superimposed gratings mutually oriented at 60° to each other will produce a hexagonal pattern of six diffraction spectra, the distance of the three pairs of spectra from the central light spot being related to the fineness of the rulings on the three superimposed gratings. The closer together the lines on the grating are, the further apart will be the resulting diffraction spectra. Only regular repeating fine structures such as the parallel engraved lines on diffraction gratings will produce diffraction spectra. Therefore, if a beam of light is made to traverse a transparent object and it produces a diffraction spectrum pattern similar to the foregoing, we can work backward and deduce that the regularity producing it consists of one or more sets of parallel lines, depending on the number of pairs of spectra, their orientation can be determined from the angles between the pairs of spectra, and the distances apart of the lines can be calculated by measuring the distances of the spectra from the central light spot. The study of diffraction patterns can therefore be used to elucidate the structure of transparent objects with regularly repeating fine structure.

X-RAY DIFFRACTION BY CRYSTALS

A crystal is, in effect, a three-dimensional diffraction grating because it consists of many parallel layers of regularly spaced atoms (in the case of simple salts like sodium chloride) or of molecules. This regular pattern or arrangement of atoms or molecules is called the space lattice or crystal lattice. In some cases, the regular repeating structural unit consists of two or more molecules, in which case the structural unit is called a unit cell. The structure of a crystal is therefore one of regularly arranged unit cells in the three dimensions of space.

The spacings between the layers of atoms, molecules, or unit cells in a crystal are much closer together than those of the lines of the finest artificial diffraction grating and are too fine to diffract light of visible wavelengths, but they will diffract X-rays because the X-ray wavelengths are much shorter than those of visible light and are in the region of 0.1 to 150 Å. If a narrow collimated beam of monochromatic X-rays is passed through a crystal and a photographic plate is placed on the opposite side (X-rays are invisible to the human eye), a complex diffraction pattern of dark spots of various intensities and distances from the central undeflected spot will be seen. If the crystal is placed in different orientations and the experiment is repeated, a different diffraction pattern is produced each time. From the number and relative intensity of the spots and their distance from the central spot—and knowing the orientation of the crystal producing a given pattern—the arrangement of the atoms and molecules of the crystal may be determined.

Soon after X-ray diffraction was applied to the study of crystals, Sir W. H. Bragg and W. L. Bragg made the suggestion that the crystal should be used as a reflection grating instead of as a transmission grating. In a crystal the atoms or molecules are arranged in parallel planes in a pattern that is determined by the symmetry of the crystal. When a narrow beam of X-rays is passed through a crystal, it is reflected by the electron shells of atoms, the multiple layers of atoms or molecules giving rise to multiple reflections in many directions. These may interfere with each other and produce an interference pattern of light spots on the photographic plate set up to receive them. This pattern is characteristic of the crystal.

The study of protein molecules by X-ray diffraction dates from 1934, when J. D. Bernal and D. Crowfoot noted the highly detailed X-ray diffraction pattern produced by crystals of the enzyme pepsin. That such diffraction patterns contained information from which the structure of the protein molecule could be deduced was not fully realized until M. Perutz in 1954 introduced the technique called multiple isomorphous replacement (explained later), which would provide important phase information; the significance of this will be apparent further on. Later, J. C. Kendrew used the method to determine the structure of the myoglobin molecule, and Perutz used it in the determination of the structure of hemoglobin. In 1965 the first enzyme structure (lysozyme) was reported in detail by D. C. Phillips.

CRYSTAL STRUCTURE DETERMINATION BY X-RAY DIFFRACTION

When a narrow beam of monochromatic X-rays impinges on a crystal at an angle θ, there will be multiple reflections from the numerous parallel layers of atoms and molecules (Fig. 6-4), and the angles of reflection in each case will be equal to the angle of incidence (θ), a well-known law of optics.

In Fig. 6-4, if we consider just two such beams, it will be evident that the one reflected from the lower layer of atoms (and deeper ones) will have a greater distance to travel than the beam reflected from the upper layer. The distance is related to the distance (d) between the crystal layers, and the angle of incidence of the X-rays is equal to $2d \sin \theta$. If, now, the lower reflected ray is exactly in step with the upper (i.e., if the crests of the waves in this beam coincide with the

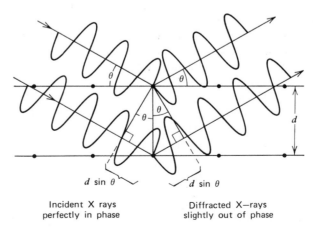

Incident X rays
perfectly in phase

Diffracted X—rays
slightly out of phase

Figure 6-4 Diffraction of X-rays by a crystal (see text for explanation).

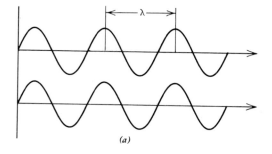

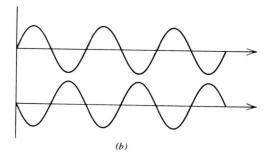

Figure 6-5 Interference of wave trains. (*a*) Waves in phase. (*b*) Waves completely out of phase.

crests of the waves in the other and likewise the troughs) (Fig. 6-5), the two wave trains are said to be in phase. The two sets of waves will therefore reinforce one another, producing what is called constructive interference; the result is a wave having a much greater amplitude (Fig. 6-5*a*). If they are completely out of step (i.e., out of phase) and the crests of one wave train coincide with the troughs of the other, they will cancel out and there will be no wave. This is called destructive interference (Fig. 6-5*b*).

The extra distance that the lower set of waves has to travel determines what type of interference will occur. If this extra distance is equal to an integral number of waves, there is constructive interference. This situation is expressed in the first of what are called the Bragg equations.

$$n\lambda = 2d \sin \theta \qquad (1)$$

where

n = a whole number

λ = wavelength of X-rays used

When the extra distance is equivalent to $(n + \frac{1}{2})$ wavelengths, there will be destructive interference.

$$(n + \tfrac{1}{2})\lambda = 2d \sin \theta \tag{2}$$

By varying the angle of incidence (θ) and keeping the wavelength (λ) of the X-rays constant, the different values of θ that produce constructive and destructive interference can be measured and noted; this permits calculation of the distance (d) between the layers that make up the crystal.

If it were possible to see atoms, on examining a crystal from different angles we would notice that it seems to be made up of many different parallel intersecting layers of atoms or molecules; this is similar to seeming to view many different rows of trees while passing by a regularly planted forest in a train or car. The X-ray diffraction pattern of a crystal that is recorded on a photographic film is made of numerous discrete interference maxima produced by reflections of the X-ray beam as it passes through different groups of imaginary equidistant layers of regularly arranged atoms or molecules in the crystal.

THE PHASE PROBLEM

The pattern of interference maxima (diffraction spots) produced by a crystal and their positions and intensities constitute the information from which the atomic structure of the crystal may be worked out. The electron shells of atoms or molecules are, in effect, three-dimensional waves in which the electron density varies from place to place. The electron density can be represented as a series of contour lines, in much the same way as the heights of land above sea level are represented on maps. In the simple case of the hydrogen molecule, the electron density has the form of a sinusoidal wave; therefore the electron density in the hydrogen molecule varies in a sinusoidal manner.

The wave describing the electron density of more complicated molecules such as those of proteins are much more complex. Such complex sinusoidal waves may be thought of as being compounded from or being the resultant of several simpler waves or elements. The X-ray diffraction pattern of a crystal tells us what the simpler wave elements are from which the complicated electron density wave form of the molecules in the crystal is compounded. The mathematical technique of deducing the elemental wave composition of the complex wave is known as Fourier analysis; this is what the X-ray diffraction "picture" amounts to. Hence, by determining the nature of these simpler waves, they can be combined to give the complex wave from which they were derived—Fourier synthesis—and the electron density distribution of the molecules of the crystal can be determined; things are not as simple as this, however. Three pieces of information are needed to perform the Fourier synthesis. First, the frequency and direction of a unit wave is needed; this is provided by the position of a diffraction spot. Second, the amplitude of the wave must be known; this is given by the density of the diffraction spot. Third, the phase of the unit wave must be known, and herein lies the problem: the electron diffraction pattern of a crystal provides no information on this point, or at least it is ambiguous; several unit waves of identical length and amplitude may be combined to give many different complex wave forms, depending on the differences in the phases. The phase must be known before the Fourier synthesis can be carried out, so the task of the X-ray crystallographer is to select the right one. This is what is known as the "phase problem."

Fortunately, there is a chemical trick that enables a virtually unambiguous choice to be made of the many theoretical possibilities. The procedure is called isomorphous replacement; it was introduced into protein crystallography by Bragg, Perutz, and Kendrew of the University of Cambridge, England. Proteins may be treated with certain reagents that alter their molecular structure slightly; this causes subtle alterations in the X-ray diffraction pattern of the protein crystal. Heavy metal atoms such as mercury can be introduced into specific positions in the unit cells of a protein crystal without significantly distorting their structure or arrangement. It is best if the inserted atom replaces a lighter molecule, such as water. The pattern of spots constituting the X-ray diffraction picture of such a treated protein is essentially the same as that of the untreated protein, but the spots differ slightly in intensity from those given by the un-treated one. If a series of such derivatives are prepared with different reagents, a technique called multiple isomorphous replacement, and the X-ray diffraction patterns of each are analyzed, one unambiguous value of the phase for each unit wave can be assigned. Fourier synthesis—the reconstruction of the molecule from the X-ray diffraction data—is relatively easy in the case of simple crystals and only involves choosing the right structure from a number of possible alter-natives. Experience and intuition on the part of the investigator play a significant role here. In the case of protein molecules, the task is enormously more compli-cated; if it were not for the development of high-speed computers, the mathe-matical calculations needed to work out the structure of a relatively small globular protein molecule such as myoglobin would have been virtually im-possible. Nevertheless, the usual methods are long-winded, expensive, and inefficient. In late 1975, D. M. Collins et al. described two major improvements at important stages in the technique of determining protein molecule structure from X-ray diffraction data. One of these permits resolution to be improved by simply collecting more data from the native protein without the need for mul-tiple isomorphous replacement data in the same range. The second makes use of an interactive computer graphics system in the construction and fine adjustment of molecular models. The computer graphics system is employed in the efficient fitting of the molecular models to the electron density sections obtained by experiment at a resolution of 1.5 to 3.0 Å. One of the great advantages of this method is that it permits quick and accurate atomic coordinate readouts and provides high-quality stereoscopic diagrams.

THE MYOGLOBIN MOLECULE

Myoglobin is an important heme-containing globular protein; it is an oxygen-storing protein in vertebrate muscles. It is especially abundant in the muscles of diving mammals such as whales, porpoises, dolphins, and seals and in diving birds like penguins; in these animals it stores oxygen when the oxygen circulating in the bloodstream has been used up during prolonged dives.

The first protein to have its molecular structure worked out was sperm whale myoglobin. The molecular weight of this myoglobin is 17,000, and the molecule consists of a single polypeptide chain of 153 amino acid residues. A heme group is bound by hydrophobic forces to the protein and combines reversibly with molecular oxygen.

J. Kendrew and his co-workers carried out the X-ray diffraction analysis of the three-dimensional structure of sperm whale myoglobin at the University of Cambridge, and the detailed structure was published in 1961. The crystal of myoglobin is made up of unit cells containing two myoglobin molecules; the crystal lattice is monoclinic. In order to secure sharp diffraction patterns, crystals must be kept in the supernatant liquid in which crystallization occurs (the "mother liquor") so that the water of crystallization is included in the crystal lattice. This water may occupy up to 50% of the total volume of the crystal. J. D. Bernal and D. Hodgkin showed that such hydrated myoglobin crystals would give sharp X-ray diffraction patterns, with more than 20,000 separate reflections (Fig. 6-6).

Myoglobin derivatives obtained by multiple isomorphous replacement were used in the X-ray analysis. Four different heavy metal-containing reagents were used that, when added to the liquid in which the myoglobin was crystallizing, had the effect of combining with a few specific chemical groups in the myoglobin molecule. Each of the different derivatives therefore had various heavy metal atoms inserted at precisely known positions in the myoglobin molecule, and these scattered X-rays strongly.

Comparison of the X-ray diffraction pattern of the four isomorphous derivatives and the untreated protein enabled Fourier synthesis to be performed and the three-dimensional electron density within the unit cell to be calculated, a formidable task that was greatly facilitated by use of high-speed computers. The electron density was to be mapped at different levels within the unit cell. A series of such maps resembling the contour maps of geography books, were drawn on plastic sheets; each sheet corresponded to the level within the unit cell of the particular electron density distribution. When these were stacked together, the result was a greatly enlarged three-dimensional map of the electron density distribution. It resembled the technique of reconstruction of small organisms or embryos by serially sectioning them at various levels and then transferring enlarged drawings of the sections in proper sequence to glass or transparent plastic sheets. These are then stacked in correct order so that an enlarged three-dimensional representation of the organism or embryo can be seen.

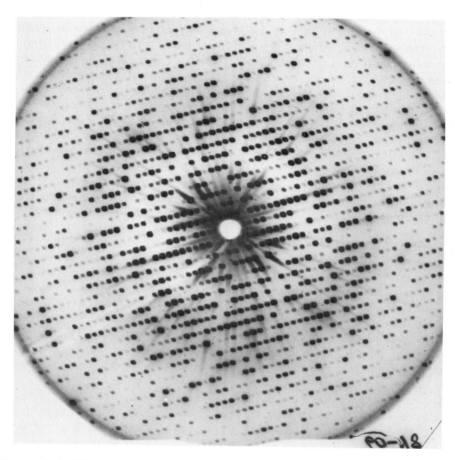

Figure 6-6 X-ray diffraction pattern of myoglobin. (Courtesy of Dr J. C. Kendrew.)

At the commencement of the work on myoglobin, resolution was limited to 6 Å. This gave a coarse reconstruction of the myoglobin molecule but, even for this, it was necessary to analyze approximately 400 reflections. It enabled the pattern of twists and bends of the α-helical segments of the molecule to be made out. When the resolution was taken to 2 Å, the number of reflections to be analyzed rose to 10,000. At 1.4 Å, the number was 20,000. The 2 Å analysis enabled the secondary structure and even the primary structure, to be worked out in detail. The 1.4 Å analysis was necessary for certain identification of all the amino acid residues, although some could be made out at 2 Å.

The myoglobin molecule, whose shape is variously described as being like a triangular prism or an oblate spheroid, measures about 44 Å × 44 Å × 25 Å. Of the 153 amino acid residues, about 118 are involved in α-helical secondary structure, so that about 75% of the myoglobin polypeptide exists in α-helical form. There are eight α-helical segments, designated A to H beginning from the N-terminal; these alternate with uncoiled (random) segments. The lengths of the coiled segments vary from 26 amino acid residues, as in segment H to only 7 residues as in segments C and D. The tertiary folding that gives the molecule its compact roughly globular shape is made possible by bending of the polypeptide chain in the random regions. Most of the hydrophilic amino acid side chains are directed toward the external aqueous environment, whereas most of the hydrophobic groups are directed toward the inside of the molecule, so that the myoglobin molecule is an example of the "solubilized oil drop" structure.

Some of the hydrophobic groups line the inner surface of a "cleft" or "groove" in the molecule; in this cleft the heme group (Fig. 6-7) is situated. This hydrophobic cleft is found in many globular proteins, especially in enzymes, where it is usually associated with the active site. The iron atom in the heme group of myoglobin is held by six coordinate covalent bonds, four of them to the four nitrogen atoms of the four pyrrole rings of the heme and a fifth to one of two histidine side chains on opposite sides of the cleft. The sixth bond is to the oxygen molecule in oxymyoglobin. The second histidine residue is indirectly bonded to the iron atom by this oxygen molecule, since the histidine is too far away to bond directly with the iron. In metmyoglobin a water molecule replaces the oxygen molecule and the histidine residue bonds with the oxygen atom of this, which bonds to the iron atom. In deoxymyoglobin there is nothing bridging the gap between the heme iron atom and the second histidine residue (i.e., the oxygen binding site is empty, even of a water molecule).

The orientation of the heme group in the cleft is automatically determined by the distribution of the hydrophilic and hydrophobic substituents at the periphery of the tetrapyrrole nucleus. Hydrophobic groups are located on one part of the

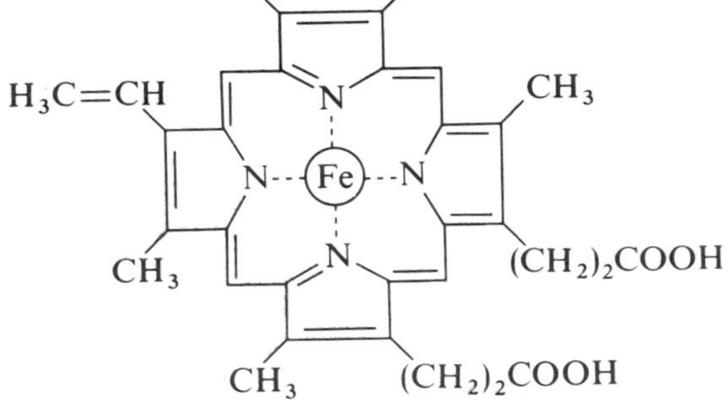

Figure 6-7   The heme molecule.

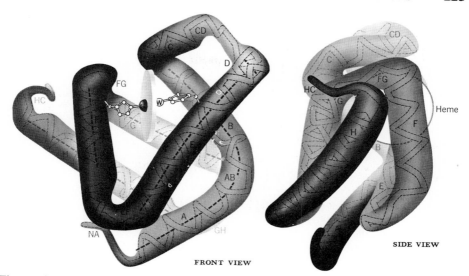

Figure 6-8   The myoglobin molecule (two views). (Reprinted from *The Structure and Action of Proteins* by R. E. Dickerson and I. Geis. W. A. Benjamin Inc., Menlo Park, California, Publisher. Copyright 1969 by Dickerson and Geis.)

periphery of the heme molecule, and so this region of the molecule buries itself completely within the hydrophobic environment of the cleft. That part of the periphery with the two hydrophilic propionic acid groups projects outward. The presence of the heme group probably stabilizes the entire myoglobin molecule.

The three-dimensional structure of the myoglobin molecule is shown in Fig. 6-8.

CHAPTER SEVEN
Enzymes

In philosophical discussions about the nature of "life," claims are often made regarding the primal importance of different types of cellular biomolecules; most discussion usually centers around proteins versus nucleic acids. Perhaps such discussions might be more rational if the relative importance of various life functions were considered instead. Catalysis and catalyzed chemical reactions are of primary importance. Without the many catalyzed reactions that go on in living cells, life as we know it would be hard to imagine. Even the replication of nucleic acids and of cells themselves would be difficult to imagine in the absence of suitable catalytic machinery.

CATALYSTS AND CATALYSIS

Ostwald's classical definition of a catalyst is that it is "an agent which affects the velocity of a chemical reaction without appearing in the final products of the reaction." Negative catalysts slow down reactions and positive catalysts speed them up. Usually we are concerned only with positive catalysts. One should realize that catalysts influence the speed of chemical reactions that are already inherently possible, but do not initiate new reactions of their own that do not occur in their absence.

Very few of the multitude of chemical reactions going on in living cells would proceed fast enough or in a sufficiently orderly manner within the ranges of temperature and pH encountered in living cells without the aid of suitable biological catalysts, or enzymes, as they are called. How do enzymes—and catalysts generally—speed up chemical reactions? Chemical reactions proceed spontaneously in a direction leading to greater overall stability and loss of energy (i.e., they occur in the direction of thermodynamic probability) and not the reverse. However, because a given reaction is thermodynamically probable and therefore spontaneous, it does not follow that it will proceed quickly; it may not even occur at all. Thermodynamics tells us nothing about reaction *rate*.

A good illustration of this is the combination of hydrogen with oxygen; when ignited, the combination gives water. The reaction yields a good deal of energy (the gas mixture explodes), light and heat and the product (water) represents a much more stable energy state than the mixture of hydrogen and oxygen. Yet, if these two gases are mixed together, they do not show any tendency to react. If a lighted match is applied, the two gases combine explosively, and water is formed. Therefore, it appears that the individual molecules of hydrogen and oxygen at ordinary temperatures are not sufficiently reactive to combine and form water. The covalent bonds of the hydrogen and oxygen must first be broken and then rearranged to form the covalent bonds of the water molecule. Increasing the temperature causes molecules to move faster, and their covalent bonds "bend" and "stretch" more violently so that the molecules become more reactive. The lighted match (heat) supplies the energy necessary to get some of the gas molecules into a reactive state so that they combine much more readily to release energy, which activates other hydrogen and oxygen molecules. The lighted match therefore effectively starts a chain reaction and an explosion results, the initial input of energy starting the reaction that sustains itself by its own energy release. The situation is as if a cannonball was situated on top of a hill in a small hollow. Theoretically, the cannonball would spontaneously tend to

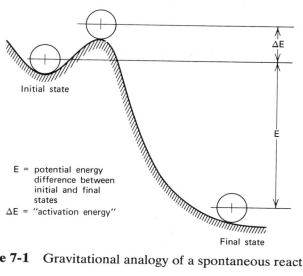

E = potential energy difference between initial and final states

ΔE = "activation energy"

Figure 7-1 Gravitational analogy of a spontaneous reaction.

achieve a more stable state of lower energy by rolling down the hill, yet as long as the barrier is there it will not do so. An initial input of a little energy is required to get it over the hump; after this it will roll down the slope under gravity without any further help (Fig. 7-1).

In the case of oxygen and hydrogen and other reactions involving energy yield and greater stability of the products, an initial input of energy is needed to get the reactants over an "energy hump" before they will react spontaneously. This energy hump is the extra energy that is required to get the molecules into a sufficiently reactive state. Elevated temperatures and/or extremes of pH usually do the trick. However, extremes of either temperature or pH could not be tolerated by any cell, so that cells have evolved enzymes that speed up the reactions at the "mild" temperature and pH conditions of cells. The question is, how do they do it?

CHEMICAL NATURE AND PROPERTIES OF ENZYMES

Before the mechanism of enzymatic catalysis can be properly understood, the chemical nature of enzymes must first be described. Enzymes are nondialyzable; this fact suggested to early workers that they are large molecules. The properties of enzymes and the effects of pH and various chemical and physical agents on them strongly indicate that they are proteins.

(i) Influence of pH on enzyme activity

Most of the physical properties of proteins in aqueous solution, such as solubility, viscosity, and electrical conductivity, exhibit a maximum or minimum value at one particular pH value. Solubility of a given protein, for example, is

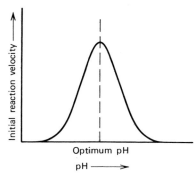

Figure 7-2   Graph illustrating the effect of pH on enzyme activity.

minimal at the pH, where the number of ionized acidic sidechain groups balances the number of ionized basic groups so that the protein molecule as a whole is electrically neutral. This pH is the so-called isoelectric point, and the protein exhibits special properties at this particular pH value. The catalytic activity of enzymes exhibits a maximum at one particular pH, known as the optimum pH; the enzyme is also most stable at this point. Activity falls off rapidly as the pH is raised or lowered above or below the optimum value (Fig. 7-2). In this respect, enzymes exhibit a property characteristic of proteins with respect to the effect of pH changes on a specific property (i.e., the catalytic activity).

(ii) Influence of heat and protein precipitants

The catalytic activity of most enzymes is destroyed by temperatures over 60°C, although a few are known that survive higher temperatures (e.g., certain enzymes from thermophilic bacteria). Protein precipitants such as strong acids, alkalis, and heavy metal salts (heavy cations) and the heavy anions of acids such as picric, phosphotungstic, and trichloracetic acids inactivate enzymes. In this respect, enzymes behave like typical soluble proteins, which are usually rendered water-insoluble by heating or by treatment with protein precipitants, a phenomenon called denaturation. There is simultaneous loss of biological properties. Even merely shaking a solution of an enzyme may denature and hence inactivate it. Almost always, denaturation is irreversible. Exceptions are the reversible denaturation of the enzyme ribonuclease with $8M$ urea described earlier, and a few enzymes that are inactivated by temperatures over 60°C, but that recover activity on cooling; however, such cases are very unusual.

The effect at various pH values of low concentrations of heavy anions and cations provides further evidence of the protein nature of enzymes and, in addition, indicates that catalysis is not a property of the enzyme molecule as a whole. If enzymes are proteins, they should be positively charged in acid solution and therefore should be especially sensitive to heavy, negatively charged ions (anions). In alkaline solutions they will be negatively charged and sensitive to heavy cations. Experiments with the enzyme yeast saccharase have, in fact, shown that the effect of gradually increasing concentrations of silver ion (Ag^+) on the activity of the enzyme is more marked on the acid side than on the alkaline side of the optimum pH, whereas the reverse is true when the experiment is repeated with phosphotungstic acid. *The concentrations of these reagents needed to inhibit or to inactivate the enzyme are much lower than those required to denature the protein.* Therefore, the effect of silver ions under these conditions evidently does not involve all of the negatively charged groups on the enzyme protein, but is localized and specific for certain negatively charged groups involved in catalytic activity. This suggests that catalytic activity is not a property of the entire molecule, but that there is a localized, catalytically "active site" or sites.

ENZYMES AS PROTEINS

All enzymes so far isolated and purified have been shown to be proteins. Many enzyme molecules have a nonprotein component more or less firmly attached to the enzyme protein. This may be a metal ion or an organic molecule of relatively low molecular weight; it is known as a prosthetic group or cofactor, so that such enzymes are conjugated proteins. The presence of the cofactor is essential for full enzymatic activity. The enzyme protein is often referred to as the apoenzyme, and the complete enzyme, consisting of apoenzyme plus the cofactor, is called the holoenzyme.

Like other proteins, enzymes are denatured by heat and certain chemical reagents so that tertiary and secondary structure is destroyed and the protein assumes the random coil form, which is catalytically inactive. Hydrophobic groups are exposed that previously were hidden within the native or natural active enzyme, and so the denatured form is much less soluble and may precipitate out of solution. Alternatively, it may form a layer at the surface of the water, or the interaction of denatured molecules of different types may form insoluble complexes. That enzymes are proteins is hardly surprising; of all the biological macromolecules, only proteins possess a sufficient inherent complexity of structure that can provide the basis for their remarkable binding and catalytic specificity.

In some cases the primary, secondary, and tertiary structures of enzymes have been determined; in every instance, the enzyme molecule has a compact and roughly spherical shape (i.e., they are globular proteins). Several hydrolytic enzymes have had their three-dimensional structures worked out. Examples are lysozyme, pancreatic ribonuclease, carboxypeptidase A, α-chymotrypsin, subtilisin, and papain. All of them differ markedly from myoglobin in their secondary and tertiary structures. The proportion of α-helix is quite low in these proteins, and segments of extended polypeptide chain in an approximately β-pleated configuration is a prominent structural feature of them, as the following brief description will show. All of these enzymes consist of a single polypeptide chain.

(i) Lysozyme

(Molecular weight = 14,600)

This has 129 amino acid residues; it contains six very short α-helical regions that take 55 residues, so that the α-helix content is 42%. Residues 45 to 60 are in a nonhelical part of the chain that is turned back on itself in an antiparallel, pleated sheet configuration. The hydrophobic side chains of lysozyme are directed into the interior of the molecule, and hydrophilic groups are mostly on the surface.

(ii) Pancreatic ribonuclease

This is a kidney-shaped molecule measuring $38 \times 28 \times 22$ Å; it has an even lower α-helix content; only 22 (16%) of the 124 amino acid residues are involved in two short α-helical segments. There are four intrachain disulfide groups. Two of these are located at opposite ends of the molecule; between the two ends, 45 consecutive amino acid residues are arranged in three approximately antiparallel chains. The ribonuclease molecule has a somewhat "open" structure, and some of the hydrophobic residues are found near the surface; for the most part, the "hydrophobic in, hydrophilic out" rule is followed.

(iii) Bovine carboxypeptidase A

This is a compact molecule measuring about $52 \times 44 \times 40$ Å with a molecular weight of 34,600. It has one disulfide bridge. Thirty-five percent of its 307 amino acid residues are involved in eight α-helical segments. Twenty percent of the residues form a large pleated sheet structure consisting of eight antiparallel chains; the whole is twisted so that the first chain makes an angle of 120° with the last. The α-helices are situated on both sides of the pleated sheet. In this respect (α-helices outside and pleated sheet inside), carboxypeptidase A differs markedly from the other enzymes described here. Another 20% of the amino acid residues form an irregular coil at one end of the molecule.

(iv) α-Chymotrypsin

This contains 241 amino acid residues and has five disulfide bridges. There are only two very short α-helical segments formed from a total of 8 amino acid residues near the C-terminal end of the chain, most of which is fully extended and folded back on itself, forming antiparallel segments; most of the molecule is in the β-pleated form.

(v) Subtilisin

In the roughly spherical 275-residue molecule of this enzyme, there are eight α-helical segments that take up 30% of the total number of residues. Of these, 7 residues are roughly parallel and have the same N-terminal and C-terminal orientation. A twisted β-pleated sheet runs through the molecule and is formed from the five sections of extended chain, each containing 5 or 6 residues each.

(vi) Papain

Papain has four short α-helical segments formed from about 40 (20%) of its 211 residues. The molecule is mostly irregular in shape.

One should be careful not to consider enzyme molecules as "odd" when comparing them with myoglobin just because of their low α-helical content and the conspicuous part played by the β-pleated, antiparallel sheet in their structure. For many years, myoglobin was the only globular protein molecule whose detailed structure was known; its possession of 75% α-helical structure and the absence of antiparallel extended chains were possibly unconsciously regarded as "normal" or "typical," whereas it would now appear that myoglobin is unusual because it possesses such a high helical content and lacks an antiparallel extended chain structure. So far, the collagen triple helix has, so to speak, been conspicuous by its absence from enzyme molecule structure. There may still be surprises in store as more globular protein molecules have their three-dimensional structure worked out; the presence in them of any type of protein secondary structure cannot be certainly excluded.

ENZYME SPECIFICITY AND THE ENZYME-SUBSTRATE UNION

Unlike inorganic catalysts, enzymes exhibit a remarkable specificity regarding the reactions that they catalyze. Some enzymes will catalyze only a few closely related reactions, such as the hydrolysis of ester or amide bonds; this is known as group specificity. Not infrequently, an enzyme will catalyze only one particular reaction and no other, such as the enzyme urease, which catalyzes the hydrolysis of urea and no other compound. This is called absolute specificity. Some enzymes react specifically with only one of the two forms of an asymmetric substrate (i.e., a compound), the molecule of which contains at least one carbon atom to which four different groups or atoms are attached. The two forms in which asymmetric compounds exist are spatially related, as an object is to its mirror image; the usual analogy is the difference between a right and a left glove or shoe. Amino acids and sugars are examples of asymmetric compounds; enzymes for which these are substrates will usually react with the L but not the D forms. Another form of specificity is shown by enzymes that react with a substrate to give only one of two possible *cis-trans* isomeric products e.g., oxidation of succinate by succinate dehydrogenase yields fumaric acid, a *trans* compound but never the *cis*-isomer (maleic acid). Both of these forms of specificity that are determined by the precise three-dimensional arrangement of chemical groups in the substrate are forms of stereospecificity.

It is difficult to imagine how the catalytic effects of an enzyme—to say nothing of its remarkable specificity—could be manifested unless there is close contact,

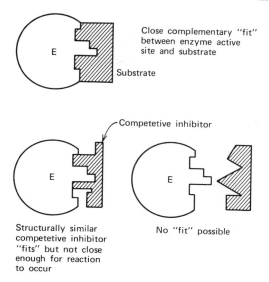

Figure 7-3 The "lock and key" model of enzyme-substrate binding (E = enzyme molecule).

even if this was only transient, between enzyme and substrate molecules. The existence of such transient enzyme-substrate "complexes" has been shown by spectrophotometric methods. In this connection we must not forget that enzyme and substrate molecules have definite and specific three-dimensional "shapes"; the specificity of an enzyme may therefore be visualized as arising from a close complementary "fit" between the substrate molecule and the active site or sites of the enzyme. This idea was first put forward by E. Fischer in 1894. Additionally, assuming this to be true, we may further speculate that a substance similar in molecular shape to the substrate may fit the active site, but not closely enough for reaction to occur, or it may not fit at all (Fig. 7-3). This idea of complementary fit between enzyme and substrate has given rise to the well-known "lock and key" or template analogy of the specifity of the enzyme-substrate union. The substrate is analogous to the lock that fits a key (the enzyme) and is opened by it (reaction occurs). Another lock may fit the key, but not closely enough to open the lock (no reaction); a third similar key may not fit at all (no union possible between enzyme and substrate).

The union of enzyme with substrate is reversible, so that as well as reaction products being formed with the release of free enzyme again, the enzyme-substrate complex may alternatively dissociate to give free substrate and enzyme. These phenomena are summarized in the following equation.

$$E + S \rightleftharpoons ES \longrightarrow E + Products$$

where

E = enzyme molecule

S = substrate molecule

ES = enzyme-substrate complex

COMPETITIVE ENZYME INHIBITION

Evidence for the reversible complementary fit between enzyme and substrate is provided by the phenomenon of competitive enzyme inhibition. The best-known example of this is the competitive inhibition of the enzyme succinate dehydrogenase by malonate. Succinate dehydrogenase catalyzes the removal of a pair of hydrogen atoms from succinate, one hydrogen atom coming from each of two adjacent carbon atoms (never two hydrogen atoms from the same carbon atom in this and other dehydrogenations). The product of this reaction is fumaric acid.

$$
\begin{array}{ccc}
\text{COOH} & & \text{COOH} \\
| & & | \\
\text{H---C---H} & & \text{C---H} \\
| & \longrightarrow & \| \quad +2\text{H} \\
\text{H---C---H} & & \text{C---H} \\
| & & | \\
\text{COOH} & & \text{COOH} \\
\text{Succinic acid} & & \text{Fumaric acid}
\end{array}
$$

If increasing amounts of malonate are added (other things being kept constant) to a mixture of succinate and succinate dehydrogenase, the rate of dehydrogenation of the succinate gets progressively slower and, finally, almost ceases if enough malonate is added. The inhibition can, however, be reversed by adding more succinate, and the dehydrogenation reaction begins again. Careful studies reveal that the degree of inhibition is not related to the absolute concentration of the malonate, but to the ratios of the concentrations of both malonate and succinate; that is, the degree of inhibition is proportional to the ratio concentration of malonate/concentration of succinate. The explanation of this effect is that malonate is structurally very similar to succinate.

$$
\begin{array}{c}
\text{COOH} \\
| \\
\text{H---C---H} \qquad \text{Malonic acid} \\
| \\
\text{COOH}
\end{array}
$$

The malonate molecule can therefore fit into the active site of succinate dehydrogenase; however, no dehydrogenation can occur because, in the malonate molecule, there is only one carbon atom carrying hydrogen atoms instead of the two, as in succinate, so that removal of a pair of hydrogen atoms, one from two adjacent carbon atoms, cannot occur. Malonate, therefore, inhibits succinate

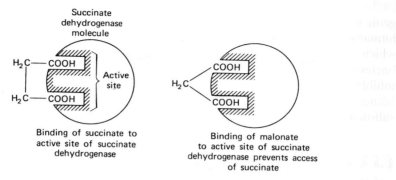

Figure 7-4 Competitive inhibition of succinate dehydrogenase by malonate (shematic).

dehydrogenase because it prevents access of the substrate to the active site (Fig. 7-4). If the amount of succinate is held constant, increasing the malonate concentration enables malonate molecules to "crowd out" succinate molecules from the active site, so that progressive inhibition is observed. If more succinate is now added, the succinate molecules can now crowd out the malonate molecules, and the reaction begins again. Thus, the rate of the dehydrogenation reaction in the presence of malonate is a function of both the malonate and succinate concentrations. It is a situation in which the substance in higher concentration has the better competitive advantage in vying for the active site. For this reason, this type of enzyme inhibition is called competitive inhibition and, as we have seen, it can be reversed by adding more substrate. Structurally, competitive inhibitors always closely resemble the substrate in their molecular shape.

Certain biologically active compounds owe their effects to competitive enzyme inhibition. A well-known example is *p*-amino benzene sulfonic acid amide

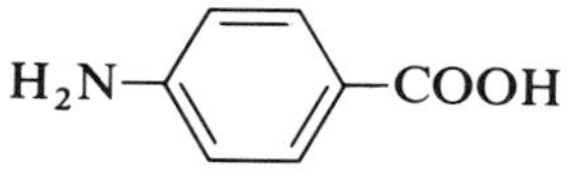

$$H_2N-\langle\ \rangle-COOH$$

p-Amino benzoic acid

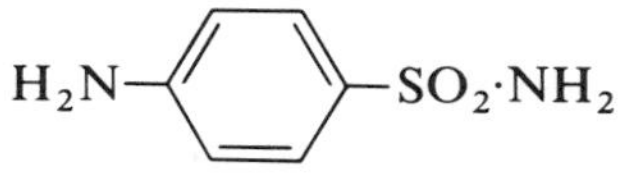

$$H_2N-\langle\ \rangle-SO_2\cdot NH_2$$

p-Amino benzene sulfonamide

Figure 7-5 Molecular structures of *p*-amino benzoic acid and *p*-amino benzene sulfonamide.

(sulfonamide) which is selectively toxic toward certain bacteria that are pathogenic to humans, but is relatively harmless to humans. The bacteria differ from humans in that they have a dietary requirement for *p*-amino benzoic acid (PAB), which is an essential metabolite precursor; it can therefore be regarded as a bacterial vitamin. The enzyme for which PAB is the substrate is competitively inhibited by sulfonamide, the molecule of which has a close structural resemblance to that of PAB (Fig. 7-5), and this is the basis of the toxicity of the sulfonamides toward bacteria that require PAB.

ELUCIDATION OF ACTIVE SITE CONFIGURATION

As we have seen, the substrate molecule fits closely into the complementary surface of the enzyme active site much as a hand fits a glove; the phenomenon of competitive inhibition lends strong support to this concept. By studying the rate of reactions of various substrates when they are acted on by a given enzyme, and by noting the effects of various competitive inhibitors, it should therefore be possible to build a conceptual picture of the actual dimensions and configuration of an enzyme's active site. As well as being of theoretical interest, the elucidation of the configuration of active sites and identification of catalytic groups can have great practical importance. In the case of pharmacologically active compounds, the action of which depend on reversible enzyme inhibition, such knowledge can assist in the rational design of active compounds in the search for more effective agents. A good example is the "probing" of the active site of the enzyme acetylcholinesterase, which catalyzes the hydrolysis of several esters. It is especially active toward esters possessing a cationic group situated about 5 Å from the ether oxygen atom in the ester group. An example is acetylcholine itself, the natural substrate that is hydrolyzed according to the following equation.

$$(CH_3)_3\overset{+}{N}.CH_2.CH_2O.CO.CH_3 \xrightarrow{\text{H.OH}} (CH_3)_3\overset{+}{N}.CH_2.CH_2.OH + CH_3CO_2H$$

Acetylcholine   Choline   Acetic acid

As is well known, this reaction is of great importance in neuromuscular physiology with regard to the transmission of impulses across the neuromuscular junction. The positive charge on acetylcholine is permanent and is unaffected by pH changes. Studies of the efficacy of binding of different substrates at various pH values led I. Wilson to conclude that an imidazole group was present in the active site; he proposed a structure for the active site in which there were two subsites, one anionic, which binds the substrate by electrostatic attraction, and one esteratic site or group, which splits the ester bond.

Acetylcholinesterase is reversibly inhibited by those compounds in which the molecular structure resembles the substrate, acetylcholine (Fig. 7-6). Presum-

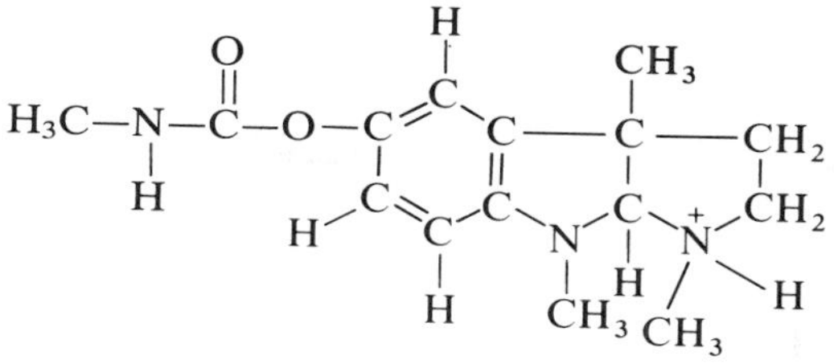

Physostigmine

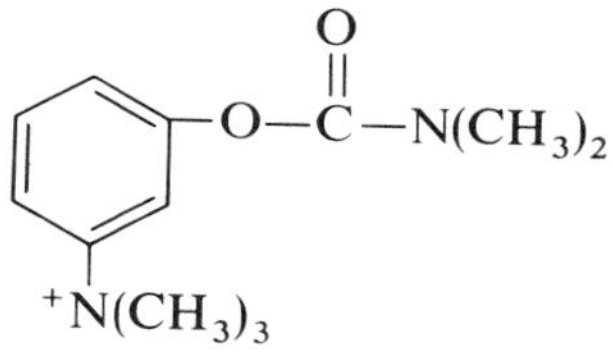

Neostigmine

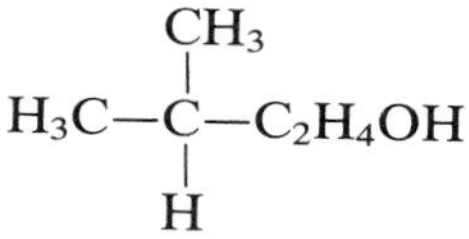

$$H_3C-\overset{\overset{\displaystyle CH_3}{|}}{\underset{\underset{\displaystyle H}{|}}{C}}-C_2H_4OH$$

Isoamyl alcohol

$$H_3C-\overset{\overset{\displaystyle CH_3}{|}}{\underset{\underset{\displaystyle H}{|}}{N}}-C_2H_4OH$$

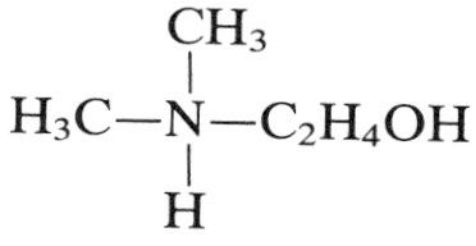

Dimethylamino ethanol

Figure 7-6 Competitive inhibitors of acetylcholinesterase.

ably these molecules bind to the active site and render it inaccessible to the natural substrate. The effects of several such inhibitors with rigid molecular structures and known dimensions were investigated by Friess and Baldridge in 1956. The most powerful inhibitors have a positively charged (cationic) group and an ester or hydroxyl group separated by a nonpolar (hydrophobic) region. If the cationic group or ester or hydroxyl groups are missing, the substance is a much less powerful inhibitor. Studies of the effects of these compounds on acetylcholinesterase led Friess and Baldridge to conclude that the anionic and

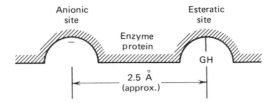

Figure 7-7   Conceptual model of the active site of acetylcholinesterase. (After I. Wilson.) ("G" originally meant glyoxaline, which is another name for imidazole.)

esteratic sites of acetylcholinesterase are separated by a distance of 2.5 Å. Figure 7-7 shows Wilson's conceptual model of the active site of acetylcholinesterase based on these data. Subsequently, it was found that the inhibitory power of the ammonium ion and the methylamines steadily increases with increasing methyl substitution. Also, 3,3-dimethylbutyl acetate, which has no cationic group, was found to be hydrolyzed by the enzyme as easily as acetylcholine, which does possess a cationic group. Hence, it was concluded that Van der Waals forces attributable to the methyl groups clustered around the cationic nitrogen of acetylcholine must be as important as the positive charge on the nitrogen atom in binding the substrate to the anionic site of the enzyme. A later, revised version of binding of acetylcholine to acetylcholinesterase with this refinement is shown in Fig. 7-8.

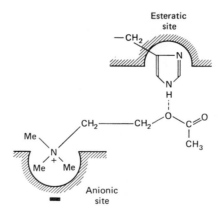

Figure 7-8   Binding of acetylcholine to the active site of acetylcholinesterase. (Redrawn from A. Albert.)

THE FLEXIBLE ACTIVE SITE CONCEPT

Insofar as the template, or lock and key, model envisions the active site as having a fixed, three-dimensional configuration, it is a "static" concept; although it explains many enzymatic reactions and inhibition phenomena quite well, it does not give a satisfactory explanation of everything, such as noncompetitive enzyme inhibition. One type is caused by the "blocking" of a catalytic group in the active site (e.g., by formation of a covalently bound metal atom, as when mercury ions combine with catalytic sulfhydryl groups).

$$\text{Enzyme-SH} + \text{Hg}^+ \qquad \text{Enzyme-S-Hg} + \text{H}^+$$
$$\text{(active)} \qquad\qquad\qquad \text{(inactive)}$$

Noncompetitive inhibition, unlike competitive inhibition, is irreversible and is not affected by the substrate concentration or reversed by increasing the substrate concentration. Now, another type of noncompetitive inhibition is known in which the inhibitor is known not to combine or bind with the active site or a group therein, but is bound to a part of the enzyme molecule remote from the active site. How, then, does such an inhibitor exert its effect? In 1952, J. Wyman and D. W. Allen proposed that protein molecules may have more than one stable, three-dimensional configuration, all with approximately similar energy. Changes from one conformation to another should therefore be relatively easy. To explain aspects of enzyme action that are not easily explainable in terms of the static template model of the active site, D. Koshland put forward the idea that the active site and, indeed, the whole enzyme molecule, probably is flexible; the approach and binding of the substrate was considered to induce the proper alignment of catalytic groups (i.e., the correct fit). Conformational changes of the shape of certain enzyme molecules have actually been detected when the substrate binds to them.

Various inhibitors may be bound, but prevent proper alignment of catalytic groups because of the geometry of the inhibitor molecule. The essence of this idea of "induced fit" is shown in simplified form in Fig. 7-9a and 7-9b. Such a model is readily able to explain the type of noncompetitive inhibition caused by binding of an inhibitor at a site remote from the active site, as shown in Fig. 7-9c, and many other phenomena not explainable in terms of a simple template model. The flexible, active site model and induced fit hypothesis are therefore dynamic concepts and represent an advance over the older static template concept of the active site of enzymes.

NATURE AND MECHANISMS OF ENZYMATIC CATALYSIS

The binding of the substrate to the active site of an enzyme appears to be mostly by short-range, noncovalent forces; the detachment of the reaction products is

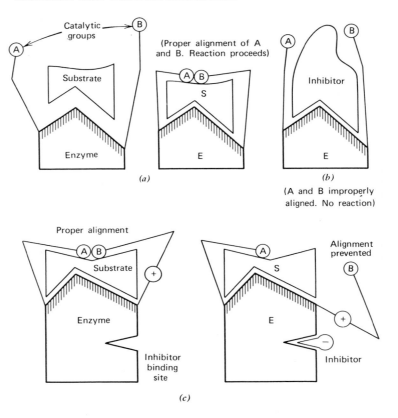

Figure 7-9 "Induced-fit" model of enzyme-substrate binding (schematic). (*a*) Binding of substrate. (*b*) Binding of competitive inhibitor. (*c*) Left, normal reaction; right, noncompetitive inhibition. (After Koshland.)

thereby made easy. If covalent forces bound substrate to the enzyme, release of the reaction products would presumably be more difficult, and the efficiency of the enzyme as a catalyst would be lower. However, transient, covalent, enzyme-substrate compounds have been shown to exist as intermediates in some enzyme-catalyzed reactions, as will be shown later.

Much experimental evidence indicates that the catalytic function is effected at least partly by electrically charged groups in critical spatial orientation within the active site. These groups are provided by the ionized side chain groups of certain amino acid residues of the enzyme protein. Because they are close together in the active site, it does not follow that the groups are close to one another in the primary structure. In fact, they are usually widely separated in the primary sequence, but are brought into close proximity in the active site by the intricate

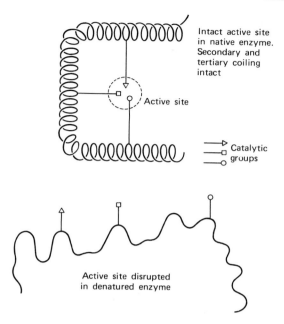

Intact active site
in native enzyme.
Secondary and
tertiary coiling
intact

Active site

Catalytic
groups

Active site disrupted
in denatured enzyme

Figure 7-10   Critical spatial alignment of catalytic groups in an enzyme active site and its disruption by denaturation (schematic).

coiling of the secondary and tertiary structures of the globular enzyme molecule. The precise positioning of catalytic groups is necessary for efficient catalysis. This spatial arrangement is destroyed when the enzyme protein is denatured and, therefore, the enzyme is inactivated (Fig. 7-10).

The process of enzymatic catalysis involves two main processes—the binding of the substrate and the action of the catalytic groups in speeding up the reaction. Actually, the overall effectiveness of enzymatic catalysis may result from the additive effect of several different phenomena, some of which may be of greater significance than others in individual cases. At least five effects have been recognized as occurring in enzymatic catalysis.

(i) Activation of substrate by "stress" and "strain" effects

The binding of the substrate to the active site of the enzyme "activates" the substrate (i.e., makes it more highly reactive). The stress could be brought about by the effect of critically placed charged groups in the active site, causing redistribution of electrons in the covalent bonds of the substrate. It could also

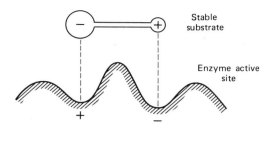

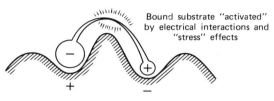

Figure 7-11   Activation of an enzyme-bound substrate molecule by "stress and strain" effects (schematic).

occur by the induction of mechanical stress in the covalent bonds of the substrate molecule (Fig. 7-11). Both of these effects would make the substrate much more reactive; hence it is said to be "activated." The activation energy hump of the reaction is lowered; so less activation energy is required, and the reaction proceeds more quickly.

A nonenzymatic example may help in understanding how strain within a molecule makes it more reactive. The compound dimethyl phosphate undergoes hydrolysis according to the following equation.

$$H_3C-O-\overset{\overset{\displaystyle O}{\|}}{\underset{\underset{\displaystyle O^-}{|}}{P}}-O-CH_3+-OH \longrightarrow H_3C-O-\overset{\overset{\displaystyle O}{\|}}{\underset{\underset{\displaystyle O^-}{|}}{P}}-O^-+CH_3OH$$

Dimethyl phosphate

Ethylene phosphate, which can be regarded as the cyclic analog of dimethyl phosphate, undergoes hydrolysis 10^8 times more rapidly.

Ethylene phosphate

This can be attributed to the strained covalent bonds in the five-membered ring of the latter compound; these bonds are much more reactive toward water than the unstrained bonds in the molecule of dimethyl phosphate.

(ii) Enhanced collision frequency and orientation effects

Enzymes are remarkably good catalysts, far better than any inorganic catalyst. What it is about enzymes that makes them so efficient in speeding up chemical reactions can be explained partly by reference to a well-known organic reaction and the effect of nonenzymatic catalysts on it. This example provides insight into the nature of enzymatic catalysis and its efficiency. The reaction chosen is the mutarotation of tetramethyl-β-D-glucose. In aqueous solution, glucose and its simple derivatives undergo this reaction; when pure tetramethyl-β-D-glucose is dissolved in water, the pyranose ring reversibly opens and closes by the breaking and remaking of the covalent bond from C_1 to the ring oxygen atom. The ring may close to regenerate a tetramethyl-β-D-glucose molecule, or it may close with the C_1 hydroxyl group in the other position (because of free rotation around the C_1—C_2 bond), thus giving tetramethyl-α-D-glucose. Ultimately, an equilibrium mixture of the α- and β-isomers is formed (Fig. 7-12). In the isomerization reaction, the ring oxygen accepts a proton, and the C_1 oxygen loses a proton. Attainment of equilibrium is hastened by adding phenol to the reaction mixture because it donates protons. Pyridine also accelerates the reaction because it is a proton acceptor. Addition of a mixture of both phenol and pyridine makes the reaction proceed faster than either phenol or pyridine used separately. (Phenol is such a weak acid and pyridine is such a weak base that no appreciable neutralization occurs when they are mixed). 2-Hydroxy pyridine, which combines both structures in one molecule, accelerates the reaction most of all, speeding it up to no less than 7000 times faster than a mixture of phenol and pyridine. The reason for this enormous acceleration is twofold. First, using 2-hydroxy pyridine means that only two molecules need collide in the reaction, whereas three are required to do so when phenol and pyridine are used separately. Statistically, the former is much more probable and therefore more frequent than the latter. Second, the two reactive groups in 2-hydroxy pyridine are just about the right distance apart for them to effect the proton transfers simultaneously and most efficiently. The catalysis of mutarotation of tetramethyl-β-D-glucose by 2-hydroxy pyridine provides a nonenzymatic model of enzymatic catalysis in that increased collision frequency, and several catalytic groups arranged in the correct complementary spatial arrangement with regard to groups on the substrate bring about the observed highly efficient catalytic effect (Fig. 7-12).

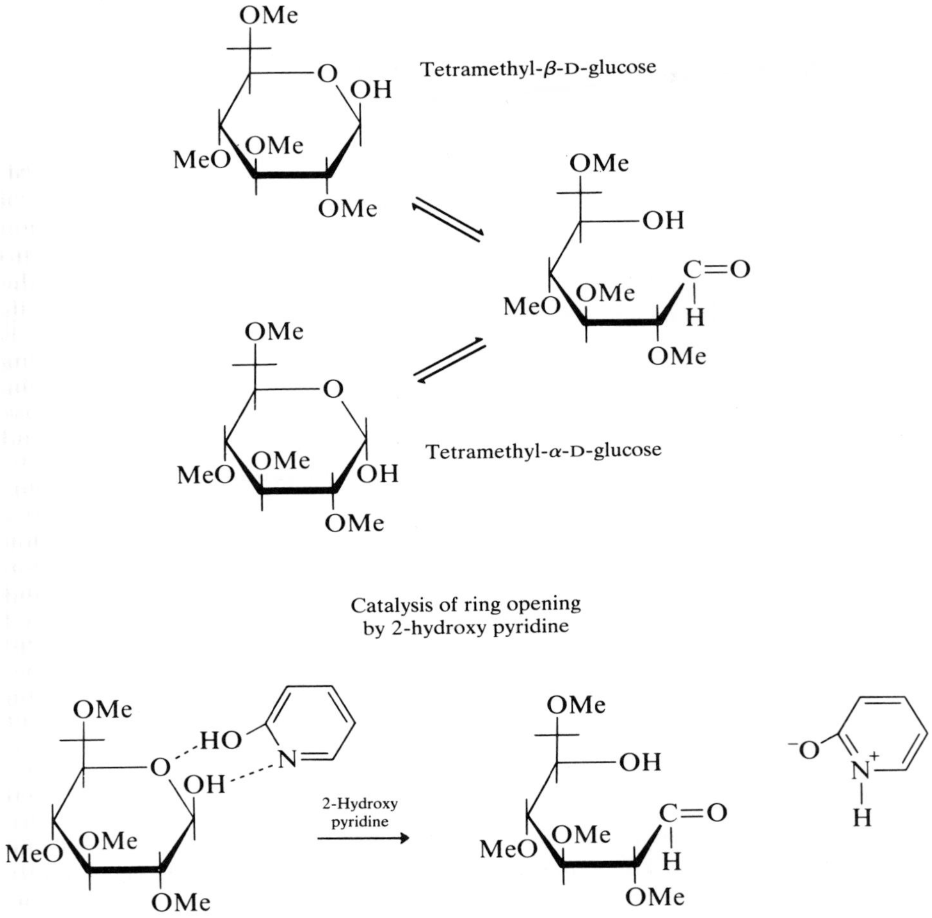

Tetramethyl-β-D-glucose

Tetramethyl-α-D-glucose

Catalysis of ring opening
by 2-hydroxy pyridine

2-Hydroxy
pyridine

Figure 7-12 Mutarotation of tetramethyl-β-D-glucose.

The analogy with enzymes is further illustrated by noting that the proton accepted by the nitrogen atom on the catalytic molecule is transferred to the doubly bonded oxygen atom, so that the original 2-hydroxy pyridine is regenerated. The catalytic molecule has therefore speeded up the reaction without appearing to have taken part in it and it is recoverable, unchanged, at the end of the reaction. The type of catalysis exemplified in this reaction and involving simultaneous proton addition and removal is known as concerted acid–base catalysis; it is a commonly encountered mechanism in enzyme-catalyzed hydrolysis.

(iii) Unusual reactivity of catalytic groups

Not only may catalytic groups be arranged in a critical pattern in the active site, but the catalytic groups may also be unusually reactive. Chymotrypsin, which is a digestive enzyme secreted by the pancreas, provides an example of both the importance of critically oriented catalytic groups and of unusually reactive catalytic groups in catalyzing a hydrolysis reaction.

Chymotrypsin catalyzes the hydrolysis of peptide bonds of protein molecules, especially those whose carbonyl groups are adjacent to an amino acid side chain containing an aromatic ring such as tyrosine or phenylalanine. The protein-digesting enzymes, as a group (peptidases), provide a good example of enzyme specificity. This type of bond specificity shown by chymotrypsin differs from that of another digestive enzyme, such as carboxypeptidase A, which acts preferentially on C-terminal peptide bonds adjacent to side chains containing six-membered carbon rings; trypsin acts only on peptide bonds, whose carbonyl groups come from an amino acid with a positively charged side chain.

The specificity of the digestive peptidases was elucidated by their ability to catalyze hydrolysis of a range of low-molecular weight, synthetic peptide substrates. The fact that these peptidases would act on such small peptide molecules also showed that the large size of the natural substrate molecules (proteins) played no part in the enzyme-substrate union or in the catalytic reaction and its specificity.

The susceptibility of peptide bonds to hydrolysis by various peptidases is therefore influenced by their immediate chemical environment. Chymotrypsin is usually described as a peptidase with a preference for peptide bonds, the carbonyl groups of which have been contributed by amino acids with an aromatic group in the side chain. However, it not only catalyzes hydrolysis of peptide bonds, but also of several types of amides and esters. It also catalyzes transfer of acyl groups to acceptors other than water, such as certain amino acids and alcohols. An aromatic ring is not necessary in the substrate; what seems necessary is that the ring should be hydrophobic (e.g., cyclohexyl can replace benzene). Chymotrypsin is thus more accurately described as a hydrophobic acyl group transferase. The active site contains a hydrophobic pocket, which binds the side chain, and a catalytic group which effects the hydrolysis reaction.

The active site of chymotrypsin contains a serine residue (number 195 in the primary sequence), which is involved in the catalytic process and whose side chain hydroxyl group is unusually reactive. It acts as a proton donor. Reactive serine side chains are also involved in the catalytic activity of trypsin and other hydrolytic enzymes; therefore these enzymes are all known as "serine enzymes."

A. K. Balls and E. F. Jansen showed that the catalytic activity of chymotrypsin is irreversibly inhibited by diisopropylfluorophosphate. This reagent combines selectively with the active serine 195 hydroxyl group to form a covalent derivative (Fig. 7-13). The diisopropylphospho enzyme is inactive because the catalytic

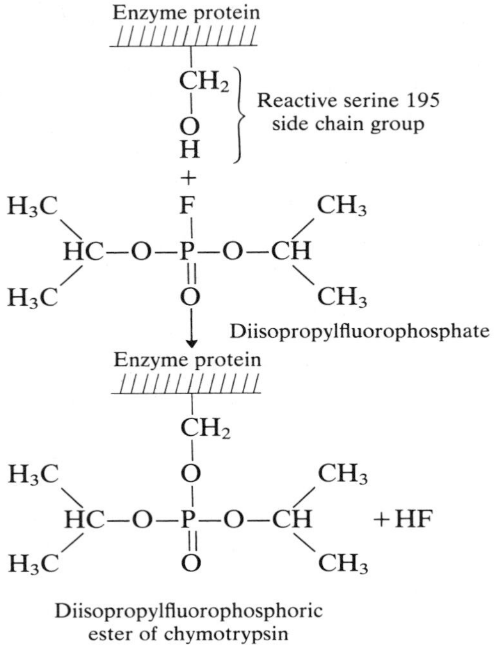

Enzyme protein
/////////////
|
CH_2 ⎫
| ⎬ Reactive serine 195
O   ⎪ side chain group
H   ⎭
+

H_3C F CH_3
    \ |
    HC—O—P—O—CH
    /   ‖   \
H_3C O CH_3

Diisopropylfluorophosphate

Enzyme protein
/////////////
|
CH_2
|
H_3C O CH_3
    \ | /
    HC—O—P—O—CH $+HF$
    /   ‖   \
H_3C O CH_3

Diisopropylfluorophosphoric
ester of chymotrypsin

Figure 7-13 Formation of diisopropylfluorophosphate derivative of chymo-trypsin.

serine group has been "blocked." The chymotrypsin molecule contains 28 serine residues, but it is only serine 195 that is blocked in this way. Evidence that this active serine residue is situated in the active site is provided by the fact that a competitive inhibitor of chymotrypsin, 3-phenyl propanoic acid, prevents the reaction of diisopropylfluorophosphate with chymotrypsin, presumably by blocking the active site and thus preventing access of the reagent to the reactive serine.

What is of interest is that not even this serine hydroxyl will react with diisopropylfluorophosphate if the enzyme is first denatured. Apparently, the immediate vicinity of the hydroxyl of serine 195 in the native enzyme, which is determined by the pattern of secondary and tertiary coiling of the enzyme polypeptide, somehow enhances the reactivity of the hydroxyl group, and dena-turation destroys this configuration so that the reactive serine becomes "ordinary." What is believed to be responsible for the reactivity of serine 195 is the close proximity in the active site of two histidine side chains (imidazole rings), which are also implicated in the catalytic reaction. The catalytic role of these imadazole rings was demonstrated by E. N. Shaw and co-workers, who exposed the enzyme to *p*-toluene-sulfonyl-phenylalanyl-chloromethylketone. This molecule resembles chymotrypsin substrates and reacts specifically and irreversibly with the active site histidines to inhibit the enzyme completely. The two imidazole groups function cooperatively with the active serine. There are other groups in the active site that are of significance in catalysis, but their roles are not yet clear.

(iv) Participation of the enzyme in the reaction by formation of transient covalent intermediates

One possible mechanism for the chymotrypsin-catalyzed hydrolysis of an ester envisions it as a concerted acid–base mechanism, one state of which involves formation of a transient covalent intermediate (Fig. 7-14). The first step is removal of a proton from the active serine residue by the imidazole group of a

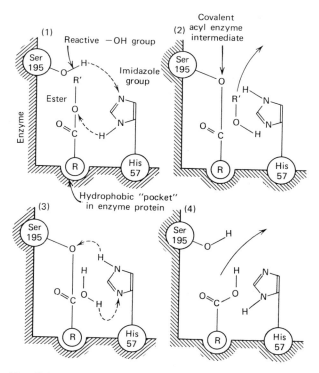

Figure 7-14   Possible mechanism of ester hydrolysis by chymotrypsin. (For descriptive simplicity one only of the two histidine residues involved in catalysis is shown.) (After M. Bender.)

histidine residue, which is a good reversible proton donor or acceptor. The resultant negative charge on the serine oxygen promotes nucleophilic attack on the carbonyl carbon atom of the peptide, amide, or ester substrate. Electron rearrangements result in liberation of an amine (in the case of peptide hydrolysis) and formation of a covalent acyl enzyme intermediate. The second step is deacylation of the enzyme, which is brought about by interaction of acyl-enzyme with an activated water molecule; this gives rise to a carboxylic acid and regenerates the original group in the active site.

(v) Hydrophobic environment effects at the active site

In the description of the structure of the myoglobin molecule in the previous chapter, it will be recalled that the heme group lies partly buried within a groove or cleft on the surface of the molecule, and that this is lined with hydrophobic groups. The interaction of the hydrophobic groups of the heme molecule with those in the cleft automatically arranges the heme group in the correct orientation with respect to the protein. Enzyme molecules often possess more or less well-defined depressions on some part of their surface, and active sites are found near or within these depressions in a hydrophobic environment. The reactivity of chemical groupings and the rate of chemical reactions in nonpolar media may differ markedly from what they are in aqueous media, because the dielectric constants of nonpolar media are very low compared with aqueous and other polar media. From elementary physics we know that the electric forces between two charged bodies is increased considerably in a nonpolar environment as compared with their magnitudes in a polar environment. When a substrate molecule becomes bound to the active site, it becomes immersed in a hydrophobic environment, and the forces that come into play between the substrate molecule and the catalytic groups will be much stronger than if the active site environment were polar. The hydrophobic environment effect may therefore be responsible for the high catalytic efficiency of chymotrypsin and other enzymes, which is enormously greater than any nonenzymatic analog.

Hydrophobic "pockets" may also be involved in the orientation of substrates on the active site of certain enzymes and hence can be regarded as determining certain types of enzyme specificity. For instance, as already noted, the peptidase chymotrypsin acts preferentially on peptide bonds, the carbonyl groups of which are contributed by amino acids with hydrophobic side chains.

M. Bender's model of the binding and hydrolysis of peptides at the active site of chymotrypsin (Fig. 7-14) envisages "anchoring" of the hydrophobic side chains of phenylalanine or tryptophane residues adjacent to the peptide bond on the —NH side of the substrate to a hydrophobic depression within or close to the active site. The orientation probably ensures optimal alignment of substrate with catalytic groups.

THE MORPHOLOGY OF ENZYME MOLECULES IN RELATION TO SUBSTRATE BINDING

R. E. Dickerson and I. Geis draw attention to interesting correlations between the molecular morphology of the hydrolytic enzymes described earlier, whose three-dimensional structures are known, and the types of substrate on which they act. In some, like lysozyme (Fig. 7-15), ribonuclease (Fig. 7-16) and papain, there is a "crevice" in the surface of the molecule (we have already noted that active sites are often located in or near these depressed areas in a hydrophobic environment). These enzymes act on substrates whose susceptible bonds are fairly well exposed, such as polyribonucleotides and polysaccharides. Dickerson and Geis liken the reaction to the cutting of a piece of wire ("substrate") in the jaws of cutting pliers. The enzymes that act on proteins with much secondary structure and hence with relatively inaccessible bonds possess a shallow depression on the surface of the molecule, which appears to be better adapted to "fitting" a larger structure than a simple extended chain. An example of such enzymes is chymotrypsin.

Carboxypeptidase, which acts on C-terminal peptide bonds of proteins, not surprisingly possesses an active site in the form of a pit in which the end of a polypeptide chain can fit.

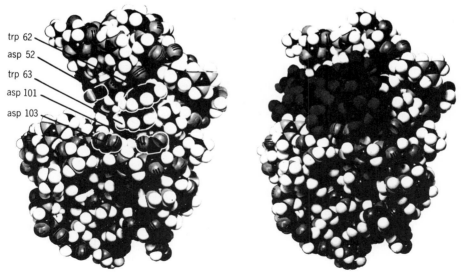

trp 62
asp 52
trp 63
asp 101
asp 103

Figure 7-15   Space-filling model of lysozyme. The left picture shows the active site crevice, and that on the right shows the substrate in the active site. (Reprinted from *The Structure and Action of Proteins* by R. E. Dickerson and I. Geis. W. A. Benjamin Inc., Menlo Park, California, Publisher. Copyright 1969 by Dickerson and Geis.)

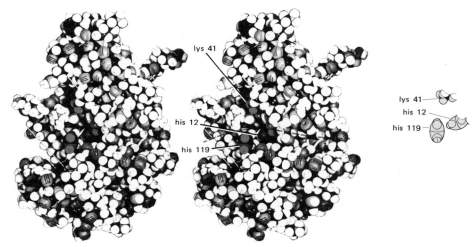

Figure 7-16   Space-filling model of ribonuclease "s" (stereoscopic pair). The catalytic groups are shown and are isolated for clarity in the drawing to the right of the stereo pair. (Reprinted from *The Structure and Action of Proteins* by R. E. Dickerson and I. Geis. W. A. Benjamin Inc., Menlo Park, California, Publisher. Copyright 1969 by Dickerson and Geis.)

ACTIVATION OF INACTIVE ENZYME PRECURSORS

Many enzymes are synthesized in the form of catalytically inactive precursors or zymogens that must be "activated" before the enzyme will function. A well-known example is the activation by enterokinase (enteropeptidase) of trypsinogen, the inactive precursor of the proteolytic digestive enzyme trypsin. Trypsin, once formed, is itself able to activate its own precursor.

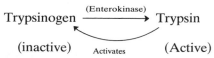

One may ask whether there is an active site already present within the precursor that is "unmasked" by activation or whether the active site is not present in the presursor but is brought into existence by activation. The following facts are known about trypsinogen, trypsin, and the activation process.

1.   Within the limits of accuracy of the experimental methods used in their determination, the molecular weights of trypsinogen and trypsin are practically identical (about 24,000). From this it may be concluded that activation does not involve extensive fragmentation of the single polypeptide chain of trypsinogen.

2. The rate of activation is correlated with hydrolysis of a single peptide bond, as shown by titration of liberated α-carboxyl groups.

3. Only one new peptide is found, as revealed by chromatography of the activation mixture after activation. It is a hexapeptide whose primary sequence is:

$$Val(Asp)_4Lys$$

4. The N-terminal primary sequence of trypsinogen is:

$$Val(Asp)_4Lys\ Ileu\ Val\ Gly\text{——}$$

5. The N-terminal residue of trypsin is Ileu.

These data indicate that the activation of trypsinogen is probably accomplished by hydrolysis of the peptide bond between lysine and isoleucine residues at the N-terminal end of trypsinogen, with release of the N-terminal hexapeptide of trypsinogen. In addition, physical measurements show that a pronounced decrease in the optical rotation of trypsinogen accompanies activation, indicating an increase in the content of α-helical secondary structure.

On the basis of these and other experimental data obtained on the nature and position of catalytic and other groups in trypsin, a tentative model of the process of activation of trypsinogen was put forward by Neurath and Dixon (Fig. 7-17). According to this model, activation generates the active site instead of unmasking a preexisting one. Presumably, the mutual repulsion of the four adjacent

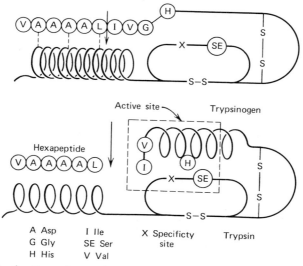

Figure 7-17 Activation of trypsinogen. (Redrawn after Neurath and Dixon.)

negative charges in the N-terminal stretch of trypsinogen provided by the 4 aspartic acid residues keeps this part of the polypeptide chain in the extended configuration. This extended region may be electrostatically attracted to another part of the molecule. Rupture of the Ileu-Lys bond in activation frees the rest of the N-terminal chain from this constraint and enables it to coil up—like a released spring—which brings about the configurational change that automatically generates the active site by bringing catalytic groups into their proper alignments.

Enzymatic removal of small peptides seems to be a common factor in other cases of activation of biologically inactive protein precursor molecules, not all of them enzymes. The molecule of the plasma protein fibrinogen, for instance, is activated by the splitting off of two small peptides by the action of the enzyme prothrombin. This gives the active protein profibrin, the molecules of which spontaneously polymerize to give insoluble fibrin fibers; this is the basis of normal blood clotting. The hormone insulin is secreted as the inactive pro-insulin, a single polypeptide chain with three intrachain disulfide bonds. Enzymatic removal of a small peptide converts the single polypeptide chain into the two separate A and B chains of active insulin that are joined together by two of these disulfide bridges.

In this context, it is interesting to note that the enzyme subtilisin catalyzes the splitting off of a 20-amino acid residue polypeptide from ribonuclease, which inactivates the latter enzyme. If the peptide is added, to the inactivated enzyme, full activity is restored. This observation led to a suggestion that has been made regarding the mechanism of action of the peptide hormones (oxytocin, vasopressin, ACTH) mentioned in an earlier chapter. It could be that these combine with inactive protein molecules in cell membranes, giving rise to biologically active proteins, possibly with enzymatic activity, that control permeability and other properties of cell membranes.

THE ROLE OF COFACTORS

The presence of cofactors is usually essential for enzymatic activity. Two broad classes of cofactors may be recognized. Some enzymes depend for their activity on the presence of the ions of certain metals such as cobalt, zinc, and manganese. These and other such metals all belong to the transition group of elements in the periodic classification and are characterized by having *d* orbitals with unoccupied quantum states. During the extraction and purification of certain enzymes, essential metal ions may be lost, and the enzyme has to be activated (note that the word "activation" is used here in a different sense from that in the last section) by the addition of the required metal ion. These types of cofactors that are only loosely bound to the enzyme protein are therefore called "activators" and are readily separated from the enzyme protein by dialysis. Relatively small organic molecules are found more or less closely bound to the protein of many

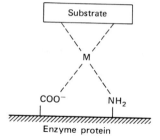

Figure 7-18 Linking of a substrate molecule to enzyme protein by a metal ion.

types of enzymes and are the second type of cofactor, usually referred to as coenzymes. When the organic molecule is strongly bound, whether the protein is an enzyme or not, it is called a prosthetic group. This type of cofactor cannot be separated from its enzyme protein by dialysis. An example from a nonenzymatic protein is the heme group of myoglobin.

(i) Metal ion catalysis

Metal ions may act as links between enzyme and substrate molecules through coordinate bonds (Fig. 7-18). That complexes are formed between metal ions bound to the enzymes and their substrates has been demonstrated by spectroscopic methods and was shown to occur by D. Keilin with the enzyme catalase, and by B. Chance with peroxidase. Certain peptidases require cobalt ions for full activity. They may function by linking an enzyme and a substrate and also facilitating the reaction by forming a strain-free, five-membered ring with the carbonyl oxygen atom and imino nitrogen atoms of peptide bonds, both of which function as electron donors (Fig. 7-19). Thus, electron withdrawal by the cobalt ion acting as an electron "sink" renders the peptide bond more susceptible to hydrolytic attack.

Formation of strain-free, five-membered rings between the substrate and metal ion may also be involved in enzymatic decarboxylation reactions, as in the case of oxaloacetate, which gives pyruvate and carbon dioxide on decarboxylation (Fig. 7-20).

Another example of metal ion participation in enzymatic catalysis is the case of the enzyme carbonic anhydrase, which catalyzes the reaction between carbon dioxide and water.

$$OH^- + O=C=O \longrightarrow O=C\begin{smallmatrix} \diagup OH \\ \diagdown O^- \end{smallmatrix}$$

Carbon dioxide   Carbonic acid

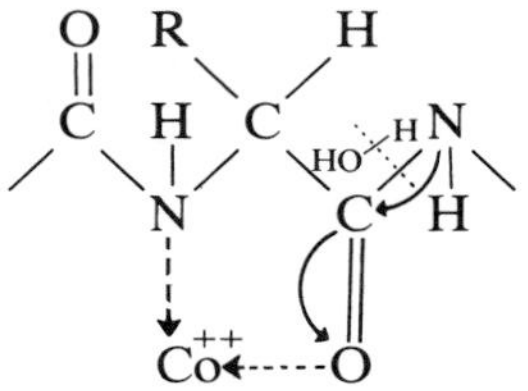

Figure 7-19   Cobalt ions in peptidase hydrolysis. (Curved arrows indicate electron transfers.)

Carbonic anhydrase contains zinc. At neutral pH, zinc can form a compound containing both carbonate and hydroxyl ions.

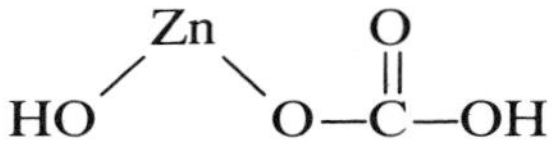

The zinc ion in the enzyme could form coordinate bonds with the substrate oxygen atoms, and this would assist the electronic movements necessary for the reaction to occur (Fig. 7-21).

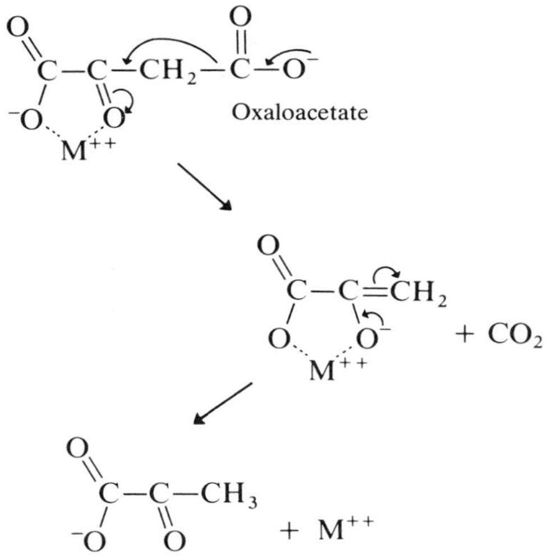

Figure 7-20   Involvement of metal ion in decarboxylation of oxaloacetic acid. (Curved arrows indicate electron transfers.)

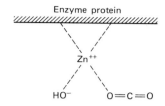

Figure 7-21   Function of the zinc ion in carbonic anhydrase.

(ii) Coenzyme catalysis (pyridoxal)

More complex examples of the importance of electron arrangements in render-ing the substrate more reactive are provided by coenzyme catalysis; a good example of this is pyridoxal-5-phosphate and its role in the many metabolic reactions that involve amino acids. The structure of pyridoxal-5-phosphate is shown in Fig. 7-22. The primary event in pyridoxal catalysis is reaction of the aldehyde group of the pyridine ring with the amino group of the amino acid substrates to give an aldimine ("Schiff base"), the molecule of which contains conjugated double bonds and hence especially mobile π electrons. Several tautomeric forms of the aldimine are therefore possible. Depending on the nature of the amino acid, the mobile electrons will be subjected to various kinds of influences, and the amino acid itself is rendered very reactive when bonded to pyridoxal in this way. There are many different types of reaction that can occur subsequent to the binding of the amino acid to the pyridoxal; they are sum-marized in Fig. 7-23, which shows a complex, generalized amino acid bound to pyridoxal. Metal ions (M^+) may also be involved in forming planar rings that facilitate electron movements. Racemization, decarboxylation, transamination, and dehydration can all occur as a result of activation of an amino acid by reaction with pyridoxal. What, then, determines which one of these possibilities will occur in a particular case?

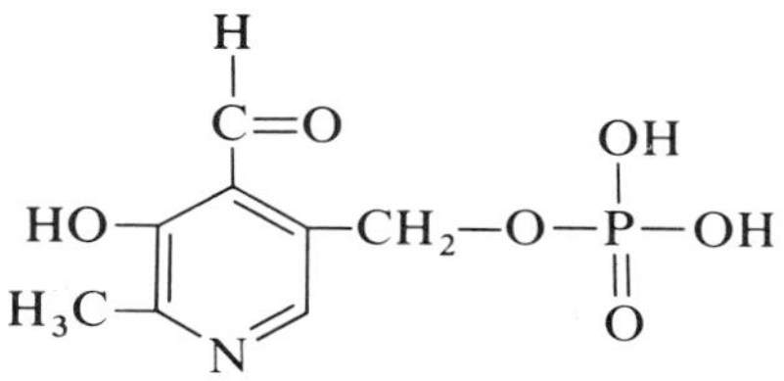

Figure 7-22   Structure of pyridoxal-5-phosphate.

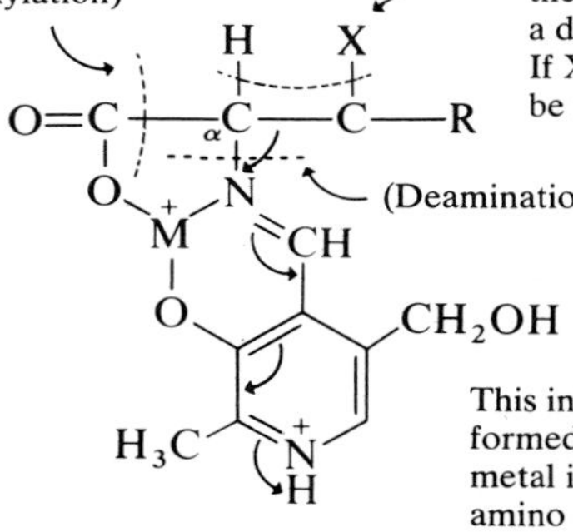

(Decarboxylation)

(If X = —OH as in serine, then this reaction would be a dehydration. If X = —SH, then H_2S would be removed)

$$O=C \overset{|}{\underset{O^-}{}} \overset{H \quad X}{\underset{\alpha}{C} — C — R}$$

(Deamination, transamination)

$$\overset{+}{M} \quad CH$$

CH_2OH

H_3C N⁺   H

This intermediate is formed from pyridoxal, metal ion, and an amino acid

Figure 7-23 Possible pathways in pyridoxal catalysis. Electron displacement toward the $\overset{+}{N}$ atom weakens bonds around the α-carbon atom. The metal ion $(\overset{+}{M})$ probably assists electron removal and maintains a planar structure, which facilitates electron transfers in the conjugated system.

THE ROLE OF THE ENZYME PROTEIN

In this (pyridoxal-5-phosphate) and other examples of substrate activation, including bonding with metal ions, there are two or more possible routes that the subsequent reaction can take. The importance of the enzyme protein is *its specific capacity to select and direct one out of several possible reactions and to facilitate it while, at the same time, other possibilities are not selected or facilitated.* Several specific transaminase and decarboxylase enzymes utilize pyridoxal-5-phosphate as coenzyme, but it is their specific protein components that promote one particular reaction—transamination or decarboxylation—with its substrate to the exclusion of others.

The enzyme protein not only exerts a highly specific directive effect on reaction possibilities, but its presence is responsible for the enormous efficiency of enzyme catalysis. This is well shown by comparison of the catalytic effects of the specific enzyme catalase and of nonenzymic catalysts on the decomposition of hydrogen peroxide. Hydrogen peroxide spontaneously decomposes very slowly, according to the following equation.

$$2H_2O_2 \longrightarrow 2H_2O + O_2$$

J. H. Wang showed that the reaction is weakly catalyzed by ferric ions and that the catalytic effect is enhanced when the iron is in organic combination, as in the heme group. The catalytic effect of heme is enormously enhanced when the heme happens to be in the prosthetic group attached to the specific protein of the enzyme catalase. The mechanism by which the protein component of enzymes brings about enormous acceleration of the catalytic effects inherent in their coenzyme is not understood. The influence of specific enzyme proteins on the catalytic properties of coenzymes is a special instance of the general principle that specific proteins can modify the properties, catalytic or otherwise, of small, nonprotein molecules when bound to them. Heme by itself is a dull, brownish compound that contains an atom of ferrous iron in every molecule. It is insoluble in water and combines with oxygen to give the ferric compound hematin. When it is combined with the right proteins, the result is bright red, water-soluble pigments (myoglobin and hemoglobin) that combine easily and reversibly with oxygen without the iron undergoing a change of valency. When bound to other types of proteins, the reversible, electron-carrying proteins known as the cytochromes are the result. Yet another protein combined with heme gives cytochrome oxidase, an enzyme that specifically catalyzes the direct transfer of electrons to molecular oxygen from one of the cytochromes as the terminal reaction of aerobic respiration; and we have just encountered catalase, a heme enzyme with yet another type of specificity.

Heme, like pyridoxal and various other cofactors, appears to possess various latent properties that are selectively "brought out" and intensified by combination with certain specific proteins.

Enzymes in "Teams"

Enzymes rarely function alone in the living cell; they almost always act in association with other enzymes. Cofactors play an important role in connecting enzymes together in teams. In the enzymatic dehydrogenation of succinate described earlier, the hydrogen removed from succinate was shown as being "free." Actually, in this and other cellular dehydrogenations, the hydrogen is passed on to a "carrier" molecule that, in its turn, may pass the hydrogen on to yet a third carrier under the influence of a second enzyme. The two reactions may be written:

$$1. \quad AH_2 + C \xrightarrow{E_1} A + CH_2$$

$$2. \quad CH_2 + B \xrightarrow{E_2} BH_2 + C$$

These equations can be written in another way that emphasizes the reversible, hydrogen-carrying function of C.

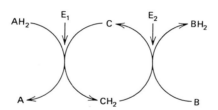

The intermediate carrier C is, in fact, functioning as a coenzyme that links together the two enzymes E_1 and E_2 into a simple enzyme system that transfers hydrogen from A to its final acceptor B. When several enzymes are linked together in this way by coenzymes, complicated multienzyme systems, or teams, are the result.

Enzymes and cofactors may be organized into solid-state assemblies, as in mitochondria, where energy released by the aerobic breakdown of carbon compounds is released and trapped as high-energy phosphate by enzymes and cofactors keyed into the inner mitochondrial membrane. Enzymes and cofactors may also be simply dissolved in the cytoplasm.

In the living cell, the activity of enzymes and enzyme systems is controlled and regulated in accordance with metabolic needs. This important topic will be dealt with in the next chapter under the section about the control of metabolic reactions.

CHAPTER EIGHT
Antibodies (immunoglobulins)

The animal body possesses several means by which it defends itself from invasion by foreign parasites such as bacteria and viruses and from nonliving substances that are alien to the body. The defenses are:

1. *Cellular.* The phagocytes of the circulating blood that ingest and destroy bacteria and other solid foreign particles.
2. *Humoral.* There are two kinds. The first is the group of substances collectively known as *complement*; this group promotes the dissolution or lysis of bacteria. The second is the group of plasma proteins known as *antibodies*, which are formed in response to foreign organisms or substances when they gain access to the body.

The antibodies are a highly specific defense mechanism, and the remainder of this chapter is devoted to their structure, function, and origin.

NATURE AND GENERAL PROPERTIES OF ANTIBODIES

The antibodies are a group of closely related soluble proteins associated with the γ-globulin fraction of the blood serum proteins. Antibody molecules are relatively compact in shape. Although they may therefore be classed as globular proteins, they differ significantly in shape and detailed molecular structure from enzymes, myoglobin, hemoglobin, and other typical globular proteins, as will be apparent later.

Antibodies are synthesized by cells of the reticulo-endothelial system, such as the spleen, the liver, and the lymphatic glands, and are formed in response to "challenge" by a foreign substance, usually macromolecular, such as proteins, polysaccharides, or other biopolymers. This is called the immune response, and antibodies are therefore often called immunoglobulins. Antibodies combine in a highly specific way with the macromolecule that elicited them, and so they are classed along with proteins like myoglobin and hemoglobin as emphores (i.e., noncatalytic proteins with the capacity to bind specifically with other molecules).

Substances capable of eliciting antibody formation are called antigens, and an antibody may be defined as a specific protective protein formed in response to an antigen. (There is, of course, a fallacy in defining one in terms of the other in a circular manner, but it emphasizes the close relationship of antibody to antigen.)

Strangely, only vertebrates exhibit the immune response that would therefore appear to be a relatively late evolutionary development. As well as being a valuable defensive response, it is a nuisance in human surgical procedures involving skin and organ grafting. The recipient's body "recognizes" the graft as being foreign ("not-self") and the immune response to it results in the graft refusing to take, a phenomenon called rejection. However, the recipient immune response can be suppressed by various means so that grafts have a better chance of being accepted; the patient is very susceptible to infection when the immune response is suppressed; special care must be taken to guard against infections. Sometimes, the normally valuable immune response can be a nuisance because it "overreacts" to a trivial antigenic stimulus, such as pollen grains, and the uncomfortable symptoms of hay fever arise. Under certain pathological conditions, antibodies to the person's own tissues may be formed; certain so-called autoimmune diseases are thought to be the result.

An antibody molecule combines specifically and most strongly with the antigen that elicited it. If the antigen was a native protein, the resultant antibody will combine with the native protein but not with the denatured protein in which the specific three-dimensional structure of the native protein has been destroyed. The resultant antigen-antibody complex is usually but not always insoluble, and therefore precipitates out of solution. This is called the precipitin reaction. Under *in vivo* conditions, a foreign soluble protein or polysaccharide will therefore be rendered insoluble and can be dealt with by the phagocytes, which ingest insoluble material but not soluble substances foreign to the body. The *in vitro* precipitin reaction is widely employed in the study of antigen-antibody reactions and is the basis of the science of immunology. Antibody production may be demonstrated by injecting, say, a rabbit with a foreign protein such as horse hemoglobin. After a few days, the rabbit's tissues and blood serum will contain high levels of antibody that combine specifically with horse hemoglobin to give a precipitin reaction. This is easily demonstrated *in vitro*. The antihorse hemoglobin antibody is absolutely specific in that it will not combine with and give a precipitin reaction with any other horse protein (i.e., with nonhomologous proteins, even from the same individual). It is very interesting to note that the antibody will combine with the homologous protein— hemoglobin—of other mammalian species to a greater or lesser extent, depending on the degree of relatedness of the species to the horse. The strongest precipitin reactions are given, in this instance, with the hemoglobins from the animal species that are phylogenetically closest to the horse. The antigen-antibody reaction therefore mirrors phylogenetic relationships between animal species. This correlates well with what we will see later (Chapter Nine) regarding hemoglobins from different species; although they are of similar physiological function, the hemoglobins all differ slightly in their primary structures, and the further apart two species are in the phylogenetic sense, the more their hemoglobins differ in their primary structures.

The two most striking characteristics of antibodies are their specificity and the astonishing versatility of the antibody response to the enormous variety of possible antigens, both natural and artificial, that may gain access to the animal body either by accident or by deliberate administration. Generally, the more foreign the antigenic substance is, the stronger and more rapid is the antibody response. The rest of this chapter will be concerned with the molecular basis of the nature and origin of antibody specificity and the mechanism of the antibody response.

THE ANTIGEN-ANTIBODY REACTION AND ANTIBODY BINDING SITES

Evidence indicates that an antigen molecule such as a protein is polyvalent (i.e., it possesses many binding sites that combine with the corresponding antibody). Since most antigen-antibody complexes are insoluble, it would seem likely that antibodies function by cross-linking antigen molecules to give complexes of indefinitely large molecular size that precipitate out of solution. If antibodies were univalent (i.e., if they possessed only one binding site), cross-linking would not be possible, although small, soluble antigen-antibody complexes would still be formed (Fig. 8.1*a*). The minimum number of antibody binding sites that would enable them to cross-link antigen molecules is two. If we imagine the rodlike antibody molecule as now having a binding site at each end, it is easy to see that cross-linking is possible (Fig. 8.1*b*).

Assuming that antibody molecules have two binding sites and that antigens have more than two, we can predict the behavior of antibodies and antigens when mixed in different ratios. If antibody is in excess, complexes of the type shown in Fig. 8.2*a* should predominate. Likewise, with antigen in excess,

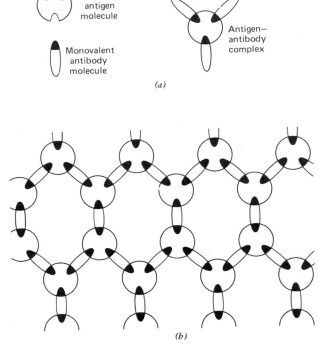

Figure 8-1 (*a*) Reaction between a monovalent antibody and a polyvalent antigen. (*b*) Reaction between a bivalent antibody and a polyvalent antigen.

complexes of the type shown in Fig. 8.2*b* should mostly be found. Somewhere in between the two extremes there will be an optimal ratio of antibody to antigen at which maximal cross-linking occurs to give a complex lattice, and a visible precipitate is therefore formed (Fig. 8.2*c*). If, in fact, gradually increasing amounts of antigen are added to a series of test tubes containing the same amount of an antibody, maximal precipitation in one of the tubes will be seen, and little or no free antigen or free antibody will be recoverable from the supernatant. Precipitation will be less in tubes on either side and least in the extreme tubes. From the supernatant in the tubes with less than optimal amounts of antigen, free antibody can be recovered. From that in the tubes with greater than optimal amounts of antigen, free antigen is recoverable. This quantitative precipitin reaction is summarized in Fig. 8.3.

Antibodies are, in fact, bivalent. In 1952, S. Singer and D. Campbell showed that if mixtures of bovine serum albumin (antigen) and the antibody to this (prepared from rabbits) were mixed in various ratios and then examined in the

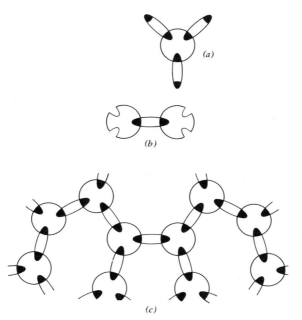

Figure 8-2 Types of complexes formed from antibodies and antigens. (*a*) Antibody in excess. (*b*) Antigen in excess. (*c*) Cross-linking of antigen by antibody when both are present in optimal amounts.

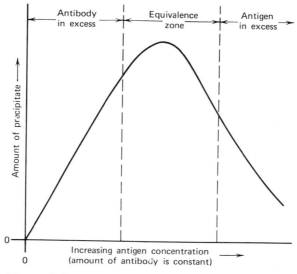

Figure 8-3 The quantitative precipitin reaction.

ultracentrifuge, the relative amounts of free antigen, free antibody, and the possible types of antigen-antibody complexes formed in these different mixtures can be explained and analyzed in terms of a reversible condensation reaction between a multivalent (antigen) and a bivalent (antibody) molecular species.

Physicochemical studies have shown that antibody molecules are asymmetrical (i.e., they differ significantly in shape from a sphere). This provisional model of an antibody molecule as being rodlike or elliptical with antigen binding sites at each end is therefore probably a fairly good first approximation to the truth.

ANTIBODY SPECIFICITY AND HAPTENS

Since the science of immunology is much older than our modern knowledge of the detailed structure of protein molecules, not much progress could be made in the study of antibody specificity and the nature of the fit between antigen (e.g., a protein), and antibody prior to the availability of this knowledge. The study of antibody specificity was greatly advanced in 1917; K. Landsteiner discovered that simple organic chemical groups of known structure were capable of eliciting antibodies specific to them if they were chemically combined with a protein molecule. Such nonprotein organic groups are called haptens. When an animal is inoculated with a protein to which a hapten has been chemically attached, the resulting antibodies are specific for the hapten as well as to other antigenic groups of the protein molecule. If the hapten is attached to a protein that is homologous to the inoculated animal's plasma proteins (i.e., its own or the plasma protein from an animal of the same species), the antibodies produced will be specific for the hapten only. The study of such hapten-specific antibodies has been immensely valuable in advancing knowledge of antibody specificity and the nature of antigen-antibody binding. In practice, haptens are usually benzene derivatives; among these, amino-sulfonic and amino-benzoic acids have been employed. The amino groups of such compounds can be diazotized and then coupled with the protein presumably by attachment to the side chain rings of tyrosine and histidine residues (Fig. 8.4). Studies with hapten groups of varying degrees of structural difference have shown that (as we saw earlier with the reaction of rabbit antihorse hemoglobin antibodies and hemoglobins of related species) antibodies do not exhibit absolute specificity and that cross reactions can occur (i.e., the antibody elicited by one hapten will react more or less strongly with structurally similar haptens, depending on the degree of similarity to the original hapten). This led to the formulation by L. Pauling of the lock and key model of antigen-antibody binding. The antigenic groups are pictured as fitting into cavities of complementary shape at the antibody binding site, where they are held by a variety of short-range, noncovalent forces. The union of antigen with antibody is therefore similar in many respects to the binding of an enzyme with its substrate.

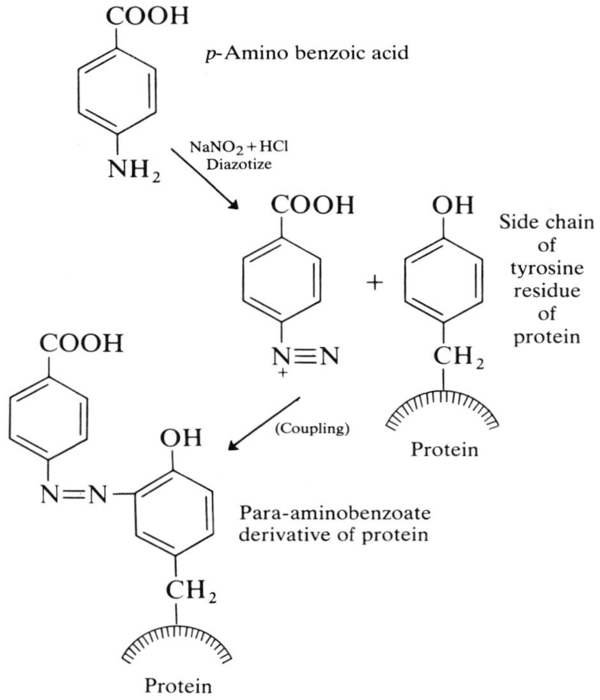

Figure 8-4 Chemical coupling of a hapten to an antigen.

THE STRUCTURE OF IMMUNOGLOBULIN GAMMA

There are many different types of human immunoglobulins, but they all have the same basic molecular structure. The most abundant immunoglobulin is known as immunoglobulin gamma (IgG). Previous chemical studies had shown that the IgG molecule consists of two identical "heavy" polypeptide chains and two identical "light" polypeptide chains held together by disulfide bonds. R. Porter showed that the molecule is split into smaller fragments by proteolytic enzymes; papain splits it into three fragments, two of which are identical and contain one antigen binding site each, designated F_{ab} ("fragment antigen binding'). Thus, the bivalent function of the antibody molecule is accounted for. The other fragment plays no part in antigen binding and crystallizes readily; hence it is called F_c ("fragment crystalline").

Porter further showed that pepsin splits the immunoglobulin molecule into an F_c fragment and a bivalent antigen binding fragment, designated $(F_{ab}^1)_2$, which consists of the two F_{ab} fragments united by a disulfide link. Since so few of the

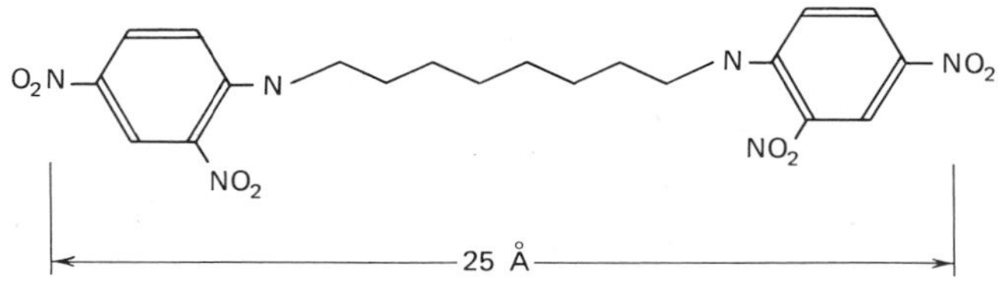

Figure 8-5   A bivalent hapten.

peptide bonds of the immunoglobulin molecule are susceptible to enzymatic attack, it would appear that those cleaved in these enzyme treatments are situated in exposed flexible regions of the polypeptide chains; the remainder of the polypeptide chains are tightly and compactly coiled so that the peptide bonds are inaccessible.

Antibody-hapten complexes have enabled the actual shape of antibody molecules to be visualized in the electron microscope. M. Green and R. Valentine carried out experiments in which a homologous series of bivalent hapten molecules of increasing molecular weight and carrying two dinitrophenyl groups were caused to react with the antibodies specific to them. The smallest of these that was capable of cross-linking two or more antibody molecules was a chain of eight carbon atoms with a dinitrophenyl group at each end (Fig. 8-5). When free antibodies are examined in the electron microscope, nothing definite can be made out but clear images were observed when the antibody-hapten complex obtained with the above dinitrophenyl compound and the corresponding specific antibodies was examined (Fig. 8-6). Structures of triangular and polygonal shapes were observed with "knobs" at the corners. These were interpreted as consisting of groups of three or more antibody molecules cross-linked by the hapten molecules; the hapten molecules were too small to be visible in the electron microscope (Fig. 8–7). When antibodies were treated with pepsin and then caused to react with the hapten, similar geometrical structures were seen in the electron microscope, but without the knobs at the corners. The knobs would therefore appear to be the F_c moiety of the antibody molecule; the remainder that is still capable of combining with antigen would be the $(F_{ab}^1)_2$ structure. All this information is summarized in the structure of the immunoglobulin gamma molecule as shown in Fig. 8-8, which also shows the primary structure peculiar to immunoglobulins of two light and two heavy polypeptide chains.

THE PRIMARY STRUCTURE OF ANTIBODY MOLECULES

The amino acid sequences of antibodies are of considerable interest because in these sequences would appear to lie the molecular basis of their enormously variable specificity. The determination of primary structure of a protein must, of

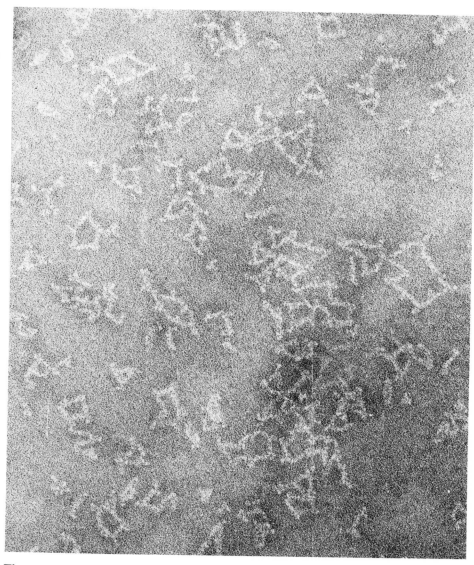

Figure 8-6 Electron micrograph of an antibody-hapten complex. (Courtesy of Dr N. M. Green.)

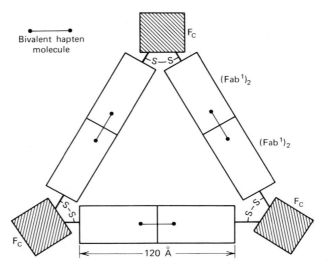

Figure 8-7 Linking of antibody molecules by a bivalent hapten. (Redrawn after R. R. Porter.)

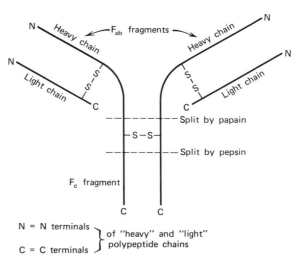

Figure 8-8 Structure of the gamma immunoglobulin molecule.

course, be performed on a chemically pure sample of the protein un-contaminated with any other protein. Such determinations in the case of antibodies is complicated by the fact that immunoglobulin gamma itself is made up of many different types of antibody molecules. In an *immunologically* pure sample of an antibody (i.e., one in which all the antibody molecules have precisely the same antigen binding specificity), the antibody molecules are therefore not all *chemically* identical. The molecular composition is heterogeneous and the antibody molecules, all with identical antigen binding specificity, differ slightly in their amino acid composition, so slightly that it is virtually impossible to separate them by present methods. Attempts to correlate amino acid sequence with antibody specificity would therefore seem to be hopeless.

Fortunately this barrier has been largely overcome; in the malignant disease called myelomatosis, which is a cancer of immunoglobulin-synthesizing cells (plasma cells), there is excessive production of one type of immunoglobulin whose molecules are chemically identical. This apparently happens because a solitary cell that synthesizes one type of immunoglobulin becomes malignant and proliferates so that large amounts of this one homogeneous immunoglobulin are produced. These proteins are known as myeloma proteins. In about 50% of myelomatosis patients, large quantities of an abnormal protein are excreted in the urine, and the presence of this protein is practically diagnostic of the disease. The abnormal proteins in the urine of myeloma patients were first described by H. Bence-Jones in 1847, and they became known as Bence-Jones proteins. Chemically, they are the homogeneous light polypeptide chains of the myeloma proteins, as shown in 1962 by G. Edelman and J. A. Gally. At about the same time, F. W. Putnam independently made the same discovery.

The interesting and important thing is that a given patient excretes only one type of light chain, and this differs from the homogeneous light chains excreted by other myelomatosis patients, all of which differ from each other. Strictly speaking, the myeloma proteins are not antibodies in the sense that they have been induced selectively by the presence of an antigen. However, they have the same structure as normal immunoglobulin, and each one has the appearance of being an individual member of the heterogeneous population of normal immunoglobulins.

Myeloma proteins are obtainable in quantity from the urine of myelomatosis patients; primary sequence determinations have been performed on these purified immunoglobulin light chains. All myeloma proteins were found to consist of a linear sequence of about 214 amino acid residues. From studies of human and animal (mouse) myeloma proteins, the remarkable fact emerges that in all light chains of the same type, the amino acid sequences of their C-terminal halves (from about position 190 on) are all practically identical, whereas the other (N-terminal) halves (residues 1 to 108) all show considerable variation in amino acid sequence. A similar situation has been found in the case of various heavy chains. The light and heavy chains thus have "constant" (C) and "variable" (V) regions. This unusual feature of constant and variable regions in two polypeptide chains making up a protein molecule is a feature unique to immunoglobulins.

G. Edelman, M. J. Waxdal, and W. H. Konigsberg set out to determine the amino acid sequence of a whole immunoglobulin molecule; the source of material was blood plasma from a myelomatosis patient. The type of globulin they studied was the most abundant, immunoglobulin gamma. Cyanogen bromide was employed to split selectively the polypeptide chains at sites where the amino acid methionine occurred. Methionine residues are relatively sparse in the immunoglobulin molecule, and this treatment yielded two of each of 10 different peptide fragments (suggesting that the intact molecule consists of two identical halves). These were separated and purified, and each peptide was broken into smaller portions by selective enzymatic hydrolysis with two different proteolytic enzymes. The primary sequences of the two different resulting series of peptides were than determined by Sanger's method. By using the method of overlapping sequences (Chapter Three), the biochemical "jigsaw puzzle" was put together, and by determining the positions of the disulfide bonds occurring in various parts of the molecule, the complete primary structure was elucidated.

As noted earlier, the immunoglobulin molecule was found to consist of two identical light and two identical heavy chains; the entire molecule has bilateral symmetry, as shown in Fig. 8-9. The chains are held together in their three-

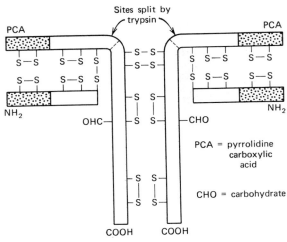

Figure 8-9 More detailed structure of the gamma imunoglobulin molecule. Shading indicates variable regions of light and heavy chains. (Adapted from *The Structure and Function of Antibodies* by Gerald M. Edelman, copyright 1970 by Scientific American, Inc.)

dimensional configuration by disulfide bonds and by weak, noncovalent inter-actions between the chains. The existence of variable and constant regions in the two polypeptide chains is the most extraordinary feature of the molecule; it is now definitely known that the amino acid sequence differences in the variable regions is the basis of the different configurations of antigen binding sites in antibodies of different specificities.

The arrangement of the intrachain disulfide bonds, which looks a little repeti-tive, is actually a reflection of a definite structural repetitiveness or periodicity within the C regions. There are three such regions or domains within the C region of the heavy chain designated C_H1, C_H2, and C_H3 in which the amino acid sequences are so similar that the three domains must be considered homologous. They are also homologous with the C region of the light chain (C_L). The V regions in the light and heavy chains are also homologous. The heavy chain is therefore described as having four domains; three are constant (C_H1, C_H2, and C_H3) and the fourth (V_H) is variable. The light chain possesses two domains: one is constant (C_L) and one (V_L) is variable (see Fig. 8-10).

Apparently, the V domains are concerned with specificity and antigen binding and the C domains are concerned with other immune, response-related functions not connected with antigen binding. The C_H2 domain exemplifies such a function because it is here that complement binds, initiating a chain of reac-tions that can cause lysis of bacterial or other cells. (It now seems that antibodies are concerned with specificity and "recognition" of invading microorganisms, whereas the complement system, which is a complex of several different pro-teins, is concerned with the destruction or lysis of the invader.)

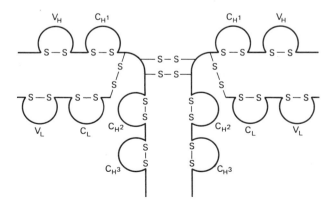

Figure 8-10   Domains of the gamma immunoglobulin molecule. (Redrawn and modified after D. R. Davies, E. A. Padlan, and D. M. Segal.)

THE CLASSES OF HUMAN IMMUNOGLOBULINS

The human immunoglobulins are a complex heterogeneous group of proteins. All immunoglobulins, as already noted, have a structure consisting of two heavy and two light polypeptide chains.

There are five main classes of immunoglobulins, three major ones designated IgG, IgA, and IgM and two minor ones designated IgD and IgE. These are distinguishable on the basis of general properties such as molecular weight, amino acid composition, carbohydrate content, and the classes of their heavy chains. The five different types of heavy chains are designated γ, α, μ, δ, and ε, respectively. In any one of these five classes of heavy chains a number of subclasses are found. For example, within the IgG immunoglobulins, there are four immunologically distinguishable subclasses reflected in differences of amino acid composition of the C terminals of the heavy chains. However, among the five immunoglobulin classes, there are only two types of short chains; these are designated κ and λ. Within the constant regions of the light or heavy chains of any one subclass, minor variations in amino acid sequences may be found that are genetically determined and are known therefore as allotypic variants or allotypes. It is important to realize that these structural differences between the constant regions of the five subclasses and their allotypic variants are not in any way related to their specificity.

The following is a brief summary of the structures and characteristics of the five main classes of immunoglobulins.

IgG.   These are by far the most abundant of the five immunoglobulin classes. They represent about 80 to 85% of the total population of antibodies. The chain structure is $\gamma_2\lambda_2$ or $\gamma_2\kappa_2$, and the molecular weight is about 150,000. The heavy chains each consist of about 420 to 440 amino acid residues; the light chains each consist of about 210 to 230 residues, depending on the source of antibody. Carbohydrate is present to the extent of 2.9%. Most of the immune capacity is referable to this antibody class.

IgA.   The chain structure is $\alpha_2\lambda_2$ or $\alpha_2\kappa_2$ and often occurs in the dimeric forms $(\alpha_2\lambda_2)_2$ and $(\alpha_2\kappa_2)_2$. The monomers are joined in end-to-end fashion by the F_c fragments. It comprises about 10% of the total serum antibodies. It has a molecular weight of about 180,000 to 500,000, and the carbohydrate content is 7.5%. It is the most abundant antibody in tears, nasal secretions, saliva, milk, and colostrum; it may be significant in conferring immunity when ingested in milk by offspring of certain mammals in which the maternal γG immunoglobulins cannot enter the fetal bloodstream through the placenta. It is one of the few proteins known to survive digestion in the alimentary canal.

IgM.   The chain structure is $(\mu_2\lambda_2)_5$ or $(\mu_2\kappa_2)_5$, the molecular weight is 950,000, and the carbohydrate content is 11.8%. This class of antibody is usually the first to appear in the bloodstream after immunization, especially if the antigen is of large dimensions. Bacteria are particularly good at eliciting IgM. The appearance of IgM in the blood following immunization is usually followed by IgG and then IgA. It occurs as a pentamer in which five basic units consisting of two short and two long chains are united in a ring formation by interchain disulfide bonds. In this immunoglobulin and also in IgA, a third polypeptide known as "J" has been identified in the polymeric forms of the immunoglobulins but not in the monomers. It is so designated because it is thought to play a role in joining the monomers together in the polymer. Because of the large size of the IgM immunoglobulins, they are often called macroglobulins. They occur to the extent of about 5 to 10% of the total serum antibodies.

IgD. The structure is $\delta_2\lambda_2$ or $\delta_2\kappa_2$ and it comprises about 1% of total serum antibodies.

IgE. The structure is $\varepsilon_3\lambda_2$ or $\varepsilon_2\kappa_2$. The molecular weight is 196,000, and the carbohydrate content is 10.7%. Less than about 0.01% of serum immunoglobulin consists of this class. It is thought to be involved in some allergic reactions.

MOLECULAR BASIS OF THE ORIGIN AND DIVERSITY OF ANTIBODY SPECIFICITY

One entire light chain and part of one heavy chain are contained in each of the antigen binding parts of the antibody molecule. To what extent is each chain involved in the binding sites? Present indications are that the heavy chain plays the greater role and that the light chain is involved but is less important. Oddly, the α-helical content of antibody molecules is practically negligible. The structure of the compact regions of the molecule must therefore be of uncoiled polypeptide chains folded into a complex tertiary configuration. In spite of the wide range of antigen binding specificity exhibited by antibodies, no gross structural differences in the antigen binding regions of different antibodies that might account for this have ever been detected. What differences there are must therefore be relatively subtle variations in the pattern of tertiary folding within the antigen binding sites. These variations in configuration should presumably be correlated with the differences in amino acid composition and sequence variation in the V regions of the polypeptide chains.

Since those parts of the antibody molecule concerned with antigen binding and specificity (i.e., the F_{ab} fragments) are composed of the V regions of light and heavy chains, it seems certain that the differences in amino acid sequence and composition in these regions are the molecular basis of antigen binding site specificity.

An examination of the characteristics of the variability of the V regions reveals that variations in amino acid sequence are caused by mutation (as it is with other proteins), because the variations are associated with changes of single bases in the DNA base triplets that code for individual amino acids in each variable position (see Chapter Seventeen). Some amino acid sites in the V region are invariant, others are variable, and yet others are "hypervariable"; variation must therefore involve some kind of selective process. Thus, as with variation in other proteins, the variability among antibodies is attributable to mutation and selection.

There are three main theories of the origin of antibody diversity and specificity.

1. The observed variation of the different V regions in a given animal is a product of evolution, and there is a different V gene for each V region. Admittedly, this would seem to require a large amount of genetic material, but this would be a small price for an animal to pay for the obvious advantage of possessing an efficient and versatile immune system.

2. There is one V gene or very few V genes that mutate very rapidly during an animal's development, and then the mutant cells that produce a given antibody are somehow selected and proliferate when challenged with an antigen.

3. There are a few V genes that have arisen during evolution and have undergone mutation. Somatic recombination can occur among these, and this is what results in the general pattern of variation. Subsequently, an antigen selects and stimulates the cells that have the potential for producing the appropriate antibody; they proliferate while others are not so stimulated. This theory is favored by many authorities.

The attractive features of somatic recombination theories of antibody diversity are avoidance of *ad hoc* hypotheses of hypermutability, and selection of antibody genes in somatic time, and the requirement of large numbers of genes. Evolutionary time takes the place of these; the diversity of antibodies in an animal has therefore resulted from intrachromosomal recombination of relatively few genes in somatic cells over millions of years.

Several mechanisms of the somatic recombination process have been suggested. One of these views the gene for, say, a light chain as having a recognition point at the midpoint of its structure. A specific enzyme that can split the gene nucleic acid only on one side of the recognition point may attach itself here. This fragmented position is repaired under the influence of other enzymes, but the fragments become jumbled, and the original base sequence is altered; the result is many different sequences.

From his studies of several human and mouse myeloma proteins, in 1967, O. Smithies proposed a model of recombination that involves few assumptions. It accounted for the pattern of amino acid variations, linkage groups, and recombinants as found in the observed amino acid sequence data. He suggests that a given antibody polypeptide is determined by a gene pair. He calls one gene a master gene and the other a scrambler gene; the scrambler gene is similar to but not identical with the master gene. One assumption is that during evolution, the pair of genes have come to differ in a way that provides for the type of variability needed by the particular animal species. The idea is best illustrated by reference to a specific hypothetical example (Fig. 8-11). This shows a light chain master gene and its scrambler gene. The scrambler gene is an inverted copy of the master gene; it is identical with the master gene at 101 sites, but differs at six sites. These six sites in the master gene are represented by the letters ABCDEF and in the scrambler gene by PQRSTU. Figure 8-11a shows the original chromosome, Fig. 8-11b shows the intrachromosomal synaptic process, and Fig. 8-11c shows the "reshuffled" chromosome. Notice that the recombinant antibody gene differs from the original only in the variable region. If there were 20 sites at which the master and scrambler genes differed, then more than 1 million (2^{20}) recombinants would be possible.

In another model relatively few V genes are pictured as arranged consecutively on a chromosome; at a distance on the same chromosome are C genes. Segments of two adjacent V genes could be removed and become attached to and fused with a C gene (translocation). If the DNA of the V gene segments was removed in the form of a ring, this process would alter the sequence of bases and give rise to variation (Fig. 8-12). In addition to accounting for variability, this model also shows how, by the same process, V and C genes may become joined; this must occur for synthesis of the complete polypeptide.

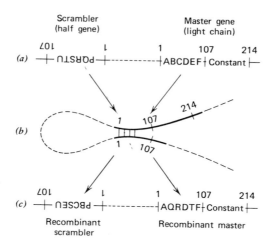

Figure 8-11 O. Smithies's theory of intrachromosomal recombination. (Redrawn after O. Smithies.)

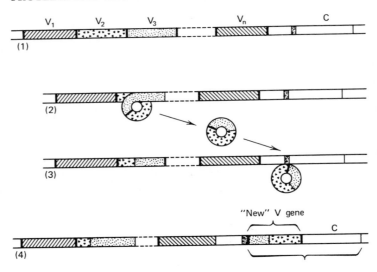

Figure 8-12 Translocation hypothesis of antibody gene variability. (1) Chromosome with "V" genes and a "C" gene. (2, 3) Removal of adjacent segments of two "V" genes as a ring and their translocation. (4) Fusion of the new "V" gene with the "C" gene. (Redrawn after G. E. Edelman.)

THE THREE-DIMENSIONAL CONFIGURATION OF ANTIBODIES

The application of X-ray crystallographic techniques to the study of four different types of immunoglobulins obtained from two different animal species has enabled their three-dimensional structure to be worked out. It appears from this that antibodies may all have a closely similar pattern of folding, since the three-dimensional structures of these four antibodies were found to be closely similar.

Chemically homogeneous immunoglobulins do not form crystals good enough for X-ray studies but fragments obtained by enzymatic digestion, such as the F_{ab} fragment of the immunoglobulin crystallize satisfactorily. The structures of F_{ab} fragments from two different sources have been investigated at a resolution of 2 to 3 Å by two groups of workers. The F_{ab} fragment from a human myeloma protein was studied by R. Poljak and L. M. Amzel. The other group, composed of D. R. Davies, E. A. Padlan, and D. M. Segal, studied the F_{ab} fragment from the antibody produced by a mouse plasma cell tumor. An F_{ab} fragment consists of four domains, two (C_L and V_L) in the complete light chain component and two (C_H1, V_H) in the half heavy chain component. The F_{ab} fragments studied at a resolution of 6 Å by Poljak and Davies were found to measure $40 \times 50 \times 80$ Å;

they consisted of two roughly spherical regions of similar size. One of these comprises the two V domains, and the other comprises the two C domains. The four domains are disposed in an approximately tetrahedral manner. More detailed studies revealed that the four domains are all cylindrical and that most of the polypeptide chain in each domain has β-pleated sheet secondary structure. Two layers of pleated sheet in each of the domains are folded so that the layers are parallel and closely apposed; one layer consists of four segments of polypeptide chain, and the other layer consists of three segments. The two layers are held together by a disulfide bridge. A molecular model of immunoglobulin gamma is shown on Fig. 8-13.

We have already noted that the variable regions of the light and heavy chains are involved in the antigen binding sites. In the variable regions of different

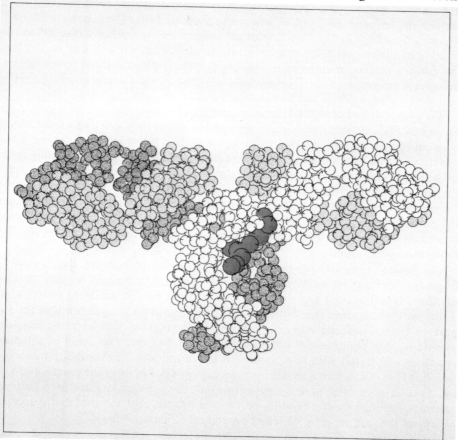

Figure 8-13 Molecular model of the immunoglobulin gamma molecule. (Courtesy of Dr D. R. Davies.)

antibodies, some sites exhibited much greater amino acid variability than others. These seem to be concentrated at sites 20, 50, and 90 residues from the N terminal, as was found by E. Kabat and Tai Te Wu, who designated these sites the hypervariable regions. The studies of the three-dimensional structure of antibodies have now revealed that the folding pattern of the variable regions of the heavy and light chains orients the hypervariable sites so as to produce a sizable antigen binding site. The manner in which the two domains in each of the V regions associate gives rise to a continuous surface whose hypervariable three-dimensional configuration can be changed by insertions, deletions, and substitutions of various amino acids in the parts of the V regions of the light and heavy chains that are hypervariable. All of an animal's immunologic potentialities are probably provided by these hypervariable surfaces in the complete immunoglobulin supply. The hypervariable amino acid sites are mostly situated outside the parts of the domains where the β-pleated sheet structures are found. We noted earlier that the binding of an antigen by an antibody resembled the binding of a substrate by an enzyme. It is therefore highly interesting to find that the antigen binding regions of antibodies possess β-pleated sheet structures (so do several enzymes, as noted in an earlier chapter), whereas it is absent from the molecules of myoglobin and hemoglobin.

We also noted that while free immunoglobulin molecules do not give clear images in the electron microscope, they do so when they are cross-linked by a bivalent hapten to give small-sized complexes. It is possible that the reason for this may be that antibodies undergo a configurational change when binding to an antigen; possibly they change from a relatively loose form to a more compactly folded or condensed form when they do so. Evidence for such a configurational change has been obtained by I. Z. Steinberg and J. Schlessinger, who found that the circular polarization of fluorescence of antibodies changed when they were bound to an antigen. The effect was not observed, however, when smaller molecules were bound.

The flexible, hingelike region of the antibody molecule between the F_{ab} and F_c fragments permits movement of these parts of the molecule relative to one another. It is believed that the antibody without bound antigen is somewhat T-shaped, but assumes the Y-shaped configuration when antigen binds to the F_{ab} portion by a swivelling movement at the hinge region. This exposes the previously hidden complement binding site (Fig. 8-14), and the chain of immune reactions associated with complement is set in motion. The antibody molecule thus behaves rather like a switch and can turn a series of immune reactions on or off. One is reminded of the configurational changes undergone by allosteric enzymes induced by the binding of various ligands and the metabolic control processes that are thereby switched on and off (Chapter Nine).

The building blocks of immunoglobulin molecules are the domains, but X-ray studies make it plain that the fundamental structural units are pairs of homologous domains that show strong interactions. The immunoglobulin molecule may

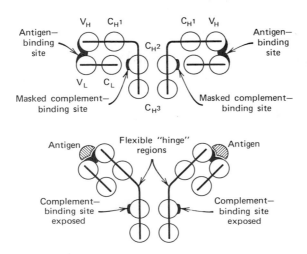

Figure 8-14   Shape change in an antibody molecule when antigen is bound. (Adapted from *The Structure and Function of Antibodies* by Gerald M. Edelman, copyright 1970 by Scientific American, Inc.)

therefore be regarded as a set of structural units that are loosely connected; there are two units in each F_{ab} fragment and two or three units in the F_c fragment.

THEORIES OF ANTIBODY ORIGIN AND SYNTHESIS

(i) The instructional or template theory

At about the time that scientists were beginning to understand that the structure of proteins consists of long, flexible chains of amino acid residues that could be folded into various configurations, the first of the two best-known theories of antibody synthesis and specificity was put forward. Around 1930, several publications appeared of theories of antibody synthesis in terms of the flexible chain concept of protein molecules and the three-dimensional characteristics of the antigen binding site. The idea put forward was that antibody molecules are synthesized in close contact with antigen molecules (the latter therefore serving as templates), so that the final folding pattern of the antibody molecule was complementary to that of the antigen. Thus, antibody specificity is determined by external information, or by "instructions" supplied to the cell (Fig. 8-15). In 1940, L. Pauling postulated that both ends of the antibody polypeptide chain could become folded into an infinite number of three-dimensional patterns that were complementary to and specific for any antigen. The presumption was that

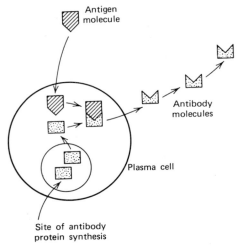

Figure 8-15   Instructional theory of antibody synthesis.

the antibody molecule was bivalent (subsequently experimentally proven) and the specificity of the antigen-antibody combination was a result of the antigen's direction of the folding pattern and did not stem from differences of amino acid sequences.

The instructional or template theory is now known to be wrong; one of the main objections to it is that antibodies do have differing amino acid sequences and that these sequences are correlated with specificity.

(ii) The clone selection theory

This was put forward by M. Burnet in the late 1950s. The theory states that stem cells, which are the precursors of the antibody-producing plasma cells, contain all the information needed to synthesize antibodies against any antigen with which it may be challenged, even without ever coming into contact with an antigen. The antigen merely selects and stimulates proliferation of the plasma cell (or cells) that is capable of producing the appropriate antibody (Fig. 8-16). Scientists believe that there are many different kinds of plasma cells and that each one can synthesize one or at most a very few different specific antibodies. All of the cells synthesizing a given antibody in an animal that has received an antigenic stimulus are believed to be a clone (i.e., the lineal descendents of a single plasma cell that contained the genetic information needed to synthesize that antibody).

This production of one specific antibody, or at most a very few, by plasma cells in response to an antigenic stimulus suggests that the activity of the genes

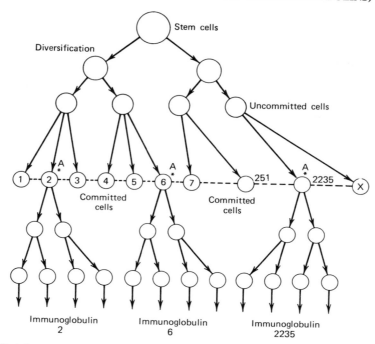

Figure 8-16 Clone selection theory of antibody synthesis. A committed cell produces one or very few antibodies. The antigen "A" may be "recognized" by more than one of the committed cells, and they are stimulated to proliferate clones that produce antibodies to the antigen "A." (Redrawn and modified after G. E. Edelman.)

controlling production of antibodies is mutually exclusive (i.e., the activity of one particular gene in a plasma cell precludes the activity of all the others). This observation is strongly reminiscent of the sequential production of non-α-polypeptides of the various human hemoglobins (see Chapter Nine). As production of the ε chain declines, the synthesis of the γ chain begins and, when this begins to decline shortly before birth, there is a simultaneous starting up of synthesis of β chains. The activity of the genes governing biosynthesis of ε-, γ-, and β-polypeptides therefore also appears to be mutually exclusive.

D. Kabat considers that the selective activation and deactivation of antibody and human non-α-hemoglobin polypeptide synthesis may involve similar mechanisms. (The reader may wish to skip the next paragraph or two if unfamiliar with the mechanism of genetic control of protein synthesis; refer to Chapter Eighteen.) The genes governing these two processes are known to be "clustered" but, whereas closely linked genes in bacterial operons are simultaneously activated, it appears that closely linked genes in mammals are expressed in a manner that is mutually exclusive. There may therefore be a cause and effect relationship between the expression of one gene and the suppression of another.

In 1972, Kabat described a mechanism of mutually exclusive gene selection and activation involving a chromosomal "looping-out excision," which is a process occurring within a single chromosome. Briefly, the cistrons for the ε-, γ-, and $\beta\delta$-polypeptides are adjacent to one another on the chromosome, in that order, and each cistron has its own operator locus (o) at one end and a chain terminator (t) at the other end (Fig. 8-17), where transcription is terminated. There is one promoter locus (p) at which the RNA polymerase (transcriptase) enzyme attaches itself. Only the ε cistron (i.e., the promoter proximal cistron) can be transcribed, because its terminator locus stops further transcription and there is only the one attachment site for RNA polymerase. Occasionally, looping-out excision occurs between the two homologous operators nearest to the promoter, so that the promoter-proximal cistron becomes excised as an acentric ring; since it lacks a promoter locus, it cannot be transcribed. Erythropoietic stem cells are, of course, diploid, and looping-out excision takes place independently on the two relevant homologous chromosomes. Kabat points out the analogy between this and the process occurring in the *E. coli* chromosome in which lysogenic bacteriophage DNA is excised.

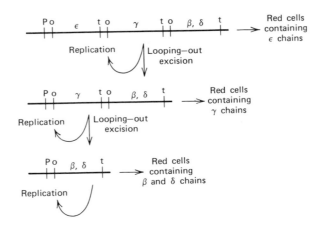

Figure 8-17 Looping-out excision theory of hemoglobin non-α-polypeptide gene activation. (There is no information regarding linkage of the ε gene, which is tentatively included in this model.) (Redrawn and simplified from "Gene Selection in Hemoglobin and in Antibody-Synthesizing Cells", Kabat, D., *Science* Vol. 175, pp. 134–140, Fig. 4, 14 January 1972. Copyright 1972 by the American Association for the Advancement of Science.)

Evidence for this model is provided by certain genetic phenomena in maize and in *Drosophila*. In addition to explaining previously unexplained findings about haematopoiesis, the evidence is also in agreement with findings regarding immunoglobulin synthesis; it could explain the "splicing" of the linked genes for the constant and variable regions of antibodies, but it could also explain the activation of the linked, variable region genes, which is mutually exclusive.

How an antigen "recognizes" a plasma cell that will synthesize the specific antibody and how the antigen stimulates proliferation of the cell are still not known.

It does, indeed, seem remarkable that plasma cells have such an astonishing capacity to provide specific antibodies against any antigen, even if these are artificial molecules totally unknown under natural conditions. However, this would seem to be no more remarkable than the fact that many enzymes catalyze reactions involving artificial substrates. In both cases, the specificity (of antibodies and most enzymes) is known to be not absolute; provided that the fit of one (enzyme or antibody to substrate or antigen) to the other is sufficiently close, reaction can occur; perhaps it will not be as strong in some cases with an artificial substrate or antigen as with the natural substrate or antigen.

CHAPTER NINE

Globular protein assemblies. Quaternary structure

We have already seen that fibrous protein molecules interact to give rise to biologically functional assemblies; examples are the triple helical molecules of collagen and the fibers that develop from them, the hydrogen-bonded assemblies of fully extended polypeptide chains in silk, and the fibrous assemblies of protein molecules in the vertebrate sarcomere. Similarly, many different types of native globular proteins consist of not just one, but of two or more globular molecules—subunits, as they are called—held together by noncovalent forces, thus giving specific functional assemblies. Such protein molecules are said to possess quaternary structure. The quaternary structure of a given protein is defined as the number, type, and spatial configuration of the separate globular subunits that make up the functional native protein molecule. A consideration of the properties of globular proteins possessing quaternary structure is again illustrative of the relation of structure to function.

Protein molecules possessing quaternary structure characteristically exhibit at least two types of novel phenomena.

1. Possession by the quaternary protein of biological properties that are not inherent in the separate subunits.
2. The ability to exert control over metabolic reactions at the molecular level.

The first type is well illustrated by vertebrate hemoglobins and other oxygen-binding proteins; the second is illustrated by multiple molecular forms of enzymes (isoenzymes) and their function in the control of metabolic reactions.

Emergent new properties and metabolic control potential often occur simultaneously.

THE HEMOGLOBIN MOLECULE

The three-dimensional structure of the molecule of horse hemoglobin was first elucidated by M. Perutz and his co-workers using the X-ray diffraction technique at a resolution of 5.5 Å; it was published in *Nature* in 1960. This work was done at about the same time that J. Kendrew was carrying out the 2 Å analysis of the whale myoglobin molecule. The hemoglobin molecule consists of four tetrahedrally arranged globular protein subunits loosely held together by noncovalent forces (Figs. 9-1 and 9-2). The four subunits, each of which is a single polypeptide chain, are of two types, designated α and β. The subunit composition (quaternary structure) of hemoglobin is therefore represented as $\alpha_2\beta_2$. In both horses and humans, the α subunit consists of 141 amino acid residues and the β of 146 residues. Both types of subunit closely resemble the myoglobin molecule in general form; the differences from myoglobin are relatively small. Each subunit possesses one heme group. The entire hemoglobin molecule is roughly spherical and measures $64 \times 55 \times 50$ Å; the combined molecular weight of the four subunits is about 64,500.

Myoglobin functions as an oxygen-storing molecule, and hemoglobin is an oxygen transporter; a comparison of the oxygen dissociation curves of the two is interesting (Fig. 9-3). Each hemoglobin molecule combines with four molecules

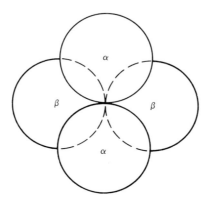

Figure 9-1 The hemoglobin molecule. A schematic diagram showing the tetrahedral arrangement of the two α and two β subunits.

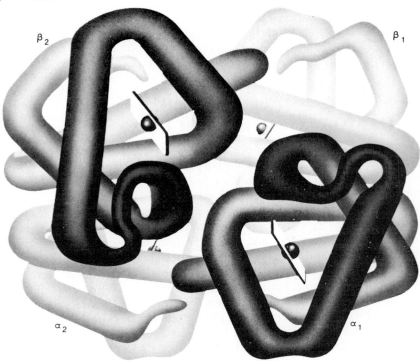

Figure 9-2   Molecule of oxyhemoglobin. The tertiary coiling of the α and β subunits is shown. The cylindrical segments are the α-helical regions. The heme groups are represented by the four planes and the iron atoms by the small spheres at their centers. (Reprinted from *The Structure and Action of Proteins* by R. E. Dickerson and I. Geis. W. A. Benjamin Inc., Menlo Park, California, Publisher. Copyright 1969 by Dickerson and Geis.)

of oxygen; the dissociation curve, instead of being hyperbolic, is sigmoid (S-shaped). At low oxygen tensions, hemoglobin combines with oxygen more sluggishly than myoglobin does (i.e., its affinity for oxygen is relatively low). Once an oxygen molecule has attached itself to a hemoglobin molecule, it appears that the affinity of hemoglobin for oxygen increases rapidly; this is progressive as more oxygen molecules are taken up by the hemoglobin. There is therefore a threshold effect; unlike myoglobin, hemoglobin does not combine readily with oxygen until a certain minimal oxygen tension is exceeded. The curve also shows that hemoglobin needs a higher oxygen tension for a given degree of saturation with oxygen as compared to myoglobin, which is another way of saying that as the oxygen tension is lowered, hemoglobin liberates its oxygen more readily (especially at very low oxygen tensions) than myoglobin. These properties are, of course, just what are required of an oxygen transporter:

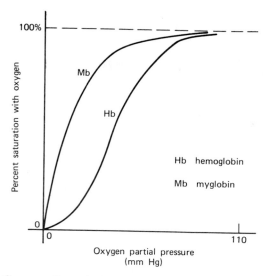

Figure 9-3   Oxygen dissociation curves of hemoglobin and myoglobin.

that it should be completely saturated with oxygen at atmospheric pressure (in the lungs) and should readily give it up where oxygen tension is low (i.e., in the tissues where it is needed). Its behavior in these respects therefore contrasts with that of myoglobin. This pattern of oxygen binding, as found in hemoglobin, is directly referable to possession of quaternary structure.

COOPERATIVE OXYGEN BINDING

The hemoglobin molecule contains four structural subunits, as we have seen, but it behaves as though it consists of two functional subunits, each of these being an $\alpha\beta$ pair. Each structural subunit attaches one oxygen molecule; therefore, a total of four oxygen molecules may be bound to a single hemoglobin molecule. One account of the binding of oxygen to hemoglobin envisages configurational changes of the structural subunits. According to this idea, the binding of the first oxygen molecule to a deoxyhemoglobin molecule is relatively difficult but, once an oxygen molecule binds to, say, the α subunit of an $\alpha\beta$ pair, this causes a configurational change in the α subunit that is transmitted to the β subunit of the pair (Fig. 9-4). The configurational shape in the β subunit somehow facilitates binding of the second oxygen molecule to itself so that the binding of the first oxygen molecule assists the binding of the second. The configurational change undergone by this $\alpha\beta$ subunit induces a similar configurational change in the other $\alpha\beta$ subunit; this increases the affinity of this second functional subunit for the two other oxygen molecules needed to saturate the hemoglobin molecule. These last two bind most readily of all; the fourth binds hundreds of times more

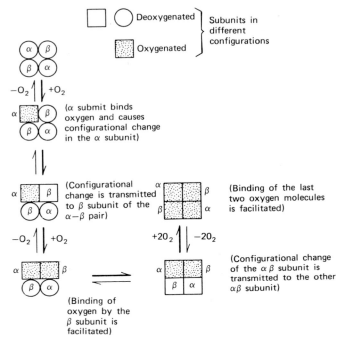

Figure 9-4   Cooperative oxygen binding by hemoglobin. A possible model. The descriptions are to be read in the direction of the heavier arrows.

readily than the first. The reverse occurs when the oxygen tension is lowered; two oxygen molecules are fairly readily released from each oxyhemoglobin molecule, the third more easily, and the fourth most easily.

This account of oxygen binding by the hemoglobin molecule seems difficult to reconcile with the fact that the secondary and tertiary structures of the individual α and β subunits do not appear to undergo detectable change when oxygen is bound. However, there is a change in the positions of the subunits relative to one another when oxygen is reversibly bound, as shown by M. Perutz in 1967 (Fig. 9-5); therefore, the hemoglobin molecule *as a whole* undergoes a configurational change. As one author picturesquely puts it, the molecule "breathes." The individual α and β subunits do not, therefore, function independently, but cooperate with and influence each other in binding oxygen. Their interaction is an example of a cooperative effect, and the sigmoid oxygen dissociation curve of hemoglobin is a direct consequence of this.

The configurational change undergone by a protein molecule as a result of binding a smaller, nonprotein molecule is a phenomenon of considerable biological importance. Configurational shape changes are suspected to occur in

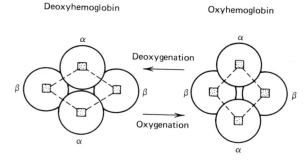

Figure 9-5 Configurational change of the hemoglobin molecule associated with binding of oxygen. When the hemoglobin molecule combines with oxygen, the distance between the iron atoms of the heme groups of the α subunits increases by 1 Å and the distance between the iron atoms of the heme groups in the β subunits decreases by 6.5 Å. This results from changes in the relative positions of the subunits when oxygen is reversibly bound.

many enzyme molecules whose activities, as we will see later, are regulated by reversible binding of small molecules, known as allosteric effectors, at sites on the enzyme molecule other than the active site. The case of the hemoglobin molecule is the first such configurational change to be actually demonstrated, and it is likely to remain so for some time; the technical difficulties involved in proving the existence of configurational shape changes in protein molecules are formidable.

THE BOHR EFFECT

The affinity of hemoglobin for oxygen is lessened if the pH is decreased, that is, if conditions are somewhat acid, such as those found in rapidly respiring tissue or vigorously exercising muscle. This acidity is caused by release of lactic acid and carbonic acid; the latter is derived from carbon dioxide and water:

$$CO_2 + H_2O \rightleftharpoons H_2CO_3 \rightleftharpoons H^+ + HCO_3'$$

Carbonic acid

Oxyhemoglobin tends to lose protons more readily than deoxyhemoglobin (i.e., it is more strongly acidic). Hence, in accordance with Le Chatelier's principle, an acid medium would tend to suppress loss of a proton from oxyhemoglobin, thus shifting it to the more weakly acidic deoxy form and thereby encouraging liberation of oxygen.

The decreased affinity of hemoglobin for oxygen in the presence of increased hydrogen ion concentration is called the Bohr effect. Again, this is a useful attribute of the oxygen-transporting protein; the slight acidity of the muscles (due to formation of lactic acid and CO_2) and other actively respiring tissues further facilitates unloading of oxygen from the oxyhemoglobin molecule, and the released oxygen is mostly used directly by the tissue. The Bohr effect again is an example of a new property emergent from possession of quaternary structure by hemoglobin; myoglobin, lacking quaternary structure, does not show the effect.

OXYGEN BINDING BY "ARTIFICIAL" HEMOGLOBIN MOLECULES

Hemoglobin molecules exposed to pH values over 11 or below 5 dissociate (i.e., the subunits separate from each other); each hemoglobin molecule gives two α and two β subunits. The resultant solution of myoglobinlike molecules still binds oxygen, but the dissociation curve is hyperbolic, like that of myoglobin. If physiological conditions are restored, the subunits reassociate spontaneously to give the native $\alpha_2\beta_2$ tetramers. The profound influence that quaternary structure and subunit composition has on oxygen affinity is vividly illustrated in experiments where conditions for the spontaneous reassociation of hemoglobin subunits are so manipulated that abnormal "unphysiological" hemoglobin molecules are formed. H. A. Itano and E. Robinson showed that if a mixture of dissociated human and dog α and β subunits were allowed to reconstitute, two dog α units might associate with two human β units, and vice versa, so that "hybrid" hemoglobins resulted with the compositions α_2 man β_2 dog and α_2 dog β_2 man, respectively. Furthermore, assemblies consisting of four β subunits from one species could be formed, so that the composition of the resulting tetramer would be β_4. Such abnormal hemoglobins have a much greater affinity for oxygen than the normal hemoglobin and do not exhibit a Bohr effect. They would therefore not be of much use as oxygen-transporting proteins, because the oxygen would not be readily liberated in the tissues. The properties of these artificial hemoglobins is a good illustration of the influence of quaternary structure and subunit composition on biological function.

NORMAL AND ABNORMAL HEMOGLOBINS

Although somewhat of a digression from the main theme, which is the relation of quaternary protein structure to function, a discussion of hemoglobin would be incomplete without mentioning certain genetically determined hemoglobin variants; some variants are associated with disease states; also, other normal human hemoglobins and their possible evolution are discussed.

(i) Hemoglobin S

Perhaps the best-known pathological hemoglobin variant is the one that is characteristic of and the molecular basis for the condition called sickle cell disease. (The term sickle cell anemia, often used as a synonym, is inappropriate, since anemia is only one of the many signs and symptoms of sickle cell disease.) The red blood cells (erythrocytes) of affected people assume unusual and bizarre shapes when the blood is exposed to low oxygen tensions. Often the cells are crescentic or sickle shaped (Fig. 9-6), hence the name of the condition. As a result, the sickle cells agglutinate readily and cause blockages (thromboses) in blood vessels, causing more or less serious pathological effects. Sickle cells are prone to more rapid destruction (hemolysis) than normal erythrocytes, and so anemia results.

The abnormal shapes of the erythrocytes are caused by the peculiar property of sickle cell hemoglobin molecules (hemoglobin S, HbS for short) of aggregating into long, needlelike paracrystalline structures called tactoids, which distort the red blood cells.

In 1949, L. Pauling and H. Itano showed that carbonmonoxy sickle cell hemoglobin differed from normal carbonmonoxy hemoglobin in its electrophoretic behavior. It moved in the opposite direction in the Tiselius apparatus, thus proving that the net electric charge of the two hemoglobins was different in sign and that the sickle cell hemoglobin must therefore differ chemically from normal hemoglobin. The prosthetic group (carbonmonoxy heme) from both hemoglobins were identical, and so they concluded that the difference was in the protein (globin). Because of this, Pauling called sickle cell disease a "molecular disease."

The nature of the abnormality in hemoglobin S was pinpointed in 1957 by V. M. Ingram, who found that the β subunits of hemoglobin S differ from the normal subunits with respect to 1 amino acid residue. The α subunits were normal. Ingram separately subjected normal and sickle cell hemoglobin to tryptic digestion, which splits specific peptide linkages. He then subjected the resulting hydrolysates to filter paper electrophoresis, which partially separated the peptide mixture. The electrophoresis papers were then turned through a right angle and chromatography was performed, which separated the peptides as discrete spots in a two-dimensional "spread." The pattern of spots is distinctive for a given protein; the technique is called "fingerprinting." Ingram observed that one of the peptide spots in the sickle cell hemoglobin hydrolysate occupied a slightly different position from that on the fingerprint of normal hemoglobin. The spot was isolated, the peptide extracted, and the amino acid sequence determined by Sanger's method. The only difference between this peptide and the corresponding normal one was in a single amino acid residue—the glutamic acid residue at position 6 in the normal β-polypeptide was replaced by a valine residue. The substitution is symbolically represented as:

$$\text{Glu}_6^\beta \longrightarrow \text{Val}$$

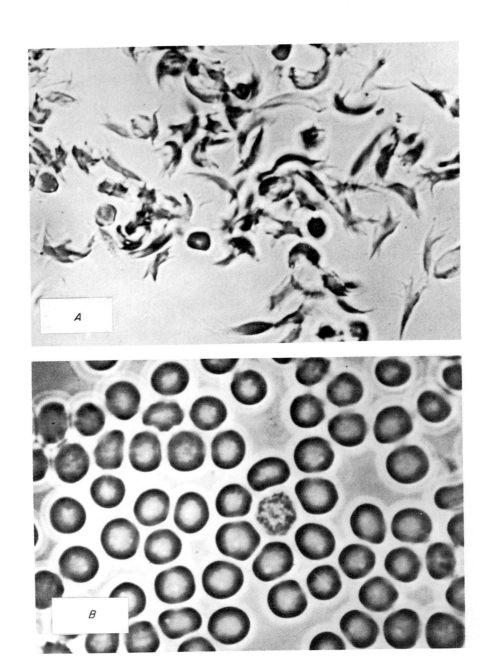

Figure 9-6 Phase contrast micrograph of human red blood cells. (*A*) From person with sickle cell disease. (*B*) Normal. (Courtesy of Dr A. C. Allison.)

Why such a seemingly slight variant from the normal should have so profound an effect on the physicochemical properties of hemoglobin S is not yet clear. The abnormal β-polypeptide is regarded as resulting from a mutant form of the gene that determines the primary structure of the normal β-polypeptide.

Sickle cell disease exists in two forms—the not too severe sickle cell trait and the more severe sickle cell disease. In the former, Itano found that 65% of the total hemoglobin was normal and 35% was hemoglobin S. In the latter, the hemoglobin consists completely of the sickle cell type. The explanation is that hemoglobin S is inherited by a simple Mendelian mechanism. If the gene determining normal hemoglobin is represented by H and the one determining hemoglobin S is represented by S, three genotypes and three phenotypes are possible.

Genotype	Phenotype	Hemoglobin(s) present
HH	Normal	HbA
HS	Sickle cell trait	HbA + HbS
SS	Sickle cell disease	HbS

A person with sickle cell trait is therefore a heterozygote; one with sickle cell disease is homozygous. (Hereditary mechanisms and genetic terminology are described and explained in Chapter Fourteen.)

Heterozygotes are able to live fairly normal lives as children and adults, provided they avoid conditions leading to low oxygen tensions (i.e., too violent exercise or climbing high mountains). Homozygotes usually do not survive to reproductive maturity, and one might wonder why this apparently "bad" gene has survived as it has done. A clue is provided by the geographical distribution of sickle cell disease. The areas primarily affected are Central Africa, the countries in Europe and North Africa along the Mediterranean, and parts of Saudi Arabia and India. These are also areas where malaria is common (or was common until recently); interestingly children with sickle cell trait (i.e., heterozygotes) have a higher resistance to a first infection with malaria than unaffected children. Therefore, in those areas of the world, the sickle cell gene in heterozygotes can be considered to have some survival value, which may account for its persistence in the population. Apparently the interior of sickle cells is not conducive to the development of that stage of the malarial parasite's life cycle that occurs within red blood cells; it may be that the tendency of hemoglobin S to crystallize is responsible. (A few years ago, I attended a lecture on sickle cell disease given by Linus Pauling. With characteristic humor, the speaker suggested that the failure of the malarial parasite to develop within sickle cells was because the long, thin "crystals" of hemoglobin S stabbed the parasites to death).

(ii) Other abnormal hemoglobins

HbH.   This hemoglobin is found in a disease in which there appears to be an inability of the afflicted individual to synthesize α subunits. β subunits are present, and they associate to give β_4 tetramers.

HbM.   This type of hemoglobin does not easily bind oxygen; the characteristic outward sign in people with HbM disease is cyanosis (i.e., a bluish-gray color of the skin because of the low oxygen content of the circulating blood). This hemoglobin differs from hemoglobin A in 3 amino acid residues. Using the same symbolism as before, the substitutions are:

$$His_{58}^{\alpha} \longrightarrow Tyr \quad His_{63}^{\beta} \longrightarrow Tyr \quad Val_{67}^{\beta} \longrightarrow Glu$$

These occur in the neighborhood of the heme group. Possibly, the hydroxyl and carboxyl groups of the abnormal amino acid residue substituents interfere with the oxygen-bonding power of the central iron atom.

Hb Bart's.   In this variant the hemoglobin molecule is a γ_4 tetramer (see later regarding γ subunits).

HbGH.   An abnormal hemoglobin, rejoicing in the name of Hemoglobin Gun Hill, was described in 1967 by T. B. Bradley et al., who discovered it in the blood of a man and in one of his three daughters. The man had had mild jaundice since adolescence, was in a compensated hemolytic state, and his spleen was enlarged. The defect in the molecule of this hemoglobin resides in the β-polypeptide, which lacks a linear sequence of 5 amino acid residues in the region concerned with heme binding. The abnormal globin cannot bind heme, so that HbGH has only half of the heme content of HbA. The α-polypeptide is normal. The conformation of the β subunits does not seem to be seriously affected by this deletion, and they are able to associate with the α subunits to give HbGH. T. S. Bradley et al. suggest that the defect might have arisen by unequal crossing over in meiosis.

Not all abnormal hemoglobins give rise to clinical symptoms, but they can be detected by chemical and electrophoretic methods. Since they do not have any detectable effects in the organism, the mutations responsible for them are said to be "silent."

(iii) Other normal hemoglobins

Beside the normal adult hemoglobin (HbA) with the composition $\alpha_2\beta_2$, there are others that occur in humans but may have only a short existence in the fetus or may be present in only small quantities in the adult (i.e., they are minor components).

Kunkel and Wallenius and also Morrison and Cook showed in 1955 that normal adult human hemoglobin is heterogeneous. HbA is present to the extent of about 80 to 90% of the whole, the remainder consisting of several minor components.

The blood of the human fetus contains a type of hemoglobin not present in adults; it is therefore known as fetal hemoglobin (HbF). The fetal hemoglobin molecule consists of two α subunits, as in HbA, but differs from it in that the β subunits are replaced by two γ subunits, which differ from the β primary structure with respect to 37 amino acid residues. The composition of the fetal hemoglobin molecule is therefore represented as $\alpha_2\gamma_2$. Fetal hemoglobin is heterogeneous; its major component is designated HbF and the minor component is HbF$_1$. Both components have the same amino acid structure, but HbF$_1$ differs from HbF in that it possesses an acetyl group on the glycine residue at the N terminal of one of the γ subunits. The functional significance of fetal hemoglobin is that its affinity for oxygen is greater than that of adult hemoglobin; this is hardly surprising, since the human fetus of necessity acquires its oxygen secondhand from the maternal blood through the placenta. The oxygen dissociation curve of fetal hemoglobin is therefore shifted to the left of the curve for adult hemoglobin (Fig. 9-7).

In the very early fetus, there is transitory production of another type of subunit designated ε (epsilon), but these are soon replaced by the γ subunits. The primary sequence of the ε subunits has not yet been determined in detail.

In the adult, there is a minor component HbA$_2$, present to the extent of 2.5 to 3.5% with yet another type of subunit δ (delta), which replaces the β subunit. The composition of HbA$_2$ is therefore $\alpha_2\delta_2$. The δ subunit differs from the β in 10 amino acid residues.

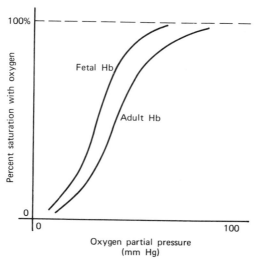

Figure 9-7   Oxygen dissociation curves of adult and fetal human hemoglobin.

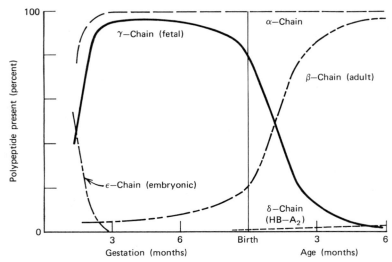

Figure 9-8 Changes in the relative amounts of human hemoglobin subunits during early fetal life until soon after birth. (Courtesy of Dr E. R. Huehns. Cold Spring Harbor Symposium on Quantitative Biology, *29*:329, Cold Spring Harbor Laboratory, 1964.)

During the lifetime of an individual, therefore, no less than five different types of hemoglobin subunits (α, β, γ, δ, and ε) are synthesized. The changes in the relative amounts of hemoglobin polypeptides present in human blood from early fetal life until a few weeks after birth are shown in Fig. 9-8.

There is yet another minor component known as HbA$_3$, which itself consists of three components (HbA$_{Ia}$, HbA$_{Ib}$, HbA$_{Ic}$); these components really need not concern us here, but one of them (HbA$_{Ic}$) is worth mentioning in some detail. It is present to the extent of 5 to 7% and is the most abundant of the minor components. Chemically, HbA$_{Ic}$ is odd in that at the N terminal of one of its β-polypeptide chains and attached by Schiff's base linkage is an aldehyde or ketone of molecular weight about 280. Although not proven, the indications are that it is a long-chain, aliphatic lipid. W. A. Schroeder and W. R. Holmquist suggest that its function may be other than as an oxygen carrier. Although it would be inappropriate here to go into their findings and discussions in detail, we may note briefly some of the suggestions made by Schroeder and Holmquist regarding possible functions of HbA$_{Ic}$.

1. *As a determinant of the tensile strength of the red cell membrane.* The low surface tension of the red cell membrane is comparable with that of other cell types. This may be due to presence of HbA$_{Ic}$ in the red cell membrane, because its structure is similar to that of other surface active agents [i.e., it has a long-chain aliphatic group and a polar group (HbA)].

2. *As a membrane buffer.* Schroeder and Holmquist calculated that there are about 5×10^6 HbA_{Ic} molecules in the red cell membrane, and each contains 38 histidyl residues. If the pH of the membrane is taken to be about 7.4, there will be about 7×10^7 hydronium ions within it. Assuming that half of the histidyl residues are ionized, the HbA_{Ic} contributes 9.5×10^7 (i.e., practically all of the hydronium ions in the membrane). Clearly, as Schroeder and Holmquist remark, a red cell membrane without HbA_{Ic} would be a very different membrane.

3. *As a "transition" molecule.* With regard to the efficiency of passive transport (e.g., of oxygen), the properties of a cell membrane should be as much like the cell interior as possible. It should resemble an equivalent thickness of the cell interior, yet should be thin enough to permit the other functions of the red cell membrane. The presence of a monolayer of HbA_{Ic} in the membrane would decrease the compositional difference between the interior of the red cell and the membrane. Such a monolayer would contain 50% by weight and 29% by volume of hemoglobin.

Evolution of hemoglobins

In an earlier chapter, the evolution of the cytochromes c was briefly mentioned. From a comparative study of the primary sequences of amino acid residues of cytochromes c from various species, we learned that when the cytochrome primary sequences were aligned in a certain way with respect to one another, they were closely related. In fact, the more closely related two species are phylogenetically, the more similar their cytochromes are. The only rational explanation is that present-day cytochrome c molecules have evolved from a common ancestral molecule.

Myoglobin and the hemoglobins are good material for similar comparative study, because the primary structures of whale myoglobin and of several hemoglobin subunits have now been determined in considerable detail. When the primary sequences of human α- and β-hemoglobin subunits are placed alongside one another, they are found to correspond with respect to 64 amino acid residues but differ with respect to 77. Whale myoglobin has the same amino acid residues as human α subunits at 37 sites and the same amino acid residues as human β subunits at 35 sites. Despite the differences in primary structure, the three-dimensional configurations are strikingly similar. That the two human α- and β-hemoglobin subunits and whale myoglobin could have arisen independently with similar functions, conformation, and with so many amino acid residues in common seems too much to ascribe to coincidence, even though they differ in more amino acid residues than they agree; it seems much more likely that they evolved from a common ancestral molecule. Comparison of the primary structures of the other human hemoglobin subunits tends to confirm this idea.

The human β, γ, and δ subunits all contain 146 amino acid residues and are closely similar to one another. Between the β- and γ-polypeptides there is a difference of only 39 amino acid residues and the β- and δ-polypeptides differ in only 10 residues. The human and horse α subunits have the same number of amino acid residues (141) and resemble one another more closely than either of them resemble β, γ, or δ subunits. Whale myoglobin differs the most from all these even in its chain length (153 amino acid residues). These differences in human and horse hemoglobin subunits and whale myoglobin are summarized in Table 9-1. Clearly, all these proteins are homologous.

In comparing proteins of similar but not identical structure from different species as opposed to similar proteins from the same species, one has to be careful about deciding whether the proteins are homologous or not. Without going into too much detail, one way of differentiating homologous from non-homologous proteins is to compare their primary sequences side by side, "shift" them with respect to one another, and observe the number of identical amino acid residues in the segments of the two chains. A successful shift is one that maximizes the number of correspondences between the segments of the two polypeptides. With homologous proteins a small number of shifts reveals striking similarities in the two primary sequences; this does not happen with

TABLE 9-1   **Amino Acid Sequence Differences between Human and Horse Hemoglobin Subunits and Whale Myoglobin**

		Human				**Horse**		**Whale Myoglobin**
		α	β	γ	δ	α	β	
Total number of amino acid residues		141	146	146	146	141	146	153
Human	α	0	84	89	85	18	87	115
	β	84	0	39	10	86	25	117
	γ	89	39	0	41	87	39	121
	δ	85	10	41	0	87	26	118
Horse	α	18	86	87	87	0	84	118
	β	87	25	39	26	84	0	119
Whale myoglobin		115	117	121	118	118	119	0

With the exception of the top row of figures (total amino acid residues), the numbers indicate the number of amino acid differences between the subunit above it (in the upper horizontal column) and the subunit to its left (in the vertical column at the left).

nonhomologous proteins. By using this technique, the "degree of relatedness" of two different homologous proteins may be demonstrated. The evidence indicates that the closeness of phylogenetic relationship between animals is generally correlated with the degree of similarity of primary structure between homologous proteins from the same animals. That is, the greater the similarities between homologous proteins, the shorter the period of evolutionary time that has elapsed since they diverged from a common ancestor. We must remember that it is not the proteins themselves that evolve, but the genes that determine their structure (see Chapters Fourteen and Seventeen). Spontaneous mutations occur from time to time; these may be of the type that result in an alteration of an amino acid in the primary protein sequence. Suppose that a hypothetical ancestral vertebrate possessed a simple oxygen-transporting protein whose molecule was like that of myoglobin (i.e., consisting of one subunit). The structure of the protein would be determined by a single gene and its identical allelic partner in the diploid homozygous organism. A mutation in one of these genes that changed 1 amino acid residue would result in a defective and possibly nonfunctional protein. However, the normal protein would still be produced by the other unmutated allele, and the organism would experience no serious impairment of function. The defective protein could therefore be tolerated, and the mutated gene would be passed on to the organism's offspring. If, subsequently, a mutation occurred in the other normal allele, all the organism's respiratory protein would be defective, and death might result; the two mutated genes would be removed from the population.

Another type of genetic event involves gene duplication or doubling (Fig. 9-9). If this occurred in the organism carrying the mutant gene, we would now have two identical normal genes and two identical mutant genes. If another single mutation occurs, then whatever gene is affected, there will be one normal allele to produce normal respiratory protein, and the organism will survive if the mutant proteins function reasonably well. Several duplications followed by occasional single random mutations would probably be nonlethal and would therefore eventually result in an organism with several different respiratory pigment genes, each producing (in addition to the normal protein) several proteins that all differ more or less from the normal protein and that are carried as dead weight. Over evolutionary time, further mutations and gene duplications might accumulate and eventually produce, in addition to the myoglobinlike protein, two different functional proteins that show a tendency to associate and form tetramers (i.e., hemoglobinlike structure with the type of oxygen-binding curve we find in animals with tetrameric hemoglobin molecules, which is more efficient as an oxygen carrier). These speculations are actually borne out in practice because it has been found that the circulating oxygen-carrying protein of the lamprey, a primitive vertebrate, does, in fact, consist of single, myoglobinlike molecules, whereas higher fishes have the tetrameric hemoglobin molecules.

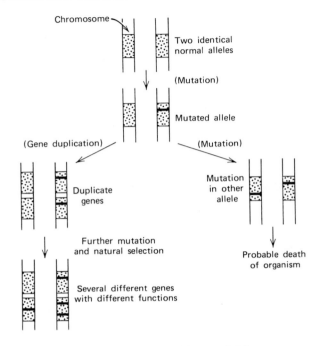

Figure 9-9   Evolution of hemoglobin genes.

In attempting to estimate the times when these mutational events occurred, biologists have to establish what is called the mutation distance between two similar proteins and a mutation rate for the genes involved. As will be shown in Chapter Seventeen, the amino acids making up an organism's proteins are ultimately determined by nucleotide triplets on the corresponding DNA molecule. Alteration of one or more nucleotides in a given triplet—as occurs in one type of mutational event—changes the triplet determining one amino acid into a triplet that determines another. Calculation of the minimum number of such changes necessary to convert a given nucleotide triplet into another triplet that determines a different amino acid is fairly easy. This number is called the mutation value of the change. If, now, in two homologous proteins the mutation values of all the dissimilar amino acid residues are added together, the mutation distance between the two proteins is arrived at. Assuming that the mutation rate for the genes determining hemoglobin synthesis has been constant during evolution, approximate dates of origin of the various subunits may be calculated. If these calculations are applied to the human and horse hemoglobin subunits and to whale myoglobin, an evolutionary tree can be worked out for them (Fig. 9-10). (In the case of the human hemoglobin subunits, the evolution has

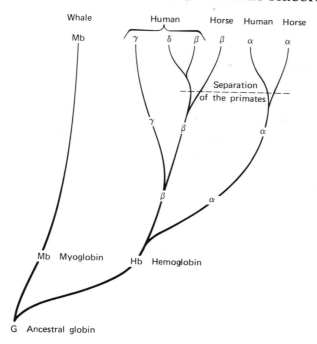

Figure 9-10 "Evolutionary tree" of human and horse hemoglobin subunits and whale myoglobin. (Redrawn after Dickerson and Geis.)

occurred during the evolutionary history of one organism.) In the case of the human, the latest evolutionary product appears to be the δ subunit. It has been suggested that the $\alpha_2\delta_2$ human hemoglobin component has properties that are better than those of the standard adult hemoglobin so that, as Dickerson and Geis speculate, here may be the appearance of a hemoglobin that may eventually supplant the $\alpha_2\beta_2$ variety.

INVERTEBRATE OXYGEN-TRANSPORTING PROTEINS

The oxygen-transporting protein pigments of invertebrates often contain copper instead of iron in their prosthetic groups and have very high molecular weights. The globin component in all known cases is much larger than that of vertebrates. Electron microscope studies reveal that they have very large molecules and possess quaternary structure so complex that the vertebrate hemoglobin molecule is dwarfed beside them and looks quite simple in comparison.

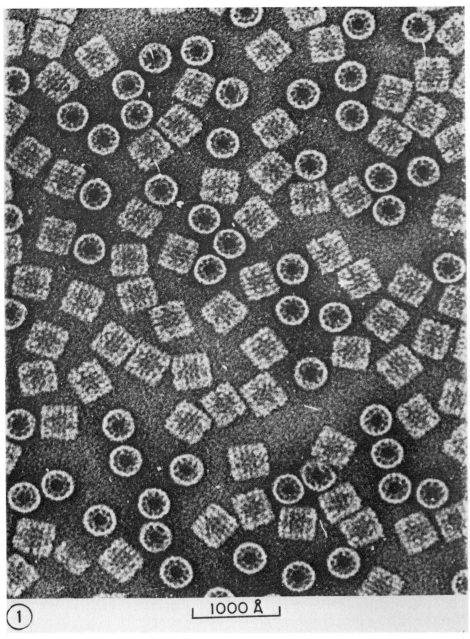

Figure 9-11 Electron micrograph of negatively stained *Helix pomatia* hemocyanin molecules showing top and side views (×260,000). (Courtesy of Dr H. Fernandez-Moran. J. Mol. Biol., *16*:191–207, 1966.)

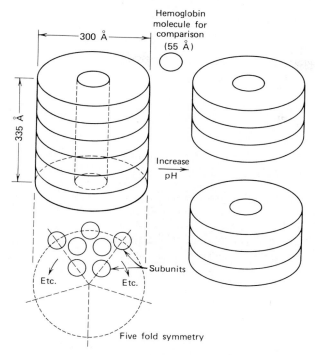

Figure 9-12   The molecule of *Helix pomatia* hemocyanin. Further dissociation appears to be sectorwise. Divalent cations are needed for reassociation (e.g., Ca^{++}, Sr^{++}, Ba^{++}, Mg^{++}).

(i) Hemocyanin

The green respiratory pigment of the snail *Helix pomatia* contains copper instead of iron and is called hemocyanin. The electron microscope studies of E. van Bruggen and E. Wiebenga showed the molecule to be a cylindrical structure measuring 300×335 Å; it consists of six layers of 25 subunits arranged in pentagonal symmetry (Figs. 9-11 and 9-12).

(ii) Chlorocruorin

Chlorocruorin is the collective name given to the green-colored, oxygen-carrying pigments found free in the plasma of the blood of certain polychaete worms. The prosthetic group is a ferroporphyrin, which is responsible for the color. Studies by E. Antonini, A. Rossi-Fanelli, and A. Caputo, using sedimentation, diffusion, and light scattering techniques, revealed that the chlorocruorin of

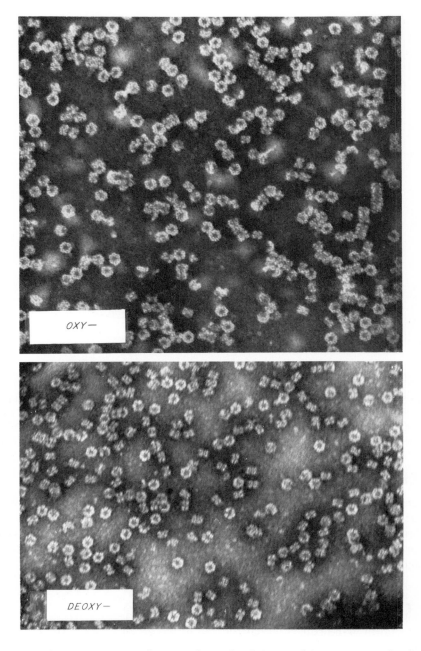

Figure 9-13 Electron micrographs of *Spirographis* oxy- and deoxy-
chlorocruorin molecules at pH 7.0 (×approx. 200,000). (Courtesy of Dr D.
Guerritore. J. Mol. Biol., *13*:234–273, 1965.)

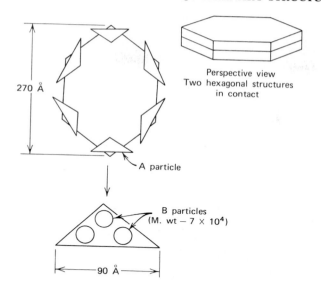

Perspective view
Two hexagonal structures
in contact

270 Å

A particle

B particles
(M. wt − 7 × 10^4)

90 Å

Figure 9-14   The molecule of *Spirographis* chlorocruorin.

Spirographis spallanzanii has a molecular weight of 2.75×10^6. The earlier electron microscope studies of J. Roche et al. indicated that the molecules of this and similar high molecular weight hemoproteins generally have a hexagonal structure. Later studies by D. Guerritore et al. showed that the *Spirographis* chlorocruorin molecule (Fig. 9-13) consists of two hexagonal discoidal structures in contact; at their corners are six identical particles (A structures) (Fig. 9-14), each consisting of three smaller structures (B structures). Apparently, each B structure contains two chlorohemes, so that the entire molecule contains 72 hemes. These investigators suggest that the cooperative interactions involved in oxygen binding occur within the A structures, which are therefore regarded as the "minimal functional units."

(iii) Invertebrate hemoglobin

The complex oxygen-carrying pigment (hemoglobin) of *Limnodrilus gotoi* has a molecular weight of 3.01×10^6. The electron microscope studies of Yamagashi et al. showed that the molecule again consists of two apposed hexagonal discs, each of which is composed of six identical submultiples. The entire molecule measures 220×160 Å. There are 108 hemes per molecule, each of which is associated with a single polypeptide chain (subunit). Yamagashi et al. suggest that each submultiple contains 9 of these heme-containing polypeptides. An electron micrograph of *Limnodrilus* hemoglobin is shown in Fig. 9-15.

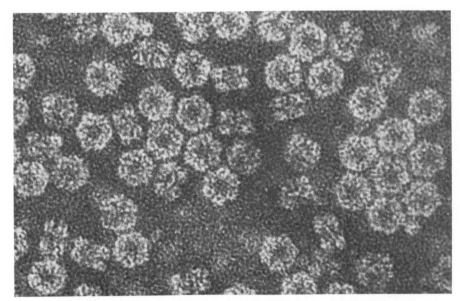

Figure 9-15   Electron micrograph of *Limnodrilus* hemoglobin molecules showing top and side views (×400,000). (Courtesy of Dr R. Shukuya. From Yamagishi, M., J. Mol. Biol., *21*:467–472, 1966.)

(iv) Erythrocruorin

The marine Lug Worm, *Arenicola marina*, contains the pigment erythrocruorin (molecular weight $= 3 \times 10^6$) which, like the *Spirographis* and *Limnodrilus* pigments, is composed of two hexagonal layers each consisting of six subunits. The protein exhibits large heme-heme interactions, like chlorocruorin, and the oxygen equilibrium curves of both these pigments are very similar.

The hexagonal chlorocruorin structure may be a fairly general "architectural theme" in invertebrate respiratory proteins. The relatively enormous size of the molecules of these invertebrate respiratory proteins ensures that they will not diffuse through the walls of blood vessels; this is probably why they are not contained within corpuscles. The vertebrate hemoglobin molecule is small enough to diffuse, but this is prevented by its enclosure within a membranous corpuscle. In addition, the hemoglobin-containing corpuscle is probably a more efficient oxygen transporter than the hemoglobin molecule would be if it were free in the plasma.

ISOENZYMES: LACTATE DEHYDROGENASE

Many enzymes exist in more than one form, all of which exhibit similar catalytic properties but dissimilar molecular structure. These structurally different but catalytically similar forms of an enzyme are called isoenzymes, or isozymes. So many enzymes are now known to exist in multimolecular forms that it now would appear that an enzyme that does not consist of several isoenzymic variants is the exception, not the rule.

The possession of quaternary structure is the basis for the existence of multiple molecular forms of enzymes. A well-known and often quoted example is lactate dehydrogenase (LDH), an enzyme that catalyses the reversible interconversion of pyruvate and lactate in vertebrate tissues. C. Markert showed that LDH molecules are tetramers consisting of two different types of equal-sized polypeptide subunits, designated A and B. Five possible quaternary structures are therefore theoretically possible (i.e., five isozymes), having the compositions A_4, A_3B, A_2B_2, AB_3, and B_4. All five have actually been identified and all have a molecular weight of 135,000. The relative amounts of each of the five LDH isozymes found in mammalian tissues vary according to the tissue type. The A_4 tetramer, for example, predominates in skeletal muscle, and the B_4 type is found principally in cardiac muscle.

PHYSIOLOGICAL SIGNIFICANCE OF LDH ISOZYMES

There is a rough correlation between the proportion of A subunits in a cell and its exposure to periods of anaerobiosis. In mammalian skeletal muscle, for instance, which often works under conditions of transient oxygen shortage, there is usually a preponderance of the A_4 tetramer, as already noted; in brain tissue and cardiac muscle, which always have an abundant oxygen supply, the B_3A and B_4 types are mostly found. The significance of this has been thought to be the sensitivity of the different LDH isozymes to inhibition by pyruvate and lactate as determined in *in vitro* experiments; the type preponderating in skeletal muscle is not inhibited by pyruvate and continues to function in the presence of even very high concentrations of lactate; this makes "metabolic sense," because high concentrations of lactate are likely to occur in skeletal muscle. On the other hand, the predominant isozymes of brain tissue and cardiac muscle are inhibited by lactate, so that if lactate did accumulate, the LDH of these tissues would be inhibited and further conversion of pyruvate to lactate would stop. Hence, by a negative feedback mechanism (product inhibition), there is selective regulation of the activity of the different LDH isozymes, and this keeps the amount of lactate in different cell types within tolerable limits.

However, E. S. Vesell doubts whether this interpretation of the physiological significance of LDH isomers is a valid one. He points out that *in vivo* concentrations of pyruvate and lactate never reach anywhere near the levels used in the *in vitro* experiments that are found to exert inhibitory effects on the LDH isozymes that preponderate in heart and brain tissue.

DISSOCIATION AND REASSOCIATION OF LDH SUBUNITS

The native tetramers of LDH isozymes can be dissociated by treatment with urea or guanidine. C. Markert demonstrated that the separated monomers can reassociate spontaneously and that the reassociation process is random. He mixed equal amounts of purified A_4 and B_4 tetramers, dissociated them, and allowed them to reassociate. The mixture was then subjected to starch gel electrophoresis. The five isozymic forms of LDH possess different electrical charges because of the differing subunit composition, and so they migrate at different rates in an electric field. Upon electrophoresis, followed by application of a suitable stain to reveal the presence of bands, the reassociated mixture of equal amounts of A and B subunits resolved into five bands; the bands corresponded to the five different tetrameric forms, and the intensity of staining of each individual band was what would be expected, knowing its monomer composition and the relative amount that would be expected from random reassociation of the monomers ($1:4:6:4:1$). See Fig. 9-16.

The electrophoretic pattern of plasma LDH isozymes has applications in clinical diagnosis. Elevated serum levels of the B_4 isozyme are found in acute myocardial infarction; elevated serum levels of the A_4 isozyme are found in muscle conditions like muscular dystrophy. Presumably, the permeability of the membranes of the damaged cells of the diseased tissues is increased, and the

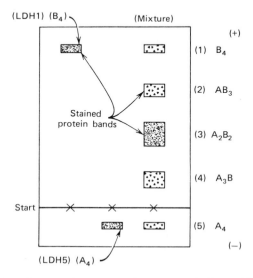

Figure 9-16 Markert's experiment on the reassociation of LDH subunits and their separation by electrophoresis.

LDH isozymes characteristic of the tissues leak into the bloodstream. The general LDH serum level is therefore elevated; by noting which specific isozyme is abnormally elevated, as shown by electrophoresis, the lesion can be localized to a particular tissue. By such means, for example, chest pain caused by heart disease can be differentiated from chest pain from other causes unrelated to the heart. Changes of serum LDH isozymes are also noted in various malignancies. Serum LDH isozyme patterns by themselves are not conclusive in diagnosing disease conditions, because a given pattern might occur in different diseases and sometimes no abnormal pattern is observed in disease processes that usually show them. Always, other laboratory and clinical studies should accompany the LDH studies before making firm diagnosis.

LDH isozymes in vertebrates illustrate the principle of physiological function dependent on and emerging from quaternary structure; this is exemplified by their differing *in vitro* sensitivity to lactate and pyruvate (whatever the physiological significance of this *in vivo* may or may not be) and to differences in susceptibility to urea denaturation, their reactivity to NAD analogs, and substrate affinity and specificity.

LDH ISOZYMES OF LIMULUS POLYPHEMUS

In 1968, G. L. Long and N. O. Kaplan reported that the LDH of the horseshoe crab *Limulus polyphemus* was specific for D-lactate, unlike the vertebrate LDH, which has specificity for the L-isomer, and that the enzyme exists in six isozymic forms. These findings were confirmed two years later by R. K. Selander and S. Y. Yang, who studied the D-LDH isozymes of *Limulus* by starch gel electrophoresis and showed that the six isozymes resolved as two groups of three bands, or "triads." One triad, the "upper group," was more anodally moving than the other triad, which they called the "lower group," and it was separated from it by a wide gap. Two individuals out of the total of 64 *Limulus* studied by Selander and Yang showed variation from the usual pattern for the "upper group" in that there were six separate bands, three of which corresponded in mobility to the usual upper triad, the other three being different. The lower triad in these two individuals had the usual appearance.

Long and Kaplan had also reported that the molecular weight of *Limulus* D-LDH was 65,000, which is about half that of the vertebrate enzyme (140,000 to 150,000). The electrophoresis patterns of *Limulus* D-LDH obtained by Selander and Yang were therefore interpreted by assuming that the enzyme molecule is dimeric and that two different polypeptides are involved, one fast-moving (Y), and the other slower (Z). They consider that these are coded for by two gene loci, y and z, respectively. The usual upper triad would therefore consist of the three possible dimers Y_2, YZ, and Z_2; individuals with this pattern would be homozygous at both gene loci. The pattern of the two variant individuals exhibiting the three unusual extra bands in the upper triad was explained by assuming that the individuals are heterozygous at the z locus for the usual allele (z) and for an alternate allele (z^1) that codes for another polypeptide Z^1, which moves a little more slowly in the electric field. The heterozygous individual presumably possess LDH isozymes having the composition Y_2, YZ, YZ^1, Z_2, ZZ^1, Z_2^1, corresponding to the upper group of six isozyme bands. The lower group of isozymes are probably also dimers and are presumably under control of separate genetic determinants from those of the upper group.

The whole field of electrophoretic analysis of isozymes and genetic interpretation of the patterns revealed is a fascinating and rapidly advancing area of investigation in molecular biology.

GLUTAMATE DEHYDROGENASE

Another striking example of the dependence of physiological properties on quaternary structure is the case of the enzyme glutamate dehydrogenase (GDH). GDH catalyzes the oxidation of glutamic acid. It has a high molecular weight (2×10^6), and the enzyme molecule consists of eight separate polypeptide subunits. Treatment with various reagents causes the molecule to dissociate into four identical subunits, as shown in the ultracentrifuge, and each of these now can function as an enzyme in its own right. They catalyze the oxidative deamination of alanine, a totally different reaction from that catalyzed by the intact GDH molecule.

QUATERNARY STRUCTURE AND THE CONTROL OF METABOLIC REACTIONS

The continuous unregulated catalysis of a given enzyme reaction would be undesirable; the cell would soon have an oversupply of the metabolite(s) formed in the reaction, and this would be wasteful of both resources and energy. Likewise, unregulated enzyme activity could result in rapid depletion of a substance that might not be as rapidly replaced. Serious imbalances of various metabolites would therefore result from unregulated enzyme activity. Not surprisingly, mechanisms exist in cells for exerting control over enzymes. A commonly encountered mechanism of enzyme regulation is that of "end product inhibition." Many metabolic pathways involve synthesis of an essential metabolite; this synthesis begins with a starting material and proceeds through a series of intermediate compounds to the end product, and each step is catalyzed by a specific enzyme. As soon as a sufficient amount of the metabolite (end product) for the cell's immediate needs has accumulated, further production in excess of this would, of course, be unnecessary and wasteful. In end product inhibition, the final metabolite exerts an inhibitory effect on the first enzyme in the biosynthetic pathway, when its concentration exceeds a certain critical value, so that the other intermediates become rapidly depleted, and production of the end product ceases, As soon as the metabolite begins to be used up, the inhibitory effect lessens and synthesis starts up again. The metabolite thus automatically regulates the rate of its own biosynthesis by a negative feedback mechanism; it employs the same principle that a thermostat does to regulate the temperature of a room and maintain it constant within narrow limits.

In the early 1950s, A. Novick and L. Szilard showed that in *E. coli*, excess of the amino acid tryptophane immediately stopped biosynthesis of tryptophane by the bacterial cell, indicating that tryptophane was inhibiting its own biosynthetic enzymes. Later, Umbarger showed that in the same organism the synthesis of iso-leucine from threonine was inhibited by iso-leucine, and that iso-leucine acted as an inhibitor of threonine deaminase, and only this enzyme, which is the first in the chain of intermediates and enzymes in the pathway leading to iso-leucine (Fig. 9-17). Thus, iso-leucine, the end product in this reaction sequence, regulates the rate of its own biosynthesis.

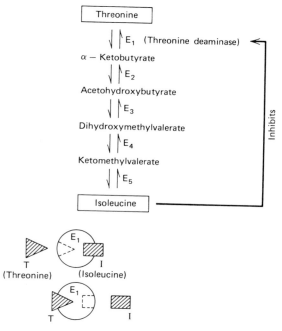

Figure 9-17 The isoleucine synthesizing system in *E. coli*. (1) Binding of isoleucine at allosteric site causes configurational change in enzyme that modifies active site so that enzyme is inhibited. (2) Binding of threonine similarly destroys allosteric site so that isoleucine cannot be bound.

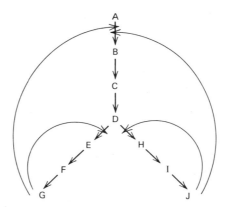

Figure 9-18   Feedback control of enzymes in branched biosynthetic pathways. The end products (G and J) might control by inhibiting the enzyme catalyzing the first common step (A ⟶ B) or by inhibiting the first enzyme in each branching pathway.

The desirability of feedback control (modulation) of enzyme activity is especially important in the case of branching biosynthetic pathways, where a single substrate is the starting material for two or more different end products. As shown in Fig. 9-18, regulation of such pathways can be achieved by feedback control of the first step common to the different pathways. Also, the first steps that diverge and that are concerned solely with biosynthesis of each separate final end product can be effectively used for control. If, however, the first common step was catalyzed by a single enzyme that was subject to feedback inhibition by one of the end products, then excess of this end product would cause a shutdown in synthesis of intermediates needed for biosynthesis of the other end products.

Among the many possible mechanisms for achieving selective feedback control of branching metabolic pathways, an effective method is one in which several isozymic variants of the enzyme catalyzing the first common step are synthesized. These all catalyze the common reaction but differ with respect to their susceptibility to specific inhibition by different end products. A simple example is the metabolism of threonine in *E. coli*. In this bacterium, threonine can follow another pathway different from the one that leads to iso-leucine; therefore it affords an example of a simple branched pathway. Iso-leucine biosynthesis occurs only under aerobic conditions, and only one enzyme is produced (threonine deaminase) susceptible to iso-leucine inhibition. In anaerobic conditions, an alternative energy-yielding pathway is taken in which alpha-keto butyrate is converted to fermentation products and a second inducible enzyme insensitive to biosynthetic end products is elaborated (Fig. 9-19).

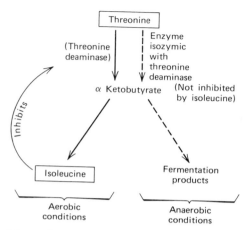

Figure 9-19 Aerobic and anaerobic metabolism of threonine in *E. coli*. An example of a simple branched metabolic pathway.

The first branched metabolic pathway that was shown to be controlled by multiple enzymes is one in which the amino acids lysine, methionine, and threonine are synthesized; their common precursor is aspartate.

CONTROL OF ASPARTATE METABOLISM

The first step in the synthesis of the amino acids lysine, methionine, and threonine is phosphorylation of their common precursor, aspartate. The reaction is catalyzed by aspartokinase. Two aspartokinases have been found in *E. coli*; one is inhibited only by lysine and the other by threonine. Knowing this, it is not immediately clear how, if the concentrations of aspartyl phosphate was low due to inhibition of one aspartokinase because one of the end products was in excess, sufficient aspartyl phosphate could be made available for biosynthesis of the other end product that was not in excess. This question was answered by G. N. Cohen and others who showed that there is a third aspartokinase enzyme that is inhibited by methionine and that there are other control points at certain steps in the branching series of reactions leading to the end products (Fig. 9-20). Note that (1) the first reaction in each branch of the biosynthetic pathway is specifically regulated by the end product of that branch; (2) there are three separate aspartokinases, all separately regulated by *three* end products—lysine, threonine, and methionine; (3) there are *two* separate homoserine dehydrogenases that are controlled separately and specifically by *two* end products—methionine and threonine; and (4) all aspartokinase activity will be inhibited only when *all three* end products are present in excess.

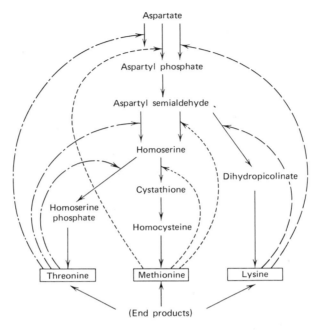

Figure 9-20   Control of aspartate metabolism in *E. coli*. Solid arrows indicate enzyme-catalyzed reactions. Dotted arrows indicate feedback inhibition by end product.

Clearly, differential control of the production of three separate end products would not be possible if only single enzymes catalyzed the first common steps in branched pathways.

MECHANISM OF END PRODUCT INHIBITION—ALLOSTERIC EFFECTS

The inhibition of enzymes by products produced by steps remote from them in a biosynthetic pathway is an established fact. Earlier, the phenomenon of competitive enzyme inhibition was described; it was shown that competitive inhibitors bear a close structural resemblance to the normal substrate and that they act by blocking the active site of the enzyme, thus preventing access of the substrate. In the case, say, of the feedback inhibition of threonine deaminase by iso-leucine in the iso-leucine synthesizing system of *E. coli*, one may ask how iso-leucine can inhibit the enzyme, since it bears virtually no structural resemblance to the substrate threonine?

In 1963, J. Monod, J. P. Changeux, and F. Jacob proposed that the inhibitor in this and similar cases binds to a part of the enzyme molecule surface remote from the active site; in doing so it causes a configurational change in the enzyme molecule that results in the active site being no longer able to bind the substrate. This idea arose from the fact that end product inhibition is quite specific; only iso-leucine inhibits the form of threonine deaminase, which catalyzes the first step in iso-leucine biosynthesis. These inhibitor binding sites are called allosteric sites (Greek. *Allos* = other, *steros* = space), and the effect is therefore called allosteric inhibition.

The basis for an allosteric conformational change in an enzyme molecule is most probably a shift in the relative positions of the subunits that make up the quaternary structure, as occurs in the case of the hemoglobin molecule when it binds oxygen. We have already seen how differences in subunit composition of an enzyme with quaternary structure affect their properties and give rise, for example, to isozymic variants; this is illustrated by the case of LDH isozymes and their different sensitivities to pyruvate and lactate and various denaturants in *in vitro* experiments. The case of GDH illustrates that changes in quaternary structure with respect to the association or dissociation of the subunits results in striking differences in substrate specificity. We have also noted the differing susceptibility of the iso-zymic forms of aspartokinase to feedback inhibition by three different amino acids. The phenomenon of allosteric inhibition shows how a subtle change in the quaternary structure, in the form of a slight shift of the relative positions of the subunits, has profound effects in that it influences the relative ease with which the enzyme binds the allosteric inhibitor or the substrate. In the case of threonine deaminase, substrate binds preferentially when the iso-leucine concentration is low. When the iso-leucine concentration rises, it has a greater chance of binding at the allosteric site, and so the enzyme is progressively inhibited; the net effect is the maintenance of an optimal level of iso-leucine.

KINETICS OF ALLOSTERIC ENZYMES

When an enzyme preparation is incubated *in vitro* with a fixed amount of substrate and the amount of product formed (or substrate that has disappeared) is measured at known time intervals, a graph similar to that in Fig. 9-21 is usually obtained. This "progress curve" of the enzyme reaction shows that the initial rapid rate of the reaction slowly declines until it finally stops. Several factors cause the progressive slowing of an *in vitro* enzyme reaction, such as a decrease in substrate concentration as the reaction proceeds, accumulation of reaction products, which may be inhibitory toward the enzyme, and slow denaturation of the enzyme. The only reliable measure of the true activity of an enzyme preparation in contact with a known amount of substrate is therefore the initial

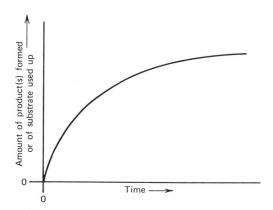

Figure 9-21 Progress curve of an enzyme reaction.

velocity before the factors tending to slow it down have had time to exert an appreciable effect. However, the amount of chemical change shortly after mixing enzyme and substrate may be so small that serious errors may creep into the estimate of the initial reaction rate. The only thing to be done is to wait until a reasonable amount of reaction product has accumulated (as the textbooks put it, without defining what "reasonable" means) that can be accurately measured so that errors from this source, and hopefully also those that accumulate as the time after mixing enzyme and substrate steadily increases, are minimized.

If the initial velocities of reaction of a fixed amount of enzyme in the presence of various concentrations of substrate under standardized conditions of temperature, pH, etc., are measured and plotted against substrate concentration, a curve like that in Fig. 9-22 is obtained. Mathematically, this curve is a rectangular hyperbola. It shows that above a certain substrate concentration, further addition of substrate causes no further increase in initial velocity. (Note that this curve is *not* the same as a progress curve.) At this substrate concentration, the enzyme is working at maximum capacity and is said to be saturated with the substrate; an increase of substrate concentration causes no further increase in initial reaction velocity. In the case of allosterically controlled (regulatory) enzymes, the graph of initial enzyme reaction velocity and substrate concentration shows that the initial velocity increases with increasing substrate concentration and is not hyperbolic, but sigmoid (S-shaped); so, also, is the graph relating end product inhibitor concentration to enzyme activity (Fig. 9-23).

These two types of curves are comparable to the oxygen saturation curves of myoglobin and hemoglobin, which are hyperbolic and sigmoid, respectively. The hemoglobin curve we have already seen is indicative of the operation of cooperative effects. In the case of enzymes showing sigmoid substrate or in-

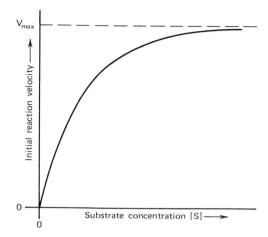

Figure 9-22   Graph of initial velocity of an enzyme reaction in presence of different concentrations of substrate under constant conditions of pH and temperature.

hibitor binding curves, cooperative effects are operative, and allosteric control is indicated. Threshold effects therefore characterize the response of regulatory enzymes to both substrates and inhibitors; the substrate-activity curve shows that the enzyme does not respond strongly until a certain substrate level—the "threshold"—is reached, which makes good physiological sense because only above a certain level of substrate does it become "worth the trouble" of processing it to form a worthwhile amount of end product. If the enzyme was unduly sensitive to low concentrations of end product and was easily inhibited, the whole point of feedback control would be nullified. The cell must have certain minimal levels of various metabolites, and only when these are exceeded is further production wasteful of energy. This accounts for the existence of threshold effects at low end product concentrations where the enzyme activity is scarcely affected but rapidly inhibited by increasing the concentration above this threshold value.

The biological significance of these sigmoid curves and cooperativity in metabolism is in these threshold effects, which prevent the cell from unnecessarily expending energy in processing uselessly small amounts of metabolites. Its significance is also in maintaining necessary levels of end products in allosterically controlled metabolic reaction sequences. This is achieved by the initial enzyme not being appreciably inhibited by concentrations below the threshold values of the inhibitory concentrations of the end products. In these respects, allosteric enzyme control mechanisms are strikingly similar to electric relays and to nerve cells, which respond only to a stimulus when a certain threshold value of that stimulus is exceeded.

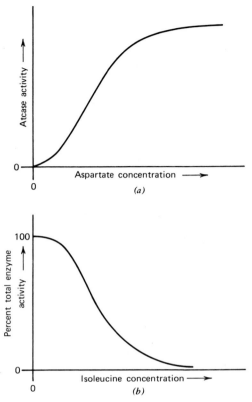

Figure 9-23   Graphs showing behavior of allosterically controlled enzymes in presence of substrates and inhibitors. (*a*) Effect of substrate concentration on activity of aspartate transcarbamylase (ATCase). (*b*) Effect of inhibitor concentration on activity of threonine deaminase.

THEORY OF ALLOSTERIC CONTROL OF REGULATORY ENZYMES

Two theoretical models of the mechanism of allosteric enzyme control have been proposed. The model of J. Monod, J. Wyman, and J. P. Changeux (the MWC model) was published in 1965; it envisages the enzyme molecule as consisting of at least two identical subunits, or "protomers"; on the surface of each subunit there are separate sites for substrate binding and the binding of various other substances or "ligands" (e.g., activators and inhibitors). The enzyme molecule can exist in two slightly different stable symmetrical conformational states that are in equilibrium; one state has a greater affinity for the substrate and is stabilized by substrate binding, and the other state has a greater affinity for the ligand and is stabilized by ligand binding. Conformational changes are viewed as occurring simultaneously, that is, in a concerted manner (Fig. 9-24).

The model of G. S. Adair, D. E. Koshland, G. Nemethy, and D. Filmer (the AKNF model) is somewhat similar, but views conformational changes caused by ligand binding as being sequential. A conformational change caused by binding of a ligand by one subunit influences the binding of the ligand by another subunit because of the interactions of the subunits. In some enzymes, binding of substrates or ligands by one subunit may not have much effect on adjacent subunits,

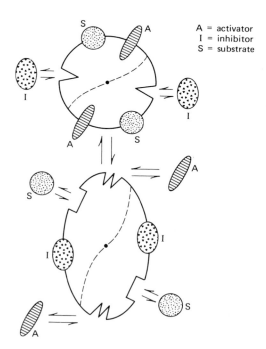

Figure 9-24   The Monod–Wyman–Changeux model of allosteric enzymes. The dimeric molecule exists in two stable states. One of these is induced by activator and substrate binding and the other is induced by inhibitor binding. The binding of substrate and activator molecules induces a configurational shape change in each subunit that alters the three-dimensional shape of the inhibitor binding sites, and this prevents binding of the inhibitor molecules. Similarly, binding of inhibitor molecules causes a different configurational shape change in the subunits that alters the three-dimensional shapes of the substrate and activator binding sites, and this prevents binding of the substrate and activator molecules. In either state, the symmetry of the enzyme molecule is preserved.

so that the binding curve will be essentially hyperbolic. At the other extreme, the configurational changes in one subunit will help to bring about configurational changes in adjacent subunits, so a strong cooperative effect will be the result. A negative cooperative effect will be produced if binding of an effector causes a similar configurational change in neighboring subunits that has the effect of slowing the reaction. Many different types of subunit interaction in enzymes are therefore represented in this model.

There are two types of ligand binding effects: (1) homotropic effects resulting from processes involving binding of identical ligand molecules; and (2) heterotropic effects resulting from processes involving different ligand molecules, such as activators and inhibitors of enzymes and their effect on enzyme activity. A feature of the MWC model is that homotropic effects are positive; the binding of one ligand molecule assists the binding of a second, so there will be a strong cooperative effect and a sigmoid curve will be given, particularly at low ligand concentrations. In the AKNF model homotropic effects may be positive or negative.

At the time that the MWC model was first proposed, there was not much concrete evidence to support it. That the molecules of regulatory enzymes possess different ligand binding sites on different protomers and that regulatory processes and sigmoid binding curves are due to interaction between the protomers was well demonstrated by the studies of J. Gerhart and H. Schachman on the enzyme aspartate transcarbamylase. Their work showed that the ideas of Monod et al. were essentially correct.

ALLOSTERIC REGULATION OF ASPARTATE TRANSCARBAMYLASE

Aspartate transcarbamylase (ATCase) has a molecular weight of 260,000 and catalyzes the reaction in which aspartate is converted to carbamyl aspartate. This reaction is the first that is unique to biosynthesis of pyrimidines. ATCase is inhibited by cytidine triphosphate (CTP), which is one of the end products in the biosynthetic pathway leading to pyrimidines. The enzyme is neither activated nor inhibited by ATP, but ATP opposes the inhibition by CTP. When the initial reaction velocity is plotted against substrate concentration for the native enzyme, the curve obtained is sigmoid, indicating that the enzyme is subject to allosteric control and exhibits cooperative kinetics. The ATCase molecule can be dissociated into six subunits, two large and four smaller, by treatment with p-mercuribenzoate. The larger subunits consist of two separate polypeptide chains; these subunits retain enzyme activity, but are no longer subject to CTP inhibition and do not bind CTP. They exhibit hyperbolic kinetics, and each has two substrate binding sites that correlate with the presence of the two polypeptide chains. The smaller subunits are noncatalytic and bind CTP. When the two types of subunits are mixed, they reassociate and regenerate the native enzyme, and cooperative kinetics are exhibited again along with CTP inhibition.

These studies prove that substrate and allosteric effectors (ligands) bind at different sites on the enzyme molecule, but they also prove that these sites are located on different subunits, as Monod et al. had predicted.

In endeavoring to unravel the mechanism of allosteric enzyme control, investigators have employed various experimental techniques, but these have yielded data only of overall changes and have given ambiguous results. Subsequent work on allosteric regulation of ATCase in *E. coli* was carried out by G. C. Hammes and Cheng-Wen Wu, who considered that the overall changes in allosteric control mechanisms could best be studied by breaking them down into their component parts. These investigators studied ATCase in *E. coli*; the methods they employed to "dissect" these very rapid reactions are known as relaxation techniques. In effect, relaxation techniques spread out the control process along the time dimension. The results of Hammes and Wu, published in 1971, indicate that the MWC and AKNF concepts of allosteric enzyme regulation are limiting cases of a more complex general model that they have put forward. Hammes and Wu consider that the entire control mechanism of ATCase in *E. coli* is much more complicated than is implied by either the MWC or AKNF models. For example, binding of a given ligand by a subunit can be associated with more than one type of conformational change, and identical end results (activation or inhibition) can be brought about by several different conformational transitions.

Many different pathways leading to the same results makes good physiological sense, because this means that the mechanism of allosteric regulation of enzymes is more flexible than the earlier models lead one to suppose, and the mechanism is thereby rendered sensitive to the influence of many different molecules.

QUINTARY STRUCTURE. THE PYRUVATE DECARBOXYLASE COMPLEX

The previous sections have dealt with the biological significance of quaternary structure in globular protein molecules. Quaternary structure arises from the association of more or less similar globular subunits that, held together by noncovalent forces, give rise to the functional molecules with new and novel physiological properties. In the case of enzymes, such as LDH, we have seen that the specific catalytic function arises directly from the quaternary association of subunits, since the individual subunits in the case of LDH do not have LDH activity.

A yet higher organizational level in the architecture of enzyme molecules has been recognized; it is called quintary structure. Such enzymes, as well as being structurally larger and more complex than quaternary enzymes, are close spatial assemblages of dissimilar molecules that themselves possess quaternary structure. In the chapter on enzymes, brief reference was made to the assemblage of enzymes into enzyme complexes. The pyruvate decarboxylase complex (PDC) of *E. coli* is an example of this, and the significance of quintary structure is illustrated by a study of this enzyme complex.

The overall reaction catalyzed by PDC is the oxidative decarboxylation of pyruvate. The reaction can be separated into three reactions, each of which is catalyzed by separate enzymes. First, pyruvate is decarboxylated, forming an ethylol group on thiamine pyrophosphate (which is attached to the decarboxylase enzyme). This group is then oxidized by a lipoyllysyl residue to give acetyl dihydrolipoyllysyl (which is attached to the transacetylase enzyme). Finally, this acetyl group is transferred to coenzyme A, flavoprotein reoxidizes the dihydrolipoyllysyl group, and NADH is formed. Thus, the end product of one reaction forms the substrate for the next enzyme with which it will be in close spatial proximity. Two of the reactions involved are transacetylation and electron transfer. The close coupling together of the enzymes involved would certainly seem to lead to a better functional performance; the biological significance of this assemblage of different enzymes and cofactors into a closely organized spatial system—quintary structure—could therefore be interpreted as being in the interests of overall efficiency in multiple enzyme systems.

PDC is inhibited by ATP, which is therefore a negative modulator. Thus, when the ATP concentration exceeds a certain level, the PDC system, which supplies acetyl coenzyme A ("active acetate") for the Krebs cycle, is inhibited.

The PDC complex has been isolated (molecular weight $= 4.44 \times 10^6$) and was visualized in the electron microscope by L. J. Reed and R. M. Oliver, who elucidated its quintary structure. It has a cubical shape, measuring 350 ± 50 Å $\times$ 225 ± 25 Å, and there are three cubically arranged assemblages of enzyme subunits. The belief is that the complex, as isolated, is the native form and is not artefactual.

The three enzymes involved are (1) pyruvate dehydrogenase, E_1, each molecule of which has a bound thiamine pyrophosphate; (2) dihydrolipoyl transacetylase, E_2, each molecule of which contains one molecule of lipoic acid; and (3) dihydrolipoyl dehydrogenase, E_3, each molecule containing one molecule of FAD. Several molecules of each of these three enzymes, plus their coenzymes and two other coenzymes are organized in a specific spatial configuration in PDC. At the center of the complex are eight trimeric E_2 subunits in a cubical arrangement; there are therefore 24 identical polypeptides. This "E_2 cube" has 24 molecules of E_1 (each of which is a tetramer consisting of two different polypeptides) arranged along its edges. On the six faces of the E_2 cube are arranged groups of two dimeric E_3 molecules in square fashion; that is, there is a total of 12 E_3 molecules (Fig. 9-25).

In the enzymatically catalyzed reactions involved in the decarboxylation of pyruvate, the lipoyl group goes through a cyclic sequence of transformations—

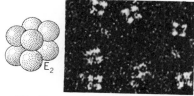

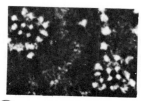

The E_2 cube of eight trimers.
Each stippled ball is 3 subunits.

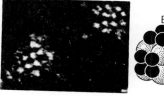

The E_2 cube with 24 molecules of
E_1 (white balls) on the cube edges.

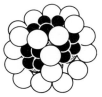

The E_2 cube with 24 molecules of
E_3 (black balls) on the cube faces.

Organization of the total complex:
E_1, E_2 (inside), and E_3.

Figure 9-25 The pyruvate decarboxylase complex. (Reprinted from *The Structure and Action of Proteins* by R. E. Dickerson and I. Geis. W. A. Benjamin Inc., Menlo Park, California, Publisher. Copyright 1969 by Dickerson and Geis.)

reductive acylation, acyl transfer, and electron transfer. This must mean that the lipoyl group attached to E_2 interacts with the coenzymes associated with E_1 and E_3. Reed and Oliver discovered that the lipoyl group of E_2 (transacetylase) is bound to the terminal amino group of a lysine side chain. Effectively, this flexible attachment provides an "arm" of the right length for the dithiolane ring that, they suggested, may permit rotation of the lipoyl group between the prosthetic groups of the other two enzymes. The precise spatial organization of the molecules in PDC is again seen to be highly effective in the efficient coupling of the sequential reactions.

The transacetylase (E_2), as well as being a catalyst, also serves a structural function. It combines with both the E_1 and E_3, but E_1 and E_3 do not combine with themselves. When all three are mixed together, the E_2 seems to organize the aggregation of the complete PDC, and one of the eight E_2 subunits can combine with E_1 and E_2 to give a functional complex that is one-eighth the size of the full assembly.

CONCLUSIONS

One aspect of the biological significance of quaternary structure, as already stated, lies in the emergence of new and novel properties in the quaternary protein molecule that are not inherent in the separate subunits. Quaternary structure makes possible allosteric conformational shape changes that are of considerable importance in the regulation and coordination of cell metabolism at the molecular level; these are fast-acting and provide means by which quick responses to the varying metabolic needs of the cell may be met. This efficient and rapid local control of metabolic reactions contrasts with the much slower-acting cellular control mechanisms that involve the genetic apparatus (see Section on "Genetic Control of Protein Biosynthesis" in Chapter Eighteen). Decentralization of metabolic control effected by allosteric proteins that rapidly and sensitively respond to changing local conditions within the cell reminds one almost irresistably of the situation in human societies in which local government is usually more rapidly responsive to immediate societal needs than the slower and more ponderous machinery of central government.

The precise spatial organization of enzyme systems as in PDC, known as quintary structure, makes possible a high degree of efficiency that would not otherwise be possible.

D. L. D. Caspar has shown that in the cellular synthesis of, say 10 polypeptides consisting of 100 amino acid residues each, there is less chance of a mistake occurring in the synthesis of any one of these than in the synthesis of one long polypeptide chain consisting of 1000 amino acid residues. A mistake (e.g., a wrong amino acid) in one position in the assemblage of a single very long polypeptide chain may render the entire molecule functionless and would be wasteful of materials and energy. If, however, the complex, functional molecule is fashioned of an association of a number of separate shorter polypeptide chains, then a single mistake would affect only one of these and only this one would have to be replaced; the others are normal and do not need to be discarded. Quaternary structure may therefore also be viewed as a device for minimizing the possible wastage of cellular energy and materials that could result from mistakes in the assemblage of large complex proteins if they consisted of only one very long polypeptide chain instead of being built from several smaller components.

Allosteric control phenomena are not necessarily confined to catalytic proteins but, as D. E. Koshland points out, many different types of biological phenomena are very probably mediated by and/or controlled by "flexible" protein molecules. Koshland considers the following phenomena to be associated with allosteric proteins that undergo configurational changes on binding with various kinds of effector molecules.

1. *Sensory Reception.* The senses of smell and taste may be mediated by configurational changes induced in protein receptor molecules by binding of the smaller molecules that are being smelled or tasted. These shape changes might trigger impulses in the associated nerve fibers. Photoreception has been definitely shown to be due to shape changes in a specific protein in the retina of the vertebrate eye.

2. *Control of Protein Synthesis.* Much of the genetic information coded by the DNA of a given cell type is "masked" by being combined with protein repressor molecules. The unmasking of genetic information can be brought about by inducer molecules that may combine with and cause conformational changes in the protein repressor molecules; this causes them to become detached from DNA, so that transcription and subsequent translation can start. Similar phenomena may be involved in cell and tissue differentiation.

3. *Immunity.* The complement fixation system destroys harmful foreign antigens that gain access to the body. The selective protective reactions of this system is induced by antibodies—specific proteins formed by the body in response to foreign antigens—and conformational changes in these proteins are believed to be involved in activating the body's protection system.

4. *Cell Membrane Transport.* Active transport of food and other materials across cell membranes may involve conformational changes in specific transport proteins in cell membranes.

PART THREE

PROTEIN-LIPID ASSEMBLIES. BIOMEMBRANES AND ENERGY TRANSDUCTION

CHAPTER TEN
Structure of lipids

The word *lipid* is not a chemical term, but a collective name given to a heterogeneous group of substances found in living cells that share two properties, but little else in common; they are insoluble in water and soluble in "fat solvents" such as benzene, chloroform, and diethyl ether. The lipids comprise a variety of chemically unrelated substances that are generally treated together because of their similar solubility characteristics. They are extracted from thoroughly dried organic material by heating with a fat solvent, preferably in a continuous extraction apparatus like that devised by Soxhlet (Fig. 10-1). After prolonged extraction, the resulting solution of lipids is heated in a distilling apparatus to remove the solvent, and the nonvolatile, oily residue is left behind and constitutes the "lipid fraction" of the material under study. It may be further fractionated by boiling with alkali; this hydrolyzes any material containing ester linkages such as triglycerides and glycerophosphatides, which are two of the three main types of lipids.

Among the products of hydrolysis of triglycerides and glycerophosphatides are carboxylic acids with rather long hydrocarbon side chains. Mixtures of the sodium or potassium salts of these that result from alkaline hydrolysis of the lipids are what are called soaps. The process of alkaline hydrolysis of these lipids is therefore sometimes called saponification, and triglycerides and glycerophosphatides are referred to as the saponifiable fraction of the total lipids. The aqueous hydrolysate is then shaken with, say, benzene, which extracts only the lipids resistant to alkaline hydrolysis; these lipids therefore belong to the nonsaponificable fraction and are called sterols. Cholesterol is the most important cellular sterol.

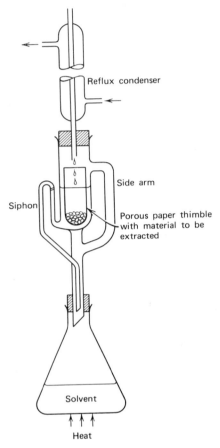

Reflux condenser

Side arm

Siphon

Porous paper thimble
with material to be
extracted

Solvent

Heat

Figure 10-1 Soxhlet extraction apparatus. Solvent vapor from the boiling flask passes through the tube in the stopper and up through the side arm into the reflux condenser. The condensed solvent drops into the paper thimble containing the material to be extracted. When it reaches a certain level, the solvent with dissolved lipids siphons over into the boiling flask. Solvent continues to drip into the extraction thimble and siphons into the boiling flask when enough has accumulated again. The filling–emptying cycle continues as long as heat is supplied to the boiling flask. The material is being continuously extracted, therefore, with fresh solvent.

Many lipids are important structural components of cell membranes. As far as cell biology is concerned, the major groups of lipids, their structures, and properties are as follows.

SIMPLE LIPIDS

(i) Neutral fats (Triglycerides, acyl glycerols)

These are carboxylic acid triesters of glycerol and have the general formula shown Fig. 10-2. The carboxylic acid side chains may all be identical, as in triolein, which is the glycerol triester of oleic acid, or they may all be different. In this case, the central carbon atom of the glycerol moiety will be asymmetric, and such triglycerides will therefore exist in D and L forms.

Triglycerides in which the side chains of the carboxylic acid residues contain only single bonds are therefore fully saturated with hydrogen and are called saturated fats. They are characteristically found in animals. Unsaturated triglycerides contain side chains with one or more double bonds and are characteristically found in plants.

The triglycerides possess no ionizable groups; therefore they are water-insoluble and cannot form hydrogen bonds. They do not play a structural role in cell membranes. Their functions are mainly as sources of cellular energy.

(ii) Fatty acids

The carboxylic acids obtained by hydrolysis of triglycerides are often called fatty acids. Many contain long hydrocarbon side chains, such as palmitic and stearic acids, which possess 15 and 17 carbon atoms, respectively, in the side chain. Almost invariably, the molecules of naturally occurring fatty acids have even numbers of carbon atoms.

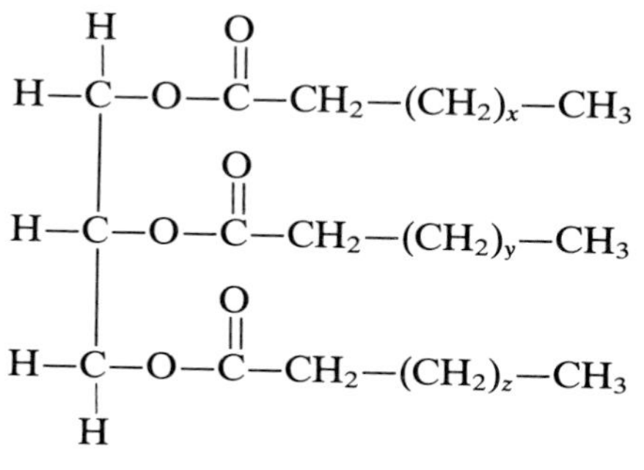

$$
\begin{array}{l}
\text{H} \quad\quad \text{O} \\
\text{H}-\overset{|}{\underset{|}{\text{C}}}-\text{O}-\overset{\parallel}{\text{C}}-\text{CH}_2-(\text{CH}_2)_x-\text{CH}_3 \\[4pt]
\quad\quad\quad\quad \text{O} \\
\text{H}-\overset{|}{\underset{|}{\text{C}}}-\text{O}-\overset{\parallel}{\text{C}}-\text{CH}_2-(\text{CH}_2)_y-\text{CH}_3 \\[4pt]
\quad\quad\quad\quad \text{O} \\
\text{H}-\overset{|}{\underset{|}{\text{C}}}-\text{O}-\overset{\parallel}{\text{C}}-\text{CH}_2-(\text{CH}_2)_z-\text{CH}_3 \\
\quad\,\text{H}
\end{array}
$$

Figure 10-2   General structural formula of a triglyceride.

One or more double bonds are found in the side chains of certain fatty acids. Oleic acid, for example, has one double bond, linoleic acid has two, linolenic acid has three, and arachidonic acid has four. These fatty acids are therefore said to be unsaturated, and their glycerol esters are the unsaturated fats. Saturated fatty acids such as stearic and palmitic acids lack double bonds and give rise to the saturated fats. Unsaturated fatty acids have a much lower melting point than the corresponding saturated fatty acids with a similar number of carbon atoms. Generally, unsaturated fatty acids are more abundant than the saturated type, especially in the cells of poikilothermic organisms that live at low temperatures.

COMPOUND LIPIDS (PHOSPHOLIPIDS, CONJUGATED LIPIDS)

These are important structural components of cell membranes. The name phospholipid derives because many of these lipids contain posphoric acid. All of them contain compounds other than glycerol and carboxylic acids. The following is a classification of the compound lipids.

(i) Glycerophosphatides

These are diesters of phosphoric acid and diglycerides; the second alcoholic component is either an amino alcohol or inositol. These compounds can be considered as esters of a parent phosphatidic acid. The general formula of the glycerophosphatides is shown in Fig. 10-3. In contrast to triglycerides, the phospholipids possess an ionizable water-solubilizing group (phosphate). This is a very important property from the standpoint of cell membrane structure. Examples of glycerophosphatides are shown in Table 10-1.

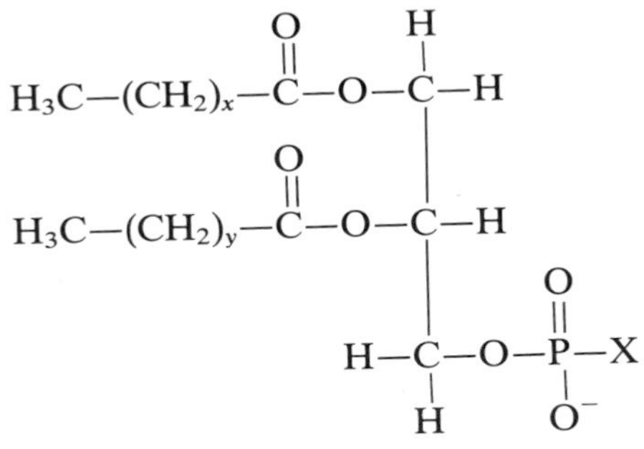

Figure 10-3 General structural formula of a glycerophosphatide. "X" can be any one of several different chemical groups (see Table 10-1).

TABLE 10-1 **Examples of Some Glycerophosphatides (refer to Fig. 10-3).**

X group	Name of Compound
—H	Phosphatidic acid
$-CH_2-CH_2-\overset{+}{N}\begin{smallmatrix}CH_3\\CH_3\\CH_3\end{smallmatrix}$	Phosphatidyl choline (lecithin)
$-CH_2-CH_2-\overset{+}{N}H_3$	Phosphatidyl ethanolamine (cephalin)
$-CH_2-\underset{\underset{COO^-}{|}}{CH}-\overset{+}{N}H_3$	Phosphatidyl serine
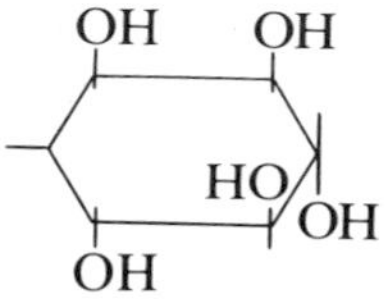	Phosphatidyl inositol

From C. U. M. Smith.

(ii) Sphingolipids

These are also phosphodiesters, but the place of the diglyceride component is taken by sphingosine, an unsaturated amino alcohol (Fig. 10-4). The second hydroxyl group of the phosphoric acid is usually esterified with choline. Examples of this type of phospholipid are the sphingomyelins (Fig. 10-5), which are found principally in the myelin sheath of nerves.

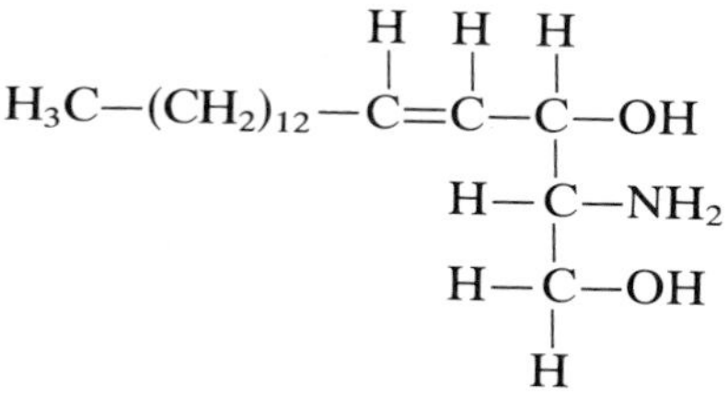

$$H_3C-(CH_2)_{12}-\overset{\overset{H}{|}}{C}=\overset{\overset{H}{|}}{C}-\overset{\overset{H}{|}}{C}-OH$$
$$H-\overset{|}{C}-NH_2$$
$$H-\overset{|}{C}-OH$$
$$\overset{|}{H}$$

Figure 10-4 Structure of sphingosine.

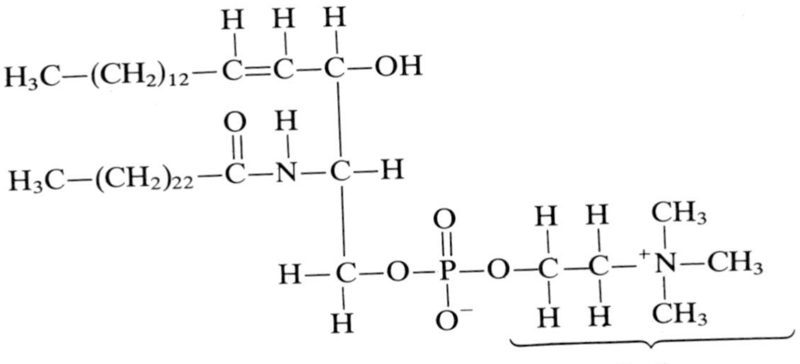

$$H_3C-(CH_2)_{12}-\overset{\overset{\displaystyle H}{|}}{C}=\overset{\overset{\displaystyle H}{|}}{C}-\overset{\overset{\displaystyle H}{|}}{C}-OH$$

$$H_3C-(CH_2)_{22}-\overset{\overset{\displaystyle O}{\|}}{C}-\overset{\overset{\displaystyle H}{|}}{N}-\overset{|}{C}-H$$

$$H-\overset{|}{\underset{|}{C}}-O-\overset{\overset{\displaystyle O}{\|}}{\underset{\underset{\displaystyle O^-}{|}}{P}}-O-\overset{\overset{\displaystyle H}{|}}{\underset{\underset{\displaystyle H}{|}}{C}}-\overset{\overset{\displaystyle H}{|}}{\underset{\underset{\displaystyle H}{|}}{C}}-\overset{+}{N}-CH_3$$

Choline

Figure 10-5 Structure of sphingomyelin.

(iii) Plasmalogens

Plasmalogens contain a vinyl ether group attached at the 1-position of the glycerol, a carboxylic acid residue, and glycerylphosphorylethanolamine or the related derivative of choline (Fig. 10-6). Plasmalogens are found in relatively large amounts in brain and nerve myelin.

(iv) Cerebrosides (glycolipids)

These are not really phospholipids but, since they are derivatives of sphingosine, they can be justifiably included here. A cerebroside can be thought of as a combination of a molecule of a sugar (e.g., galactose), a molecule of sphingosine, and a long-chain carboxylic acid (Fig. 10-7). At least four cerebrosides

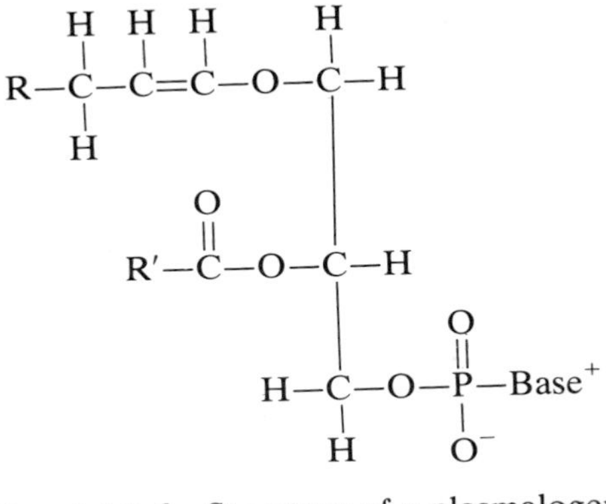

$$R-\overset{\overset{\displaystyle H}{|}}{\underset{\underset{\displaystyle H}{|}}{C}}-\overset{\overset{\displaystyle H}{|}}{C}=\overset{\overset{\displaystyle H}{|}}{C}-O-\overset{\overset{\displaystyle H}{|}}{C}-H$$

$$R'-\overset{\overset{\displaystyle O}{\|}}{C}-O-\overset{|}{C}-H$$

$$H-\overset{|}{\underset{\underset{\displaystyle H}{|}}{C}}-O-\overset{\overset{\displaystyle O}{\|}}{\underset{\underset{\displaystyle O^-}{|}}{P}}-Base^+$$

Figure 10-6 Structure of a plasmalogen.

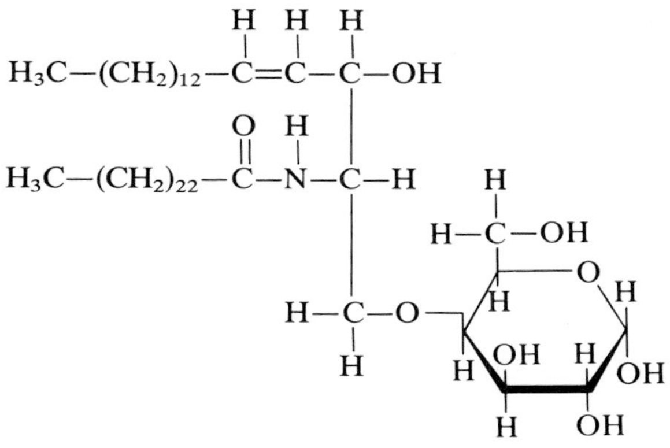

$$H_3C-(CH_2)_{12}-\overset{\overset{\displaystyle H}{|}}{C}=\overset{\overset{\displaystyle H}{|}}{C}-\overset{\overset{\displaystyle H}{|}}{C}-OH$$

Galactose

Figure 10-7   Structure of a cerebroside.

are known (nervone, oxynervone, kerasin, and phrenosin), and they occur in brain, adrenals, kidney, spleen, liver, and egg yolk, among other sources.

(v) Gangliosides

These are complex molecules and derive from sphingosine; the sphingosine is linked by galactose successively to glucose, a hexosamine and neuraminic acid. Gangliosides are present in cellular membranes and possibly have a role as receptors of virus particles. They influence transport of ions across membranes.

(vi) Sulfatides

Sulfatides are derivatives of sphingosine that contain sulfuric acid esterified to galactose.

STEROLS (CHOLESTEROL)

The steroids are a group of compounds of widely different properties that all have the steroid carbon skeleton in common. In the precise but formidable language of organic chemistry, this is the cyclopentanoperhydro-phenanthrene nucleus. This is a structure consisting of three fused benzene rings in an obtuse angular conformation (phenanthrene, Fig. 10-8a); to this is fused a five-membered carbon ring to give the steroid structure (Fig. 10-8b). The properties of individual steroids are determined by the substituent groups in the steroid skeleton and the degree of saturation or unsaturation of the carbon–carbon

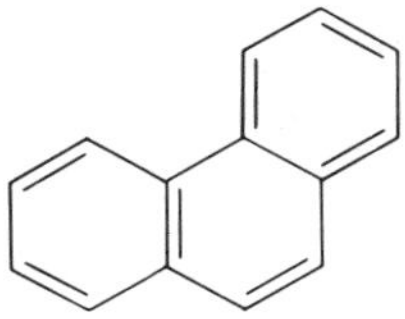

(a)

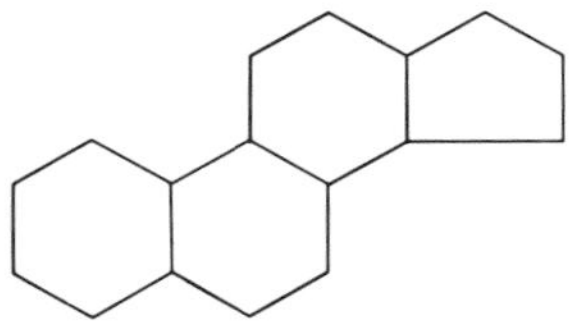

(b)

Figure 10-8 Structures of phenanthrene and the steroid skeleton. (a) Phenanthrene. (b) Steroid structure (cyclopentanoperhydrophenanthrene).

bonds. Some idea of the diversity of these substances may be gained from the list of the main classes of steroids.

Male and female sex hormones.
Bile acids.
Sterols.
Vitamins D.
Adrenal corticosteroids.
Cardiac glycosides.
Saponins.

Of these, the sterols are so-called because they contain an alcohol (—OH) group; the best-known and most important example from the point of view of cell membrane structure is cholesterol (Fig. 10-9). Cholesterol is one of the most

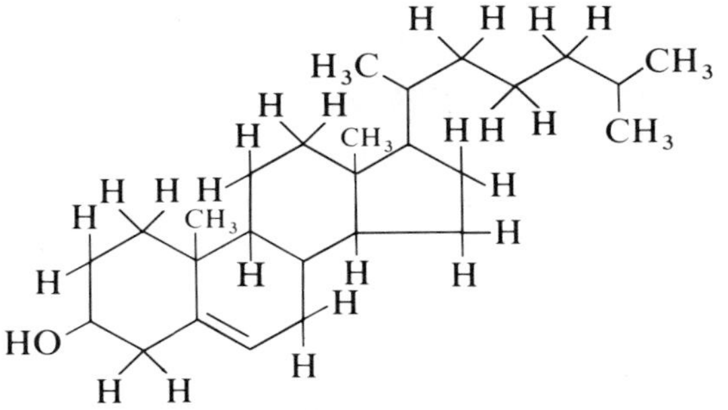

Figure 10-9   The molecule of cholesterol.

important structural molecules of cell membranes. One end of the molecule is hydrophilic because of the presence of the hydroxyl group; the other end, because of the presence of the hydrocarbon chain and the steroid ring structure, is hydrophobic, so that in these characteristics it resembles the phospholipids in being an *amphipathic* molecule.

PHYSICOCHEMICAL PROPERTIES OF PHOSPHOLIPIDS

Many textbooks show the general structural formula of a phospholipid (glycerophosphatide) with the ionizable phosphate group lying alongside the hydrophobic carboxylic ester groups. In reality, as shown in Fig. 10-3, the ionizable group and the hydrophobic groups in the phospholipid molecule repel each other. Free rotation is possible around single covalent bonds, and so the phosphate group swings round to the other side away from the carboxylic ester groups. The phospholipid molecule thus has two ends with different physicochemical properties; this is emphasized in the schematic diagram of a phospholipid molecule in Fig. 10-10.

Because phospholipids possess both hydrophilic and hydrophobic groups, they are said to be *amphipathic*. This property is of great importance from the point of view of cell membrane structure, as will be apparent in Chapter Eleven. Because they are amphipathic, phospholipid molecules have a strong tendency to orient themselves and to form organized aggregates in the bulk of a solvent or at an interface. At an air–water or an oil–water interface, for example, the phospholipid molecules at the liquid surface have a strong tendency to arrange themselves in a regularly arranged layer of molecules, with the hydrophobic

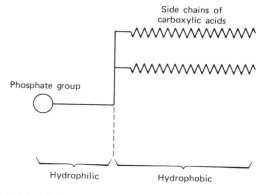

Figure 10-10   Schematic model of a phospholipid molecule.

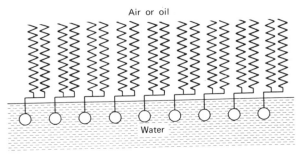

Figure 10-11   Formation of a monomolecular layer of phospholipid molecules at a water–air or water–oil interface.

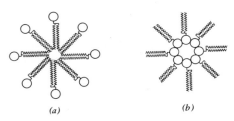

(a)   (b)

Figure 10-12   Micelle formation. (a) Aqueous medium. (b) Organic solvent (e.g., benzene).

"tails" projecting into the oil or air phase and the hydrophilic "heads" projecting into the aqueous phase. Such a film of oriented molecules at an interface is called a monomolecular layer or monolayer for short (Fig. 10-11). The formation of such sheets of oriented molecules is the basis of cell membrane structure (Chapter Eleven). In the bulk of a solvent, spherical aggregates of molecules called micelles form. In an aqueous medium, the hydrophobic heads all point outward and the hydrophobic tails point inward (Fig. 10-12a); the reverse occurs in an organic solvent such as benzene (Fig. 10-12b). This property therefore renders phospholipids soluble in both aqueous and hydrophobic solvents. Micelle formation is thought to play a role in one conceptual model of cell membrane structure (Chapter Eleven).

CHAPTER ELEVEN
Biomembranes

Much of the structure of eucaryotic cells consists of membranes and membrane derivatives. The naked cytoplasmic surface is covered by the plasma membrane, which is in structural continuity with the complex system of parallel sheet- or tubelike membranes within the cytoplasm; this system is known as the endoplasmic reticulum. The cell nucleus is enclosed in a double membranous envelope; electron microscope studies of thin serial sections of cells indicate that this, too, may be continuous with the endoplasmic reticulum and is therefore ultimately continuous with the plasma membrane. Cytoplasmic organelles such as the Golgi body (a specialized region of the endoplasmic reticulum), lysosomes, mitochondria, chloroplasts, and vacuoles are all membranous structures. Developmentally, they are believed to be derived from the endoplasmic reticulum membranes.

According to one theory, chloroplasts and mitochondria may be the descendants of what were once endosymbionts in ancestral eucaryotic cells (see Chapter Twenty). Eucaryotic cells are therefore highly compartmentalized by membranous structures.

Functions of Biomembranes

Cellular membranes should not be thought of as inert structural materials. They provide surfaces on which cellular activities proceed, and they are deeply involved in vital processes. The chapter on biological energy transducers will show, for example, that the coenzymes of cellular respiration are probably integral parts of the intramitochondrial membranes, and that the visual pigment of vertebrate visual receptors (which transduce light energy into the energy of a nervous impulse) is an integral structural component of the intricately ordered membranes of the photoreceptor cells. In animal cells the synthesis of proteins is closely associated with the endoplasmic reticulum membranes. The most thoroughly studied function of cellular membranes is their highly important role as permeability barriers. The permeability properties of the plasma membrane may be summarized as follows.

1. Substances readily soluble in lipid solvents penetrate the membrane easily.
2. Small anions, such as the chloride ion, pass readily through the membrane.
3. Certain other ions, such as sodium and potassium ions do not penetrate readily; neither do carbohydrates and proteins. There seem to be special mechanisms for aiding the penetration of these substances through the plasma membrane.

The internal chemical and ionic composition of cells must be kept constant within very narrow limits because the ionic composition, to quote one factor, of the cell interior critically influences the enormous number of interdependent reactions that go on within it. This constancy is maintained by the plasma membrane.

THE REAL EXISTENCE OF THE PLASMA MEMBRANE

At one time biologists thought that there was no actual membrane bounding the cell and that the physical properties of the cell surface were due to surface tension effects. With the introduction of micromanipulators into biological research, the existence of a definite membrane at the cell surface was clearly demonstrated.

Around 1920, R. Chambers demonstrated the existence of a membrane at the cell surface by using delicate microneedles to tear the surfaces of sea urchin eggs from which the extraneous layers of protective material had been removed. A wave, or "ripple," was observed at the cell surfaces, and the previously sharp cell boundaries became fuzzy. This appearance results from destruction of a surface film.

A different type of evidence was provided by Chambers's experiments in which harmless pH indicators were injected into living cells. The dyestuffs diffused readily throughout the cytoplasm, but did not diffuse out on reaching the boundary. Cells suspended in solutions of the indicators did not become

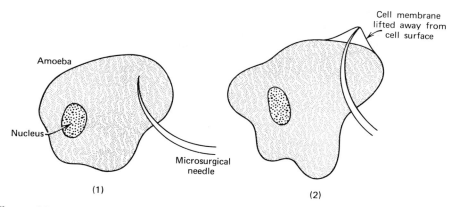

Figure 11-1 Demonstration of the existence of a membrane at the cell surface of ameoba by microsurgery. (After De Fonbrune.)

stained. Evidently, there is a boundary layer at the cell surface that is responsible for the impermeability of the cell to these substances.

Many years ago, De Fonbrune made a motion picture that showed that a definite film could be lifted off the cell surface of ameobae by penetrating them with curved microsurgical needles (Fig. 11-1). However, this surface membrane probably included the extraneous coats as well as the true plasma membrane.

STRUCTURE AND CHEMICAL COMPOSITION OF BIOMEMBRANES

As previously mentioned, permeability studies of plasma membranes indicate that lipid-soluble materials penetrate the plasma membrane quite easily. The simplest explanation of this would be that plasma membranes consist, at least partly, of lipid material.

In a classic experiment, Gorter and Grendel in 1927 quantitatively analyzed isolated red blood cell membranes ("ghosts") that had been washed free of hemoglobin. They extracted the lipid from the ghosts, and this was spread as a monomolecular film on water in a Langmuir trough. The close-packed molecules presumably oriented themselves with their hydrophilic ends immersed in the aqueous phase and the hydrophobic ends projecting out of the water. The area of the film was just about twice the surface area calculated for the amount of red cells that they had used, so they concluded that the lipid was arranged as a bimolecular layer within the red cell membrane, the hydrophobic ends facing in toward each other and the hydrophilic ends facing outward and inward. Subsequently, other investigators showed that the surface area of the red blood

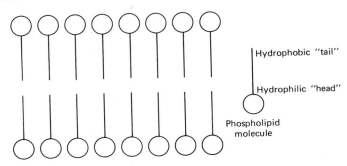

Hydrophobic "tail"

Hydrophilic "head"

Phospholipid molecule

Figure 11-2 Phospholipid bilayer of cell membranes.

cells as calculated by Gorter and Grendel was probably about 50% greater than it actually was; also, the acetone extraction method that those workers used only removed about two thirds of the lipids in the ghosts. Fortunately, these errors in Gorter and Grendel's experiments cancelled out, and the lipid bilayer model still was a valid proposal. The organization of the lipid bilayer in cell membranes that is generally accepted by biologists is shown in Fig. 11-2. Such an arrangement is well suited as a barrier between two aqueous solutions such as the cell interior and the external environment. The hydrophilic ends of the lipid molecules are directed toward the aqueous phases, and the hydrophobic ends form an "oily" interior to the bilayer. A phospholipid bilayer structure for the cell membrane would account well for the known permeability properties, because it would be relatively impermeable to charged particles and ions but easily penetrable, by fat-soluble materials. However, results of measurements of surface tension, elastic properties, and electrical capacitance could not be easily explained on the simple phospholipid bilayer model.

THE DANIELLI–DAVSON MODEL OF CELL MEMBRANES

To accomodate these facts, Danielli and Davson in 1935 proposed that the phospholipid bilayer is overlaid with a monomolecular layer of protein molecules in the extended configuration on both of its sides so as to form a protein-lipid "sandwich" (Fig. 11-3). Danielli later suggested that protein might be incorporated in the membrane in places in the form of protein-lined aqueous pores.

Since the dimensions of the molecules entering into the composition of the Danielli–Davson membrane model were known, it enabled them to calculate a minimum thickness for the plasma membrane (they had no way of knowing whether the plasma membrane consisted of a single protein-lipid sandwich or of

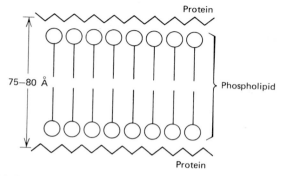

Figure 11-3 The Danielli–Davson model of the cell membrane.

several). This was estimated to be around 75–80 Å which, of course, is well below the limit of resolution of the light microscope. What relation, then, does the Danielli–Davson membrane model bear to the structure actually seen at the cell surface with the light microscope, which has been interpreted as the cell membrane? Before answering this question, we must define exactly what is meant by the cell membrane. At first, this may seem a little odd and may smack somewhat of hair-splitting about the obvious, but the concept is not as obvious as it may seem. For one thing, what biologists interpret as the cell membrane depends largely on the perfection of the instrumentation used to visualize it. In the preelectron microscope era the cell membrane, as seen in the light microscope, was defined as the thinnest dense line at the cell surface separating the cytoplasm from the exterior (Fig. 11-4a). When thin sections of cells were examined in the electron microscope more than one dense line was visible at the cell surface, and so difficulties arose as to which one was to be properly considered *the* cell membrane. Robertson showed that when thin sections of muscle fibers were examined with the electron microscopes of relatively low resolving power that were available in the early 1950s, *two* thin dense lines of unequal thickness were seen at the cell surface, the combined width of which was below the resolution of the light microscope (Fig. 11-4b). The cell membrane was then redefined as the single dense line at the cell surface, which was nearer to the cell interior. A few years later, improvements in electron microscope resolution showed that the single dense line defined as the cell membrane in low-resolution electron micrographs itself had a double structure (Fig. 11-4c). Examination of a wide variety of cell types showed that the double dense line was ubiquitous and constant in appearance, and Robertson concluded that it was the true cell membrane. The plasma membrane at the surface of a red blood cell is shown in the electron micrograph of Fig. 11-5. He called it the "unit membrane," because it appeared to be the structural unit from which complex membranous structures were built. Such a complex membranous structure is the myelin sheath of

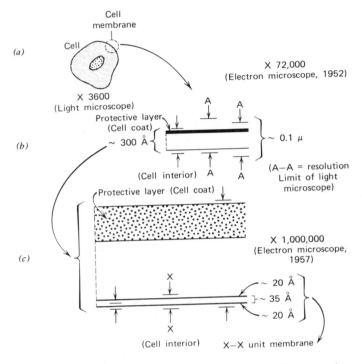

Figure 11-4   What is the cell membrane? (See text.)

medullated nerve fibers. Much of the early work on the chemical composition and molecular organization of biomembranes was done with nerve myelin which, as will be shown, consists of regularly stacked layers of unit membranes. It is a plentiful source of pure membrane material for chemical and structural analysis. However, nerve myelin is not representative of cell membrane structure and composition in general.

STRUCTURE OF NERVE MYELIN

Before the electron microscope was invented quite a lot was known about nerve myelin. Chemically, it consists of lipid and protein. X-ray diffraction analysis and the work of Schmidt on its optical properties indicated that in fresh, unfixed, myelinated nerve fibers, the myelin has a lamellar structure. The lipid molecules are radially arranged, and the protein molecules are tangentially disposed in each lamellar sheet. The X-ray diffraction studies of Schmitt and others in 1935 showed that the radial repeat distance between the lamellae was about 180 Å.

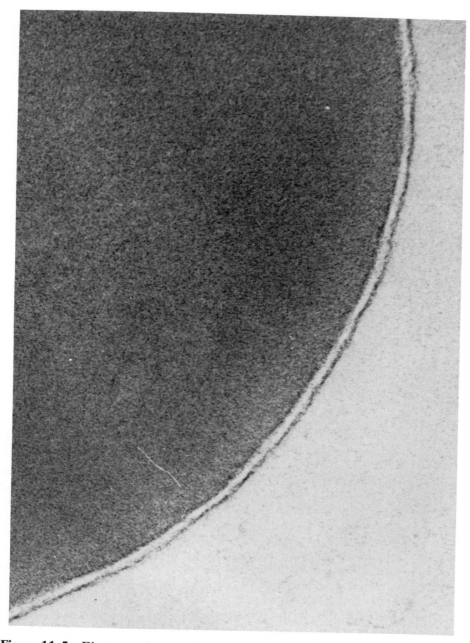

Figure 11-5 Electron micrograph showing the plasma membrane at the surface of a red blood cell. (Courtesy of Dr J. D. Robertson.)

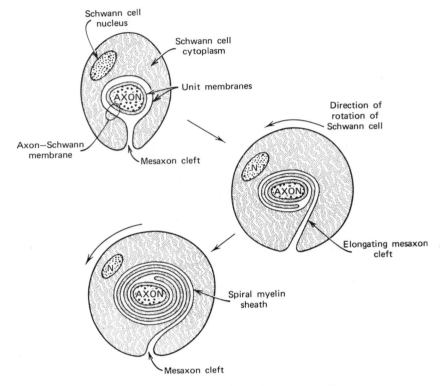

Figure 11-6 Formation of nerve myelin.

The axon of a peripheral nerve cell, which is destined to become invested with a myelin sheath, is initially associated with the so-called Schwann cells. The axon sinks into a Schwann cell, giving the appearance in section as shown in Fig. 11-6. The cleft into which the axon sinks is called the mesaxon cleft, because of its resemblance to the two sheets of mesentery in which the intestines of vertebrates are slung. The two juxtaposed layers of unit membrane forming the mesaxon cleft elongate and, at the same time, the Schwann cell rotates. This has the effect of wrapping the axon around with an ever-thickening, spirally arranged, multi-layered sheath of unit membrane material (Fig. 11-6). This constitutes the myelin sheath when the process is complete.

Examination of transverse, ultrathin sections of osmium-fixed myelin reveals the presence of concentric circular layers of dark (electron-dense) bands in regular repetitive arrangement. Another series of less dense and less distinct bands, parallel to the dense bands, divides the spaces between the dense bands. These lighter bands are designated intraperiod lines. Measurements of the repeat distance of the dense or intraperiod lines give values of about 180 Å, which is in accord with the X-ray diffraction data.

Robertson traced the two sets of lines round the spiral and down to their origin and found that the dense lines were composed of the two fused inner layers of Schwann cell membranes. The intraperiod lines corresponded to two fused outer layers of the membrane. Since the space between the two intraperiod lines contains two layers of unit membrane, and because the lines are 180 Å apart, the Schwann cell membrane is considered to be about 90 Å thick. Electron microscope studies of other cell membranes indicate that this value may be rather high, since most cell membranes appear to be about 75–80 Å thick, which is in excellent agreement with the thickness estimated for the Davson–Danielli plasma membrane model. The plasma membrane therefore appears to consist of a single Davson–Danielli type of membrane corresponding to a single Robertson type of unit membrane. The two electron-dense lines are believed to be due to protein, and the lighter space between them corresponds to the lipid bilayer.

In osmium-fixed preparations the dense and intraperiod lines have different appearances in the electron microscope, and it therefore seems that the two osmiophilic layers of the unit membrane might be dissimilar in composition. Possibly, this compositional asymmetry of the unit membrane is related to the

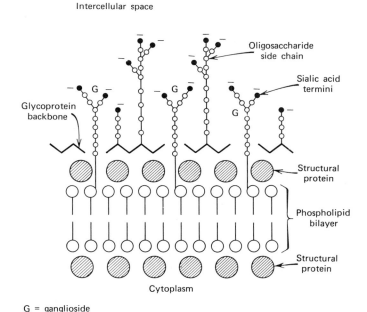

Figure 11-7 Cell coat on the outer surface of the cell membrane. (Redrawn after Lehninger *Proc. Nat. Acad. Sci. USA*, *60*: 1069–1080, 1968.)

fact that the external surfaces of various cells are often covered with a coating of polysaccharide material, such as chondroitin sulfate in cartilage cells and mucus in epithelial cells (Fig. 11-7).

THE MICELLAR MOSAIC MODEL

Certain specific permeability phenomena of the plasma membrane cannot be easily explained on the basis of the continuous phospholipid-protein bilayer model. Small molecules penetrate the membrane more rapidly than larger ones; this indicates the existence of some kind of "molecular seive." It is necessary to postulate that there are small local areas or pores that differ in permeability from the rest of the membrane.

Chapter Ten showed that as well as forming monolayers at interfaces, phospholipids form globular micelles in the bulk of a liquid phase. That the cell membrane may possess a micellar mosaic instead of a phospholipid bilayer structure has been proposed to explain these special permeability properties. Two possible micellar arrangements of the lipids in the plasma membrane are shown in Fig. 11-8. One of them possesses charged pores. It seems that these would exert a major influence on the entrance and exit of materials to and from the cell. Also, such a membrane would have essentially the same thickness as the bilayer structure. The permeability properties would be quite different, however.

The evidence for the existence of a micellar structure of the plasma membrane comes from electron microscopy, but it may well be that the observed appearances are merely due to preparative artefacts.

The sandwichlike phospholipid-protein bilayer structure gives an impression of a rigid static structure. However, it is quite possible that it could reversibly change to a micellar structure, which would result in variable permeability of the membrane. The plasma membrane may therefore now be visualized as a dynamic labile structure, and this is perhaps more realistic. The molecular architecture of this can be thought of as changing reversibly from the sandwich to the mosaic form in response to the changing permeability requirements of the cell necessary for the maintenance of the physicochemical constancy of its interior.

Even this refinement cannot easily explain why certain charged ions can penetrate the lipid layer of the membrane. This often occurs against a concentration gradient in the process called "active transport." To explain these phenomena, the existence of "carrier molecules" within the membrane has been postulated. The phenomenon of ion extraction analysis provides a nonliving analog of the manner in which carrier molecules may function in transporting charged ions through the phospholipid membrane.

Typical ionized inorganic salts such as sodium chloride are completely insoluble in lipid solvents (e.g., chloroform). If a solution of methylene blue (the

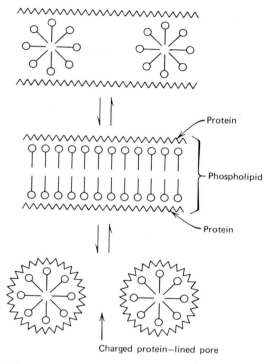

Figure 11-8 Micellar mosaic model of the cell membrane. The phospholipid bilayer can change reversibly into two types of micellar structures.

hydrochloride salt of a fairly strong organic base) is added to a solution of a well-ionized organic acid such as picric acid, a precipitate of methylene blue picrate is formed, which, unlike typical salts, is soluble in chloroform. The reason for this is that the two ions are held together not only by the electrostatic attraction, but also by extensive Van der Waals forces. The picrate anion and methylene blue cation are both fairly "flat" organic ring structures that can come to lie close together with overlapping of the ring structures. Short-range attractive (Van der Waals) forces become operative, the water of hydration that surrounds all ions in aqueous solution becomes squeezed out, and the ion pair becomes liposoluble. A. Albert suggested that carrier-aided transport of charged ions across cell membranes may involve formation of liposoluble ion pairs. Certain types of phospholipid molecules might perform the role of carriers because much of a phospholipid molecule consists of the nonpolar, fatty acid side chains. Neutralization of the negatively charged phosphate group by an oppositely charged ion would result in an electrically neutral complex that would probably be soluble in the lipid layer.

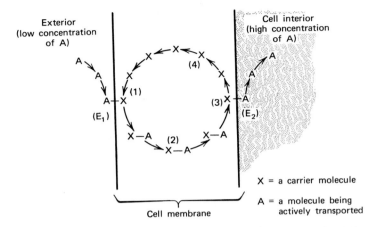

Figure 11-9   The carrier molecule theory of active transport. (1) Molecules of "A" become combined with molecules of "X" at outer cell membrane surface. Reaction is catalyzed by enzyme E_1. (2) Concentration of "A—X" builds up and "A—X" molecules diffuse to inner surface of cell membrane. (3) Enzyme E_2 catalyses dissociation of "A—X" and "A" is released into cell interior. (4) Concentration of "X" molecules builds up and they diffuse back to be used again.

A schematic model of the carrier-aided transport of materials across cell membranes is shown in Fig. 11-9. Enzymes are probably needed for catalyzing the union of the carrier and transported material and energy for transport against a concentration gradient might be provided by ATP. A second enzyme may also be required to dissociate the complex at the opposite side of the membrane.

CRITICISMS OF THE DANIELLI–DAVSON MEMBRANE MODEL

In recent years doubts have arisen regarding the validity of the Danielli–Davson concept of cell membrane structure and its refinements. One doubt is that the molecular organization of cell membranes was deduced from chemical analyses of relatively few membrane types, so that the three-layered model may not be true for all cell membranes. The relative amounts of different lipids and proteins may vary considerably from one cell type to another (see Table 11-1). It has been found in some cases that there is insufficient protein to provide a complete covering on both sides of the lipid bilayer. Analytical figures were averaged over the entire membrane, the assumption apparently being that cell membranes are

TABLE 11-1   **Protein and Lipid Composition of some biological membranes**
(Molar Ratios)

Membrane Source	Amino Acid	Phospholipid	Cholesterol
Myelin	264	111	75
Erythrocyte	500	31	31
Micrococcus lysodeikticus	524	29	0
Streptococcus faecalis	441	31	0

From C. U. M. Smith, adapted from E. D. Korn, *Science*, 153: 1496 (1966).

uniform throughout. In these respects the Danielli–Davson concept is almost certainly an oversimplification.

More serious are experimental findings that indicate that the protein molecules are not in the extended configuration, but are globular or folded. Unlike ordinary proteins, the membrane proteins are amphipathic (i.e., they possess both polar and nonpolar regions on their surfaces). Other experiments show that some of these proteins are on the surface of the lipid bilayer and that others penetrate it; some even pass right through it. It also seems very probable that these globular proteins function as carriers in the carrier-aided transport of materials across the membrane.

There is also the matter of the thermodynamic stability of the three-layered membrane, an important point that so far has not been mentioned. The maximizing of hydrophobic and hydrophilic interactions so that the lowest free energy state will be attained is important. This principle also operates in the stabilizing of tertiary structure of ordinary globular protein molecules in which hydrophobic (nonpolar) amino acid side chains interact with each other maximally and are sequestered in the center of the molecule away from the aqueous exterior. The hydrophilic (polar) amino acid side chains are oriented so that most are exposed to the aqueous exterior or interact with each other. These requirements, when applied to theoretical lipoprotein membrane models, are bound to restrict the number of acceptable structural concepts. In the Danielli–Davson structure, the bimolecular phospholipid layer maximizes hydrophobic interactions insofar as the nonpolar fatty acid chains interact with each other and are oriented away from the aqueous region; the polar phosphate groups are directed toward the water phase on both sides. However, the nonpolar amino acid residues of the unfolded surface proteins necessarily are largely exposed to water, and the protein layer itself prevents interaction of the phosphate groups of the phospholipid layer with the aqueous phase. Hydrophilic and hydrophobic interactions are therefore not maximal in the Danielli–Davson model. In a membrane model that satisfies the thermodynamic requirements of maximal stability, the nonpolar regions of the proteins should interact with the nonpolar fatty acid side chains of the lipid layer and not be exposed to the aqueous phase. Likewise, the polar phospholipid groups and polar regions of proteins should interact and/or be exposed to the aqueous phase.

THE FLUID MOSAIC MODEL OF CELL MEMBRANES

The recently proposed fluid mosaic model of cell membranes satisfies thermodynamic requirements and provides satisfactory explanations of permeability properties and other biological properties of the cell membrane; until this model was developed, adequate explanation on previous models did not exist.

The most important departure from the trilaminar protein-phospholipid sandwich is that in accordance with experimental data, most of the globular proteins actually penetrate the phospholipid layer either partially or completely. The fluid mosaic model of the cell membrane envisions the phospholipid bilayer as a discontinuous fluid film in which the globular proteins "float." Their orientation and degree of penetration into the phospholipid layer are governed by the distribution of polar and nonpolar groups on their surface, following the principle of maximal hydrophobic and hydrophilic interactions (Fig. 11-10). This

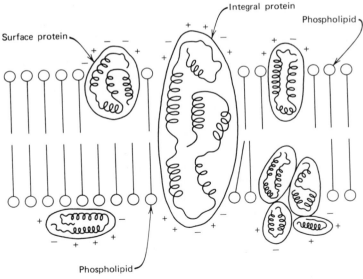

Figure 11-10   The fluid mosaic model of the cell membrane.

model receives striking support from the technique known as freeze-etch electron microscopy. In this method an aqueous suspension of cell membranes is rapidly frozen and then split with a sharp blade. Many membranes will be oriented so that they will be split through the middle of the lipid bilayer. The fracture surface has a thin film of carbon and platinum deposited on it, and it is then examined in the electron microscope. Many particles about 85 Å in diameter are seen embedded in the fractured membrane (Fig. 11-11), and these are greatly reduced in number if the fractured membranes are first treated with a proteolytic enzyme that presumably digests them. The indications therefore are that these particles are globular proteins and that they are embedded in a phospholipid matrix.

The phospholipid bilayer must be sufficiently fluid to permit free movement of the functional membrane proteins, such as the globular proteins that are involved in active transport. The degree of fluidity of the lipid layer depends on the types of fatty acid side chains in the phospholipid molecules. In a membrane containing a single type of phospholipid, the molecules would tend to be stacked in an orderly, quasicrystalline structure that would be somewhat rigid. Where there are several different types of phospholipids, especially those with unsaturated side chains, there would be a more disorderly and fluid type of

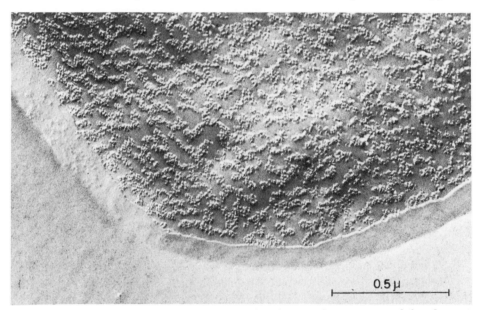

Figure 11-11 Human red blood cell showing surface prepared by freeze-etching. Note the exposed interior of the membrane showing numerous small particles (×40,000). (Courtesy of Dr D. Branton. Reprinted by Permission of Rockefeller University Press.)

arrangement, which is what would be required where membrane transport mechanisms depend on the free movement of carrier molecules within the membrane. This is, in fact, found to be the case. X-ray diffraction studies indicate that the type of fatty acids found in animal cell membranes have a liquid crystalline structure at physiological temperatures. Furthermore, the various temperatures at which different cell types grow is reflected in the qualitative composition of the cell membranes. Cells growing at relatively low temperatures have a higher proportion of the unsaturated fatty acid side chains than those growing at higher temperatures. This enables fluidity of the membrane to be maintained at the lower temperatures.

One hypothesis of carrier-aided transport suggests that the process is effected by globular protein molecules that can change their shape (i.e., they are al-losteric, acting singly or cooperatively in groups). A schematic model in which a hypothetical allosteric carrier protein has the properties of a revolving door is shown in Fig. 11-12.

Although this chapter has emphasized the Danielli–Davson and Robertson concepts of cell membranes (mainly because of their historical importance), the

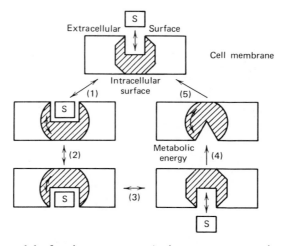

Figure 11-12 A model of active transport. Active transport may involve a *"carrier protein"* (cross-hatched) with the properties of a revolving door. Carrier protein captures substance "S" in dilute solution outside the cell and transports it into the cell where the concentration may be greater. When S is bound to protein, protein changes shape (1), which enables it to rotate (2). When S is detached and enters the cell (3), protein returns to its immobile form. Metabolic energy must be used (4) to change its shape back to the mobile form so that it can rotate again presenting its binding site to the cell exterior (5). (Adapted from *The Structure of Cell Membranes* by Fred C. Fox, copyright 1972 by Scientific American, Inc.

fluid mosaic is, in fact, now the generally accepted structure. It provides for the formulation of new experimentally testable hypotheses of cell surface phenomena such as the increased agglutinability by saccharide-binding plant agglutinins of cells that have undergone malignant transformation, the loss of "contact inhibition" of cancer cells, and cooperative phenomena in cell membranes. Details of these hypotheses may be found in the paper listed at the end of this chapter by Singer and Nicolson.

CHAPTER TWELVE

Photosynthesis, respiration and energy storage

"LIFE," ELECTRONS, AND ENERGY

One of the most characteristic attributes of living things, at least to the lay mind, is the power of spontaneous movement that is always taken as a sign of life. One may criticize this viewpoint, but at least it emphasizes energy utilization as one characteristic of living things; without energy there can be no movement. Thermodynamically, living cells are highly "improbable" because of their extreme organizational complexity. The natural tendency for all things in nature is to assume, whenever possible, a state of minimum energy and maximal entropy, or "randomness," a principle sometimes called the "law of disorder." Energy must be expended if organizational complexity is to be maintained, whether it is an organism or a cell, a nation or a dwelling house that is expending the energy. As soon as an organism dies, it begins to disintegrate (i.e., it becomes progressively more disordered), because it ceases to utilize energy.

Living things require energy for the performance of osmotic work, muscular contraction, biosynthesis of complex macromolecules, and nervous conduction. Frequently overlooked is the fact that throughout their lives, sizable organisms such as trees and elephants are constantly battling against the force of gravitation, and that energy is expended in resisting it.

The ultimate energy source for modern living organisms is solar radiation. Certain organisms utilize this radiation directly to support their vital activities

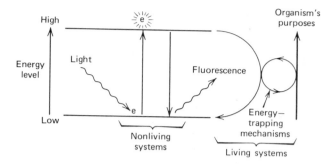

Figure 12-1   Energy levels and electron movements.

and possess specialized "light-trapping" pigments to accomplish this. When a light-absorbing molecule interacts with a photon, its electrons become "excited," and they may be promoted to higher energy levels in the molecule. The energy absorbed by the excited electron may be reemitted as the light of a longer wavelength (fluorescence) when the electron returns to its former low energy level. Alternatively, the energy of the falling electron could be imagined as being trapped and utilized to sustain life process (Fig. 12-1). This is the fundamental situation found in photosynthetic plants, and all animal life is ultimately dependent on these plants. The important point is that electron movements are the basis of the multitude of energy transactions that maintains living processes. Infinitesimally small electron movements between the covalent bonds of carbon compounds are involved; this is in contrast to the mass long-distance movements of electrons found in metals, the phenomenon called electricity.

Movement of electrons between atoms or molecules is governed by electronegativity (i.e., the relative affinities of atoms or molecules for electrons, as shown in the familiar reaction between hydrogen and oxygen to give water). When the covalent bonds of the hydrogen and oxygen molecules are broken and then rearranged to make the covalent bonds of the water molecule, energy is released, as revealed by the explosion and flash of light that is seen when a mixture of the two gases is ignited. This happens because oxygen is one of the most strongly electronegative elements (i.e., it has a very strong affinity for electrons). The electrons in the orbitals of the hydrogen molecules are associated with relatively high energies but those of oxygen are much lower in energy content. Consequently, when the electrons of hydrogen pass over into the sphere of influence of the oxygen atom, they lose their energy as the new covalent bonds are formed. The bonds may be broken to re-form hydrogen and oxygen molecules but energy must be expended, a fact demonstrated by the consumption of electrical energy that is needed to decompose water into molecular oxygen and hydrogen during the electrolysis of water.

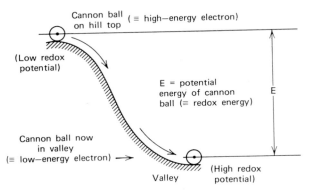

Figure 12-2 Gravitational analogy of energy levels, redox potentials, and electron movements.

    The reasons for the release of energy in this reaction and the means by which electronegativity decides the direction of electron flow can perhaps be clarified by reference to a mechanical analogy. The relatively high-energy electron of the hydrogen atom can be compared to a cannonball on top of a hill (high potential energy) that has a strong tendency to roll downhill into a valley (symbolizing the lower energy levels of the oxygen atom) and, in so doing, potential energy is lost (Fig. 12-2). The lower level of the valley (high cannonball "affinity") determines the direction in which the cannonball rolls and corresponds to the high electron affinity (electronegativity) of oxygen.

    Such reactions involving electron transfers from one atom or group to another are called *redu*ction–*oxi*dation (redox) reactions. The species donating electrons (the electron donor) is said to become oxidized, and the species that accepts electrons (the electron acceptor) is said to be reduced. Electron transfers are very often accompanied by simultaneous transfer of protons (hydrogen nuclei), so that effectively a redox reaction is often a transfer of hydrogen atoms.

$$AH_2 \; + \; B \longrightarrow A \; + \; BH_2$$

Electron or   Electron or
hydrogen donor   hydrogen acceptor

A reaction analogous to the formation of the water molecule is the formation of hydrogen chloride from hydrogen and chlorine.

$$H_2 + Cl_2 \longrightarrow 2HCl + Energy$$

The chlorine is more electronegative than the hydrogen and attracts away its electron into its own sphere of influence. The hydrogen is therefore oxidized, although oxygen itself plays no part; the same applies to all redox reactions, irrespective of whether oxygen is involved.

The electron affinities of atoms or molecules can be measured and stated quantitatively as *redox potentials*, which are expressed in volts. The *higher* the redox potential, the *greater* is the affinity for electrons. This is important to remember correctly; it sometimes leads to confusion, since in a redox reaction it is the substance with the *higher* redox potential that *gains* electrons.

ENERGY TRANSFER AND THE SYNTHESIS OF HIGH-ENERGY PHOSPHATE COMPOUNDS

The release of energy in redox reactions is the driving force of life processes. Under normal conditions, a redox reaction involving the release of energy will proceed spontaneously. The energy may be yielded in the form of light and heat, as in the case where water is formed from hydrogen and oxygen, but living cells cannot directly utilize light and heat for their vital processes.

Many years ago, a substance called adenosine triphosphate (ATP) was found to be essential for the contraction of muscle. Later the fundamental importance of ATP in the energy transactions of all kinds of cells was discovered and earned ATP the name of the "universal energy currency of life." The most important but not the only form of energy directly utilized in vital processes is the chemical energy associated with hydrolysis or synthesis of the phosphate bonds of ATP. The explanation for this virtually unique role in the life of living cells lies in the chemical structure of ATP and the magnitude of the energy changes accompanying the synthesis and hydrolysis of the covalent bonds of its phosphate groups. Adenosine is composed of a molecule of the nitrogen-containing base adenine linked to a molecule of the sugar ribose. When three linked phosphate groups are attached in the manner shown in Fig. 12-3, this gives the molecule of ATP. The biological importance, interest, and peculiarity of ATP is that hydrolysis of the two end phosphate groups (especially the terminal group, to yield adenosine diphosphate and inorganic phosphate) is accompanied by release of an unusually large amount of energy, much more than is usually encountered in the hydrolysis of phosphate bonds. Conversely, a relatively large input of energy is needed to resynthesize ATP from ADP and inorganic phosphate.

$$\text{ATP} \xrightleftharpoons[\text{Synthesis}]{\text{Hydrolysis}} \text{ADP} \quad + \quad P_i \quad + \quad \text{Energy}$$

| Adenosine triphosphate | Adenosine diphosphate | Inorganic phosphate | 12.5 Kcal/mole in the cell |

This can be explained on the basis that the three negatively charged phosphate groups repel one another strongly, and energy is required to hold them together. When one or both are removed by hydrolysis, this energy is released, and the remainder of the molecule changes to a much more stable state. Also, the energy

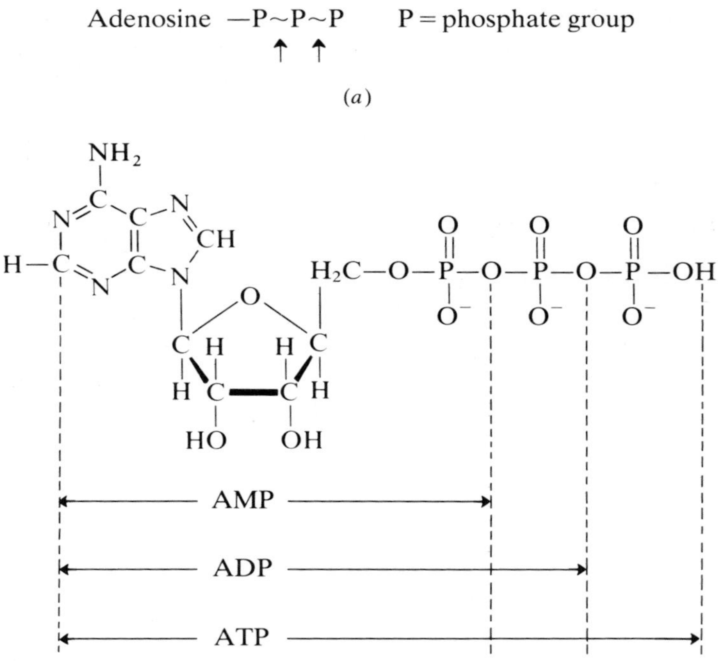

Adenosine —P~P~P   P = phosphate group

(a)

(b)

Figure 12-3 The molecule of adenosine triphosphate (ATP). (a) Schematic. Arrows indicate the so-called "high-energy" phosphate bonds. (b) Detailed structure.

of resonance of the entire ATP molecule changes, since the resonance structure of the entire ATP molecule is altered in the reaction.

The two terminal phosphate bonds of the ATP molecule are often referred to as high-energy bonds, which often gives rise to the erroneous idea that large amounts of energy are located within them. As explained above, however, the energy is not localized in these bonds, but hydrolysis or synthesis of them is accompanied by large energy changes in the ATP molecule *as a whole*. Nonetheless, the idea of high-energy bonds is a convenient fiction in ordinary conversation, provided it is remembered that it *is* only a fiction.

In order that cells can utilize redox potential energy for maintenance of their vital activities, the release of that energy must somehow be used to synthesize ATP (i.e., energy release must be "coupled" to ATP synthesis). Reactions such as the union of hydrogen and oxygen to yield water or the hydrolysis of ATP to give ADP and P_i that are accompanied by release of energy are referred to as *exergonic*, and the reverse reactions (decomposition of the water molecule to give hydrogen and oxygen; synthesis of ATP from ADP and P_i) that require input of energy are said to be *endergonic*.

Exergonic oxidation reactions provide the energy needed for synthesis of ATP from ADP and P_i and hence ATP is the primary store of useful energy in living cells.

There are two advantages in storing energy as high-energy phosphate in the form of ATP. One is that the energy is easily available and is yielded in a single-step reaction. The other is that the amount of energy released is about right for getting biochemical reactions to proceed, so that there is relatively little energy wastage; in other words, biochemical reactions in living cells are efficient. The energy change associated with hydrolysis of ATP is used to make other chemical bonds, and so the energy is transferred instead of being released. High-energy compounds therefore behave as though they "couple on" to other energy-requiring reactions. ATP is thus the link between the energy-yielding and energy-requiring reactions of cells.

As a cellular, high-energy phosphate compound, ATP is not unique; in fact, compounds like phosphoenol pyruvate and 1:3-diphosphoglycerate release much more energy on hydrolysis than ATP. What, then, makes ATP unique in cellular energy transactions? As stated above, the energy available when ATP is hydrolyzed is about the right amount for causing biochemical reactions to go; presumably much energy would be wasted if compounds such as phosphoenol pyruvate were involved, because the amount of energy associated with its hydrolysis is somewhat less than twice that of ATP, therefore, whereas it could yield the energy of one ATP molecule, the excess would be wasted, since it would be not quite enough to do the work of a second ATP molecule. ATP may therefore have been specially selected for these tasks.

THE TRAPPING OF LIGHT ENERGY

As mentioned earlier, the electrons in the covalent bonds of light-absorbing molecules may become energetically excited when exposed to light energy. This redox energy can be released again and coupled to ATP synthesis, provided that a suitable cellular mechanism is available. Development of the ability to trap and utilize solar energy was an important breakthrough in the evolutionary history of living organisms. The molecules that perform the initial tasks of absorbing light of appropriate wavelength belong to the group of pigments called chlorophylls. Structurally, a chlorophyll molecule can be considered to be derived fundamentally from porphin (Fig. 12-4). Porphin can be considered as four pyrrole rings connected together by four methene groups. It is therefore a tetrapyrrole. This ring structure is common to many important biomolecules such as the heme groups of hemoglobin, myoglobin, the enzyme catalase, and the family of electron-transporting molecules called cytochromes.

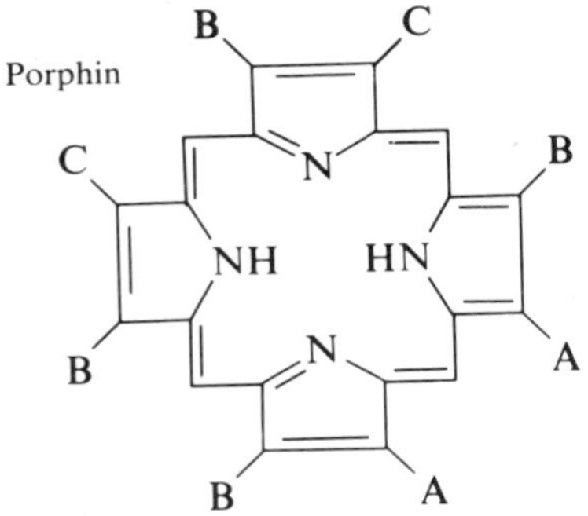

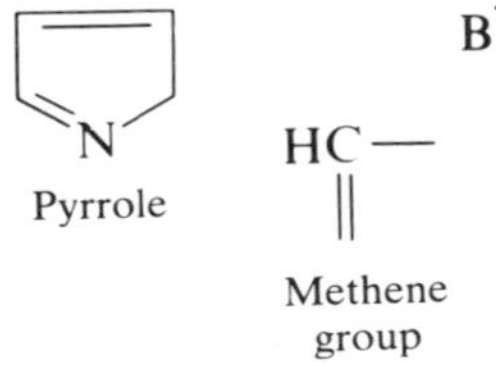

When A = $-CH_2-CH_2COOH$, B = $-CH_3$, and C = $-CH=CH_2$, this gives protoporphyrin 9, the parent molecule of heme and chlorophyll

Figure 12-4   The Porphin molecule.

The molecule of porphin is highly conjugated (i.e., it possesses many alternating single and double bonds), so that there is much electron delocalization. Because of this, the entire molecule is planar, and the possession of so many delocalized electrons is no doubt responsible for the biological properties of porphin compounds. Pullman and Pullman have shown that there are 26 π electrons in the conjugated double bonds of porphin and that these are contributed by 24 atoms.

Heme and chlorophyll can be thought of as being derived by attachment of different groups to various atoms in the porphin structure. If the chemical groups shown are attached to the positions marked A, B, and C in the porphin molecule (Fig. 12-4), this results in the structure of the molecule of protoporphyrin 9, the parent compound of both heme and chlorophyll. In the case of heme, an iron atom presumably displaces two protons from the pyrrole nitrogen atoms, and it becomes held, by four bonds (Fig. 12-5) to the nitrogens.

Chlorophyll is a more complicated derivative of protoporphyrin 9. Apart from the fact that a magnesium atom instead of an iron atom occupies the center of the molecule, there are two major structural modifications involving both of the carboxyl-containing hydrophilic side chains attached to the ring system. The one attached to the pyrrole ring marked C (Fig. 12-4) can be thought of as being modified and then interacting with the adjacent methene group to form a new five-membered ring. This has a pronounced effect on the properties of the ring system. The other carboxylic side chain becomes esterified by phytol, a long-chain aliphatic alcohol, so that the chlorophyll molecule effectively has a long

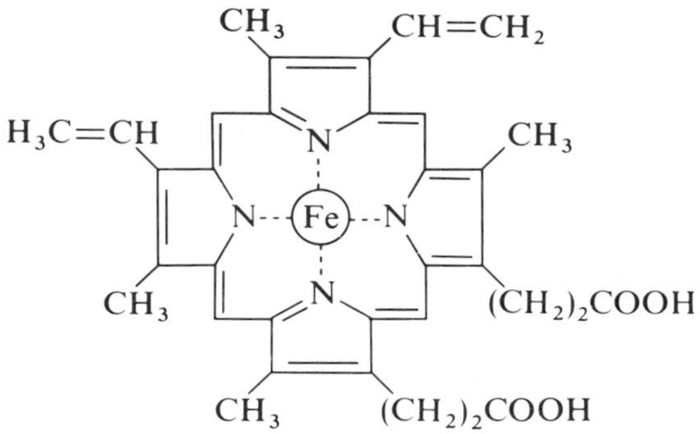

Figure 12-5 The Heme molecule.

hydrophobic tail (Fig. 12-6). The significance of this will be apparent in the section on chloroplast structure in the chapter dealing with energy transducers.

(The manner in which porphin, protoporphyrin 9, heme, and chlorophyll are derived as above, from simpler structures linking up in the manner indicated, should not be taken to imply that this is how these molecules are actually synthesized. The device is a fiction that, it is hoped, will help toward an understanding and comprehension of the structures and structural relationships between these molecules.)

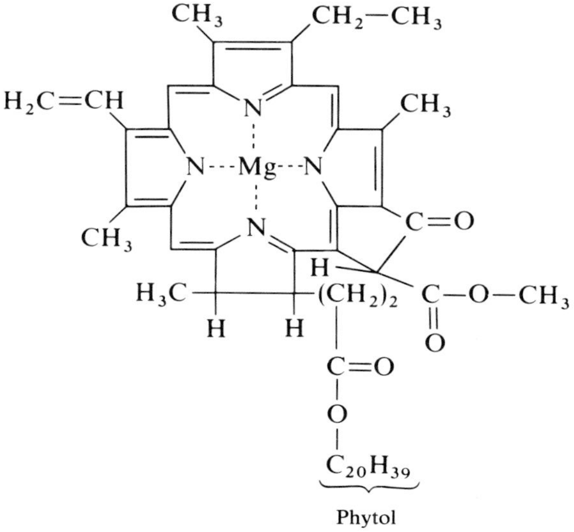

Phytol

Figure 12-6 Molecule of Chlorophyll a, the major light-absorbing pigment in green leaves.

When a solar photon is absorbed by a chlorophyll molecule, the most likely result is that one of the π electrons will become promoted to a higher energy level. The process can be symbolized by writing $\pi \rightarrow \pi^*$ or, more briefly, π, π^*. The return of the electron to its ground state would be symbolized by π^*, π. Less likely is the excitation of an electron associated with one of the nitrogen or oxygen atoms to one of the π^* orbitals, a more long-lasting type of transition. There are also other possibilities.

CYCLIC PHOTOPHOSPHORYLATION

Living cells have evolved mechanisms in which the energy emitted by an excited electron falling back to its ground state can be used to "drive" vital processes. We have already noted that cells directly utilize only the energy released when the terminal phosphate groups of ATP are hydrolyzed. Solar energy must therefore first be employed to achieve the synthesis of ATP from ADP and inorganic phosphate. In this process, an excited electron absorbs so much energy that it leaves the chlorophyll molecule; the chlorophyll molecule becomes oxidized and acquires a positive charge. Next, the electron is passed to, and traverses a series of electron carrier molecules of progressively increasing redox potential. Among these carriers are ferredoxin and the cytochromes. Each time the electron passes from one to another of these, it loses some of its redox potential energy. At certain points in the chain of electron carriers, the energy is tapped and utilized in a mechanism for synthesizing ATP from ADP and inorganic phosphate (Fig. 12-7). Finally, the electron finds its way back to the

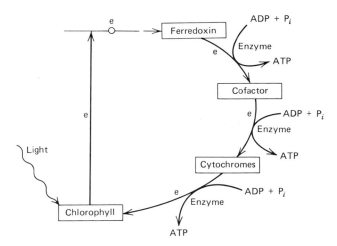

Figure 12-7 Cyclic photophosphorylation. (The "cofactor" may be flavoprotein or plastoquinone.)

chlorophyll molecule, where it will now be in its ground state again and the temporary positive charge of the chlorophyll molecule is neutralized. Because electrons are cycled to and from chlorophyll in this light-induced phosphorylation of ADP, the process is called *cyclic photophosphorylation*.

ENERGY STORAGE AND PHOTOSYNTHETIC PHOSPHORYLATION

Compared with the energy demands of an actively metabolizing cell, the amount of available energy in the ATP molecule is very small. If cells were to store chemical energy as ATP molecules, the situation would be as cumbersome as storing one's entire fortune in dimes. Clearly, a better plan would be to store energy in much larger units. The most important of these larger energy-storing units is the group of compounds called carbohydrates. The process in which solar energy is first converted into chemical bond energy as ATP and subsequently to the covalent bonds of carbohydrates is called *photosynthetic phosphorylation*.

Photosynthetic plants synthesize carbohydrates by utilizing carbon dioxide as a carbon source. Hydrogen is needed for reduction of the carbon dioxide molecule, and the biosynthetic energy is supplied by ATP. Before describing this process, we should first examine some important energy considerations.

(i) Sources of reducing power

The only way in which carbon dioxide (or, to be precise, carbon dioxide derivatives) can be reduced is by transfer of hydrogen from the reduced form of an important intracellular hydrogen carrier, nicotinamide adenine dinucleotide phosphate (NADP); the structural formula of NADP is shown in Fig. 12-8. (The reduced form of the coenzyme NADP is specifically required for this and other biosynthetic reductions. It cannot be replaced by NAD, which acts as a hydrogen acceptor in cellular respiration involving the transfer of electrons to oxygen.) The reduction of NADP can be achieved with elemental hydrogen derived from hydrogen sulfide as occurs in sulfur bacteria or, less easily, with hydrogen derived from succinate or thiosulfate.

NADP can be directly reduced by hydrogen released from hydrogen sulfide because of the high reducing capacity (low redox potential) of the donor. This does not require light energy, which sulfur bacteria need only for ATP synthesis. Thiosulfate and succinate are lower-energy hydrogen donors, and hydrogen derived from them can reduce NADP only with the aid of light energy. Electrons derived from thiosulfate or succinate are therefore first transferred to the cytoplasm, where they give up some potential energy in stepwise fashion (as in cyclic photophosphorylation); this energy is utilized to synthesize ATP. The low-energy electrons then fill electron vacancies in the chlorophyll molecule that

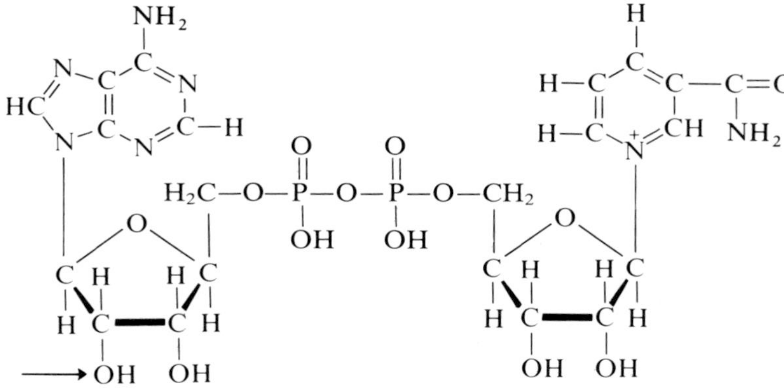

Figure 12-8 Molecule of the oxidized form of nicotinamide adenine dinucleotide, (NAD). In NAD phosphate (NADP), the hydroxyl group indicated by the arrow is phosphorylated.]

have arisen from light-induced excitation of the chlorophyll electrons. These leave the chlorophyll molecule and pass on to NADP, which therefore becomes negatively charged. Simultaneously, excess hydrogen ions (protons) originating from the removal of electrons from succinate become attracted to the negatively charged NADP, which thus becomes NADPH (Fig. 12-9).

Compounds like hydrogen sulfide, thiosulfate, and succinate are found only in localized parts of the earth's surface. Photosynthetic organisms relying on these materials as sources of hydrogen for carbon dioxide reduction would therefore be restricted to these areas. Suppose, however, that water was used as a hydrogen source. There is plenty of this available, and it is spread widely over the earth's surface, but there is one snag. The electrons of the water molecule are, as we have seen, in such a low energy state that a good deal of energy is required to detach them from the water molecule and to promote them to the much higher energy levels that they need to attain in order to reduce NADP. Some quantitative considerations will make the position clear.

The redox potentials of NADP and water are known, and so the energy difference between the electrons in the water molecule and those in the NADPH molecule can be calculated. From this can be deduced the amount of energy that electrons derived from water must absorb before they can be raised to a sufficiently high energy level for them to reduce NADP. This turns out to be 58 kcal/mole. The average energy content of the light photons absorbed by chlorophyll pigments is about 45 kcal/Einstein (An Einstein is equal to N photons where N = Avogadro's number = 6.023×10^{23}, so that an Einstein can be thought of as a mole of photons.) Obviously, the absorption of one quantum of light energy per electron will not be sufficient to raise it to the required energy

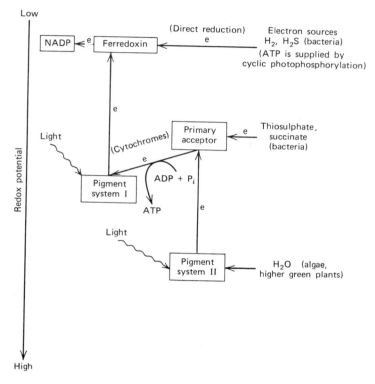

Figure 12-9   Methods of achieving reduction of NADP in different groups of organisms.

level. Clearly, a solution to this problem would be to use two photons per electron; this seems to be how this energy problem is solved in plants that use water as a source of electrons. In such plants, it appears that the existence of two pigment systems achieves the feat of energizing each electron derived from water molecules with two photons of light energy so that they can be transferred to NADP. That system, designated Pigment System I, absorbs solar radiation of higher energy content than the other, which is designated Pigment System II. Effectively, the electron is energized in two steps, and the sequence of events appears to be as follows.

Water molecules, as is well known, are dissociated to a slight extent into hydrogen and hydroxyl ions.

$$H_2O \rightleftharpoons H^+ + OH^-$$

A pair of hydroxyl groups is thought to interact to produce a molecule of water plus a "half molecule" of oxygen and one spare electron that becomes associated with Pigment System II. Subsequently, it, or some other electron in the chlorophyll molecule, becomes energized by light absorption and is promoted to a higher energy level and then passes sequentially through a chain of electron carriers that includes plastoquinone and cytochromes. In so doing some of its potential energy is given up and trapped to synthesize ATP, just as in cyclic photophosphorylation. The electron is then passed to Pigment System I, where it absorbs a second photon so that it is promoted to an energy level sufficient for it to reduce NADP through the electron carrier ferredoxin. The NADP molecule thus acquires a negative charge that is neutralized by the excess hydrogen ions remaining from the dissolution of the water molecule that gave rise to the electron we are considering and so becomes NADPH. This light-dependent (photochemical) phase of photosynthesis is called the Hill reaction, after Robert Hill who, in 1939, observed that fresh green leaves mashed in water and exposed to light evolved oxygen and reduced added hydrogen acceptors such as quinone but did not synthesize carbohydrate. The light-induced splitting of the water molecule is called photolysis. The electrons originally derived from water molecules finally end up in the NADPH molecule and are therefore not "recycled," as in cyclic photophosphorylation. The photochemical process in higher green plants is therefore called noncyclic photophosphorylation. The different methods of achieving reduction of NADP are summarized in Fig. 12-9. These light-dependent reactions therefore all have the same result: synthesis of ATP and the production of reduced NADP. These, along with carbon dioxide, are the raw materials for carbohydrate synthesis. Carbohydrate synthesis can proceed in the absence of light, and the reactions involved are therefore called the light-independent or "dark" reactions of photosynthesis. They may be summarized as follows.

$$\text{Carbon dioxide derivatives} \xrightarrow[\substack{\text{ATP} \quad \text{ADP}+\text{P}_i}]{\substack{\text{NADPH} \quad \text{NADP}}} \text{Carbohydrates}$$

(Enzymes)

(ii) Light-independent ("dark") reactions of photosynthesis

The first step in the synthesis of carbohydrate from carbon dioxide is a reaction in which carbon dioxide combines with the diphosphate derivative of a five-carbon sugar called ribulose that is already present in plant cells. The result is formation of an unstable molecule containing six carbon atoms that decomposes into two identical three-carbon fragments, phosphoglyceric acid (PGA). With the help of ATP, PGA is reduced by NADPH. The reaction is catalyzed by an NADP-specific triosephosphate dehydrogenase "working backward," as it were,

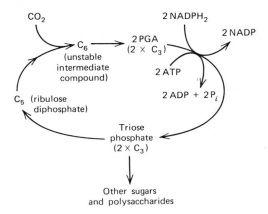

Figure 12-10   The "dark" reactions of photosynthesis (simplified).

and the result is formation of phosphoglyceraldehyde (PGAL). This is a simple sugar and is, in fact, the primary high-energy product of the dark reactions (Fig. 12-10). (PGAL is also called triose phosphate; a triose is a sugar with three carbon atoms.) PGAL is the raw material from which other sugars and amino acids, fatty acids, and glycerol are derived and, therefore, there are three major classes of cellular, energy-rich macromolecules: the polysaccharides, the triglycerides, and the proteins.

The process in which the chemical energy stored in carbohydrates is released to fulfill the energy requirements of cells is called respiration; this will be described a little later, but first some matters regarding the molecular architecture of certain energy storage compounds are important to note in relation to their suitability as cellular energy storage compounds.

MOLECULAR STRUCTURE OF ENERGY STORAGE COMPOUNDS IN RELATION TO FUNCTION

(i) The glucose molecule

The principle sugar used as an energy source by respiring cells is glucose. Glucose is one of the group of sugars called hexoses, because they contain six carbon atoms. Structurally, carbohydrates are polyhydroxy aldehydes and ketones. The simplest carbohydrates therefore contain two hydroxyl groups and an aldehyde or ketone group. Since two hydroxyl groups attached to the same carbon atom is almost always an unstable structure (one exception is the molecule of chloral hydrate), the two hydroxyl groups must be attached to different carbon atoms. The simplest sugars would therefore have the following structures [note that glyceraldehyde (but not dihydroxyacetone) and other sugars possess asymmetric carbon atoms and are therefore optically active].

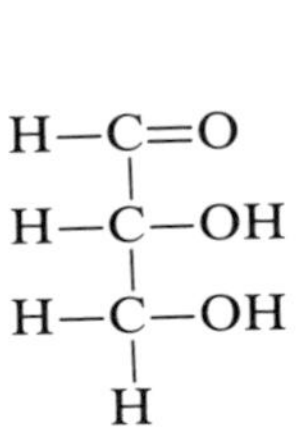

$$
\begin{array}{cc}
\text{H}-\text{C}=\text{O} & \overset{\displaystyle\text{H}}{|} \\
| & \text{H}-\text{C}-\text{OH} \\
\text{H}-\text{C}-\text{OH} & | \\
| & \text{C}=\text{O} \\
\text{H}-\text{C}-\text{OH} & | \\
| & \text{H}-\text{C}-\text{OH} \\
\text{H} & | \\
& \text{H}
\end{array}
$$

Glyceraldehyde 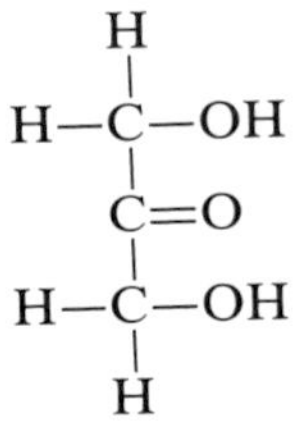   Dihydroxyacetone

Both of these substances are known and are white, water-soluble powders with a sweet taste (i.e., they have the characteristic properties of sugars). They can be considered as derivatives of glycerol and acetone, respectively. Glyderaldehyde is an example of an aldotriose, because it possesses an aldehyde group and three carbon atoms. Likewise, dihydroxyacetone is a ketotriose. Glucose is an aldehyde sugar and, since it contain six carbon atoms, it is a member of the aldohexose family. The structure of glucose and other aldohexoses may therefore be represented as follows.

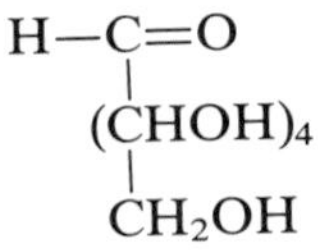

$$
\begin{array}{c}
\text{H}-\text{C}=\text{O} \\
| \\
(\text{CHOH})_4 \\
| \\
\text{CH}_2\text{OH}
\end{array}
$$

This open chain formulation accounts for some of the known properties of glucose, such as its reducing properties, which are attributable to the aldehyde group. However, aldohexoses do not exhibit certain other reactions that would be expected of an aldehyde. For example, the pentaacetyl derivative of aldohexoses does not react with hydroxylamine, and solutions of aldohexoses do not give a color reaction with Schiff's reagent. Both of these indicate that there is no free aldehyde group in the aldohexose molecule.

The properties of aldohexoses are better explained if they are formulated as ring structures that result from the aldehyde group on carbon atom 1 reacting with the hydroxyl group on carbon atom 5 and migration of its hydrogen atom to give a six-membered, oxygen-containing ring (Fig. 12-11a), which can therefore be considered as derived from the organic compound pyran (Fig. 12-11b). This structure for aldehexose sugars is therefore called the pyranose structure.

A better representation of this ring structure is the Haworth formula, which is a perspective view in which one is supposedly looking down on the ring from an

CH$_2$OH

(a)

(b) Ring structure of pyran.

Figure 12-11   Ring structure of an aldohexose molecule.

angle. The sugar represented is glucose (Fig. 12-12). The thick black edges represent the side of the hexagon nearer to the reader. The hydrogen atoms and hydroxyl groups are disposed above and below the plane of the ring, and the variations in the arrangement of these in different aldohexoses is the basis of the differences between these sugars, all of which have the same molecular formula $C_6H_{12}O_6$. Figure 12-12 shows two possible structures of glucose, the α and β forms. Note that the only differences between them is the orientation of the hydroxyl group attached to C_1. The two forms exist because the pyranose ring spontaneously opens by breakage of the bond linking C_1 to the oxygen atom, yielding the open chain form again. Because free rotation is possible about the C_1—C_2 bond, ring closure can occur in two possible ways, resulting in the two different orientations of the hydroxyl group on C_1. The orientations of the other hydroxyl groups are, of course, unaffected. An aqueous solution of glucose is therefore an equilibrium mixture of the α and β ring forms of glucose and a trace of the open chain form. Whether pure α- or β-glucose is dissolved in water, the same equilibrium mixture is ultimately attained because of spontaneous opening and closing of the pyranose ring. This process is called mutarotation. There is a much higher proportion of β-glucose than the α form in the equilibrium mixture because, as will be apparent later, the β form is thermodynamically more stable than the α form.

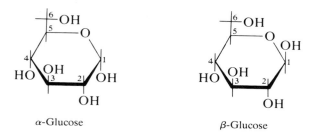

α-Glucose

β-Glucose

Figure 12-12   Haworth ring formulations of glucose.

Figure 12-13   "Chair" and "boat" configurations of the pyranose ring.

The Haworth structure, in which the pyranose ring is represented as being flat, is not strictly accurate. Because of the tetrahedral arrangement of the four valency bonds of carbon, the six-membered pyranose ring is "puckered" and exists in two different configurations, one in which two of the carbon atoms (C_1, C_4) lie above the plane of the other three and the oxygen atom (which together form a square), and the other in which one carbon atom (C_4) is above and the other (C_1) is below the plane of the square formed by C_2, C_3, C_4 and the oxygen atom (Fig. 12-13). This is much easier to understand if constructed from ball and stick molecular models. These configurations are respectively designated the "boat" and "chair" forms of the pyranose ring. The chair form is more stable than the boat form. The remaining bonds of each carbon atom in the chair form to which the hydrogen atoms and hydroxyl groups are attached are oriented either perpendicular to the general plane of the ring or roughly in the same plane as the ring. These orientations are designated as axial or equatorial, respectively. (Again, it is easier to visualize with molecular models.)

We now come to the point to which this detailed but necessary digression on sugar structure is leading. Of all the possible isomeric aldohexoses, *β-glucose is unique in that all of its bulky side groups* (hydroxyl groups and the $-CH_2OH$ attached to C_5) *are equatorially oriented*. This is the most stable of all possible arrangements; the significance of this is that β-glucose is the most stable cellular sugar energy source (Fig. 12-14). (This may also be related to the fact that as mentioned earlier, in glucose solutions there is a preponderance of the β over the α form.) Although glucose is present in the α form in storage polysaccharides such as amylose, amylopectin, and glycogen (see following sections), the β form is generated in the mutarotation reaction when the free glucose is enzymatically released from the polysaccharides.

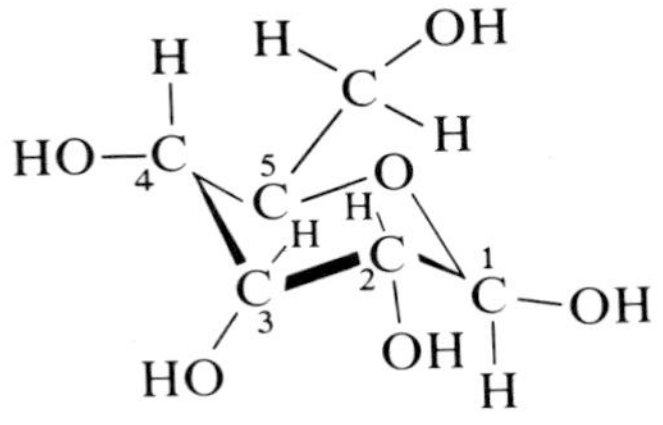

Figure 12-14   "Chair" formulation of the molecule of β-glucose.

(ii) Polysaccharides—starch and glycogen

Glucose sufficient for the respiratory needs of cells could not be stored in quantity as such because osmotic problems would soon develop as the glucose concentration increased; this would be especially acute in the case of submerged freshwater plants. Glucose is therefore stored until needed in the insoluble polymeric forms known as starch and glycogen. Starch is a mixture of two somewhat different polymers, amylose and amylopectin, and is characteristic of plants. Glycogen is virtually confined to the animal kingdom and hence is sometimes called "animal starch." (The existence of glycogen plastids in the cells of a higher plant, *Cecropia peltata*, has been reported by F. R. Rickson.) Examination of the molecular design of these polysaccharides provides a clue as to why they are distributed in this way in the plant and animal kingdoms.

Amylose and amylopectin

Monosaccharides are simple sugars, such as glucose, that consist of a single polyhydroxy aldehyde or ketone unit with three to seven carbon atoms. These units can combine chemically with one another, accompanied by elimination of water molecules to give rise to compound sugars—disaccharides, oligosaccharides, and polysaccharides. Two molecules of α-glucose can become linked by elimination of H_2O between the hydroxyl group attached to the C_1 of one glucose molecule and the hydroxyl attached to the C_4 carbon atom of a second glucose molecule (Fig. 12-15). The resulting disaccharide molecule is maltose,

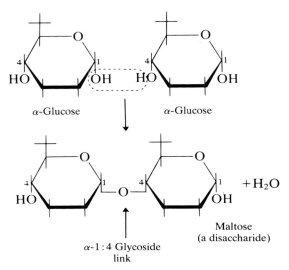

Figure 12-15 Glycoside bond formation.

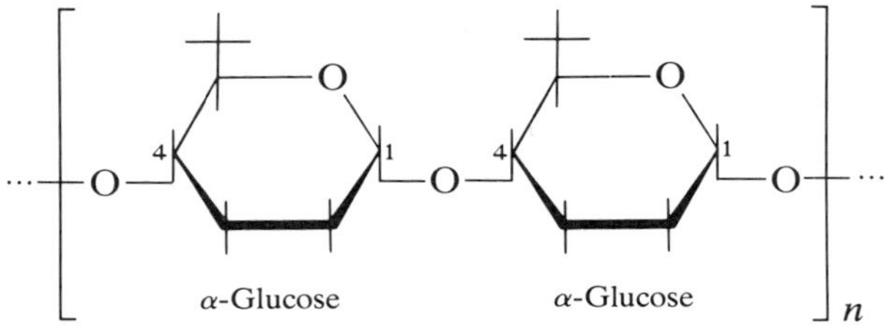

Figure 12-16 Structure of amylose.

and the oxygen atom joining the two residues in this and other di- and poly-saccharides is called a glycoside linkage. In the case of maltose in which the glucose is in the α form, and because of the locations and orientations of the hydroxyl groups involved, this is called an α-1:4 linkage. A glance at the maltose molecule will show that further α-glucose molecules can be added on at both ends by α-1:4 glycoside bond formation and that this can be repeated indefinitely. The resulting unbranched polymer consisting of α-glucose units joined by α-1 : 4 glycoside links is the structure of the molecule of amylose (Fig. 12-16).

The molecule of amylopectin is basically similar to that of amylose but, in addition to α-1:4 linkages, there are also α-1:6 linkages that result in the polymer having a repetitive branched structure. The α-1:6 linkage is shown in Fig. 12-17. On the average, there are about 12 glucose units between branch points. The structure of amylopectin is shown schematically in Fig. 12-18.

Glycogen

The molecular structure of glycogen is closely similar to that of amylopectin (glucose units joined by α-1:4 links); the only difference is in the frequency of

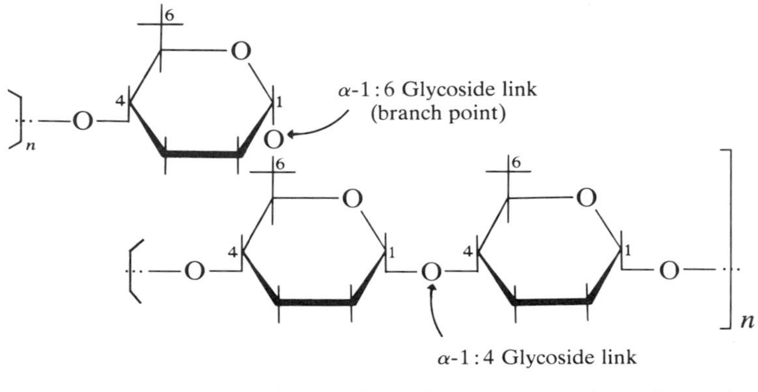

Figure 12-17 Glycoside bonds in the molecule of amylopectin.

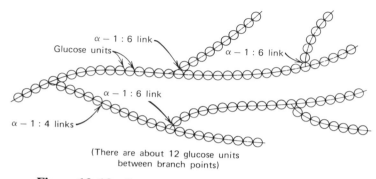

Glucose units

$\alpha - 1:6$ link

$\alpha - 1:6$ link

$\alpha - 1:6$ link

$\alpha - 1:6$ link

$\alpha - 1:4$ links

(There are about 12 glucose units between branch points)

Figure 12-18   Structure of amylopectin (schematic).

branching. In glycogen, there are about six glucose units between branch points. Glycogen is therefore a much more highly branched polymer than amylopectin.

(iii) Availability of free glucose from storage polysaccharides

Because of the more highly branched structure of glycogen, the number of terminal glucose units per unit of area of a colloidal glycogen particle is much greater than in a particle of amylopectin. Under the influence of specific enzymes, free glucose molecules required for cellular respiratory needs are hydrolytically removed from the ends of the branches, which are at the surface of the polysaccharide particles. Because of the greater surface density of terminal glucose units in glycogen, free glucose is much more readily available from glycogen than it is from amylopectin. Green and Goldberger remark that the lesser degree of branching of amylopectin and the less ready availability of free glucose from it would be no disadvantage to plants, which have lower metabolic rates than animals. The more highly branched glycogen and the corresponding easier availability of free glucose is more appropriate in an animal storage polysaccharide. Generally, animals are more active than plants (they move around in search of food or a mate) and therefore have a higher metabolic rate and a greater requirement for readily available free glucose.

Although it is broad generalization, one may suggest that this could account for the almost exclusive confinement of glycogen to the animal kingdom and amylopectin to the plant kingdom.

RESPIRATION

Living cells derive the energy necessary for their vital activities from the process called respiration. Essentially, this is the multistage oxidative breakdown of high-energy carbon compounds, especially glucose, in which the redox potential energy released is harnessed to mechanisms for synthesizing ATP. Effectively, the high-energy electrons and hydrogen nuclei (protons) associated with the covalent bonds of, say, glucose are dislodged and directed in stepwise manner through a series of hydrogen carrier molecules and electron carriers (cytochromes) until they finally combine with oxygen. The energy released during this downhill journey is coupled to ATP synthesis at certain specific points of the electron carrier chain. The electrons, with their accompanying protons, are finally passed to oxygen, the terminal electron acceptor, and a water molecule is formed.

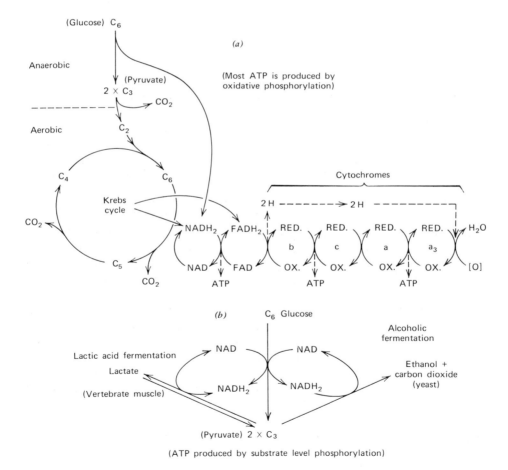

Figure 12-19 Outline of (a) aerobic respiration and (b) anaerobic glycolysis.

Part of the oxidative breakdown of glucose can occur in the absence of oxygen; this process is called anaerobic glycolysis. The glucose molecule is decomposed through several enzyme-catalyzed stages to give two molecules of pyruvate, which is further metabolized to lactic acid, as in vertebrate muscle, or to ethanol and carbon dioxide, as in yeast cells. This is as far as the process can go in the absence of oxygen. Relatively little of the total energy of the glucose molecule is released during glycolysis, which is therefore to be regarded as an inefficient process. Nevertheless, for many organisms it is the sole source of energy. The enzymatically catalyzed oxidative reactions (dehydrogenations) utilize NAD as a hydrogen acceptor, and there is a net gain of two molecules of ATP per molecule of glucose.

In the presence of oxygen, complete oxidation of pyruvate occurs in the cyclic series of reactions known as the Krebs cycle or tricarboxylic acid cycle, which are oxygen-requiring (aerobic) processes. The intimate details do not concern us here; the important thing to realize is that in this cycle the pyruvate molecule is oxidized and taken apart in an intricate series of enzyme-catalyzed reactions until it is completely broken down to carbon dioxide and water. NAD again

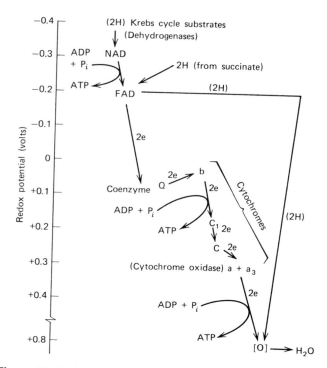

Figure 12-20   The electron transport (respiratory) chain.

serves as the hydrogen acceptor in most of the oxidative reactions (dehydrogenations). In one reaction (dehydrogenation of succinate to give fumarate), flavine adenine dinucleotide (FAD) acts as the hydrogen acceptor. The electrons are passed to the cytochrome chain from these hydrogen carriers, but they become separated from their accompanying protons, which pass into the surrounding fluid. The free electrons pass from one cytochrome to another in increasing order of redox potential, under the influence of specific enzymes, until they finally pass to oxygen and also reunite with the protons to form water molecules. At three points, the energy evolved by the electron in a "jump" between the hydrogen carriers NAD and FAD and the cytochrome molecules is coupled to ATP synthesis. The processes of respiration and electron transport are summarized in Figs. 12-19 and 12-20. There is a net yield of 38 ATP molecules from each glucose molecule when completely oxidized to carbon dioxide and water. This represents about 50% of the energy available from the glucose molecule. Therefore, the electrons that were originally associated with the covalent bonds of water molecules have traversed a complicated series of reactions and associated high-energy states and changes and finally have found their way back to the water molecule and their originally low-energy state.

CHAPTER THIRTEEN
Biological energy transducers

An energy transducer is a device for converting one form of energy into another. Energy transformations of several different kinds occur in living cells; for instance, the chlorophyll-containing cells of green plants perform the transformation of solar photon energy into the chemical energy contained in the molecule of ATP.

In both animal and plant cells, including some microorganisms, redox potential energy is converted into phosphate bond energy in the process of aerobic respiration. Photoreceptor cells in animals convert light energy into the energy of nervous impulses in optic nerves. Other forms of energy transduction are the conversion of ATP chemical energy into mechanical energy in muscle, and the production of light by fireflies.

Although some of these processes, such as the first three, seem to be qualitatively different from one another, the cell organelles in which they occur have structural features in common. These are:

1. The presence of stacks of parallel phospholipid membranes.
2. A highly regular spatial organization of the submicroscopic functional units.

The significance of these will be apparent later. The energy-transducing process on which all modern living things ultimately depend is that in which solar light energy is converted into chemical bond energy. It will be appropriate, therefore, to begin a discussion of biological energy transducers with the cell organelle in which fixation of light energy occurs—the chloroplast.

THE CHLOROPLAST
(i) Morphology and ultrastructure

Chloroplasts are the chlorophyll-containing, green-colored cytoplasmic organelles of higher photosynthetic plants. As seen in living material with the microscope, they exhibit a variety of shapes. For example, many algae contain only one or very few chloroplasts in each cell; the chloroplast in these may be like a helical ribbon, as in *Spirogyra*, or it may be star-shaped, reticulate, or bell-shaped. In the higher plants they are generally roughly discoidal or ovoid

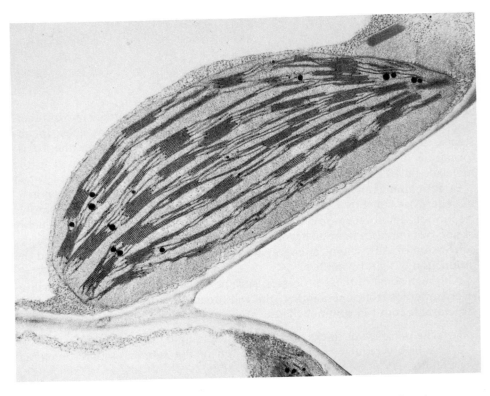

Figure 13-1   Electron micrograph of thin section of chloroplast showing grana and stroma lamellae. (Courtesy of Dr W. P. Wergin.)

bodies suspended in the cytoplasm; there may be 20 to 40 chloroplasts in each cell; in size they vary greatly, and the average diameter is 4 to 6 μ.

The chlorophyll pigment is concentrated in smaller bodies within the chloroplasts called grana (singular: granum) that are embedded within the liquid hydrophilic matrix of the chloroplast. The size of grana differs from species to species and ranges from 0.3 to 1.7 μ. They were first detected in living material by red light photomicrography of transparent aquatic plants.

When very thin sections of osmium-fixed chloroplasts are viewed under the electron microscope, a double unit membrane structure at the outer surface is evident. The interior of the chloroplast consists of a system of apparently continuous, thin, osmiophilic membranes, which form the grana structures in higher plants; these are linked together by a network of tubules or flat membranes between the grana (Figs. 13-1 and 13-2). The grana are cylindrical structures made up of stacks of circular, flattened, double membranous sacs called thylakoids (Fig. 13-3). (Some authors refer to *all* the chloroplast membranes as thylakoids and distinguish between granum and stroma thylakoids.) As many as 50 of these may be found in a single granum. Within each thylakoid sac is a closed space, or "loculus," which is shut off from the stroma. Adjacent grana are connected by thin anastomosing tubes or membranes, the stromal lamellae, that join up some but not all compartments, as shown in Fig. 13-2. Practically all of the chlorophyll within the chloroplast is contained within the grana. Developmentally, the chloroplast starts as a double membranous proplastid; the inner membrane appears to give rise to vesicles that later aggregate to form larger disks.

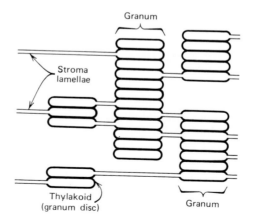

Figure 13-2 Structure of chloroplast lamellae.

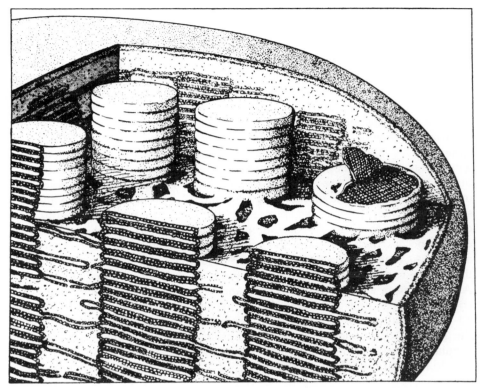

Figure 13-3   A three-dimensional reconstruction of chloroplast ultrastructure. (From *Cell and Molecular Biology* by E. J. DuPraw. Academic Press, 1968. Reprinted by permission.)

(ii) Photosynthetic reactions in relation to chloroplast ultrastructure

Both the light and dark reactions of photosynthesis can be carried out by isolated chloroplasts. The chlorophyll-containing grana have been isolated from the stromal material and found to be capable, on their own, of performing the light-dependent photolysis of water, which leads to reduction of NADP and liberation of oxygen. The unorganized stromal material, on the other hand, contains the water-soluble enzymes necessary for the dark reactions (i.e., the reduction of carbon dioxide derivatives and synthesis of carbohydrate). Evidently, there is a spatial separation of the light-dependent and light-independent phases of photosynthesis within the chloroplast.

Returning to the light reactions, one might logically ask whether whole intact grana are the smallest units capable of carrying out the light reactions or whether separate thylakoids or even fragments of thylakoids can do this; if so, is there a photosynthetic "unit" (i.e., the smallest fragment of a thylakoid that can carry out photolysis of water that could be regarded as the unit of photosynthetic activity)?

Ultrasonic disruption of chloroplasts results in fragmentation of the grana. Small fragments of granum lamellae are able to carry out the light-dependent photolysis of water, but not the fixation of carbon dioxide, which requires the soluble stromal enzymes. In an electron microscope study, R. B. Park and J. Biggins demonstrated the presence of paracrystalline arrays of globular particles on the inner surface of thylakoid membranes (Fig. 13-4). The dimensions of these particles appeared to be about $180 \times 150 \times 100$ Å, and each consisted of four smaller subunits. They have been shown to contain chlorophyll,

1000 Å

Figure 13-4   Interior view of single thylakoid disc showing paracrystalline array of quantasone particles. (Electron micrograph ×120,000. Heavy metal-shadowed preparation.) (Courtesy of Dr R. B. Park. From *Cell Ultrastructure* by William A. Jensen and Roderic B. Park. Copyright 1967 by Wadsworth Publishing Company Inc., Belmont, California 94002. Reprinted by permission of the publisher.)

carotenoids, phospholipids, and cytochromes. Fragments of thylakoid membranes consisting of as few as five of these larger particles are able to carry out the light-dependent splitting of the water molecule, resulting in the liberation of oxygen. It may be that the single large particles themselves are the photosynthetic units and, hence, they have been called *quantasomes*, but the original quantasome particle is now considered to be an artifact. Howell and Moudrianakis produced evidence that appeared to show that whatever they are, quantasomes are not Hill reaction sites, so that they cannot be regarded as units of photosynthetic activity. This was indicated in experiments with chloroplast membranes using a tetrazolium salt as oxidant in the Hill reaction. Electron microscopy of the chloroplast membranes after the reaction revealed that deposits of insoluble electron-dense formazan were uniformly present over all the membrane and not confined to quantasomes. In another experiment, apparently quantasome-free membranes were obtained by EDTA treatment of chloroplast membranes that were found to possess full Hill reaction activity when compared with untreated control preparations with intact particles. Experiments with isolated quantasome preparations showed that the quantasome does not possess photoreducing properties, but has calcium-dependent ATPase activity and is active in the dark reactions of photosynthesis.

This work has been challenged by Park and Phfeifhofer who, using freeze fracturing and deep etching techniques, showed that EDTA-treated chloroplast membranes, claimed by Howell and Moudrianakis to be particle-free and exhibiting Hill reaction activity, are not, in fact, particle-free. Park and Phfeifhofer agree that Howell and Moudrianakis appear to have removed calcium-dependent ATPase activity in particulate form by EDTA treatment, but state that their negative staining method is inadequate for revealing quantasomes. Also, in a later paper, they showed that the tetrazolium dye used by Howell and Moudrianakis can diffuse for several microns before it precipitates as the insoluble formazan.

There is no doubt that many different kinds of particles exist within the photosynthetic membranes, because they can be visualized by the freeze-etching technique. What is uncertain is how many different particles there are and what their chemical composition is. It seems uncertain whether any of these may be the ultimate photosynthetic unit or quantasome. Little else can be said at present.

The regular spatial arrangement of the particles seen on the membranes may indicate the presence of regularly arranged reactive sites to which the particles are attached, possibly as a result of some sort of interaction.

(iii) Molecular oganization in chloroplast membranes

A solution of chlorophyll in alcohol or acetone, when exposed to light, exhibits a red fluorescence. This is not observed when a fresh aqueous colloidal suspension of chlorophyll is used, and it therefore suggests that the chlorophyll molecules in the thylakoid membranes are somehow organized to "trap" the energy of solar radiation and to prevent its reemission as biologically useless fluorescent light. Spatial organization would also seem necessary when one remembers that one of the results of the light reaction is the simultaneous production of a strong oxidant and a strong reductant, which must be kept separate if the process is not to be nullified by their recombination. Compartmentalization is therefore necessary. In addition, the excited electrons from chlorophyll must pass in correct sequence from one electron carrier to another if phosphorylation is to occur, and this implies spatial organization of the components involved. One model of this molecular oganization is shown in Fig. 13-5. The structural significance of the long hydrophobic phytol "tails" of the chlorophyll molecule will now be apparent, because they interact with the hydrophobic layers of carotenoid and phospholipid molecules and "anchor" the light-trapping chlorophyll molecules in a highly ordered array.

The machinery of the dark reactions, however, would not seem to require organization, because the reactions are thermochemical. As stated, these reactions are catalyzed by water-soluble enzymes and do, in fact, occur in the liquid chloroplast stroma, which lacks the membranous ultrastructure of the grana.

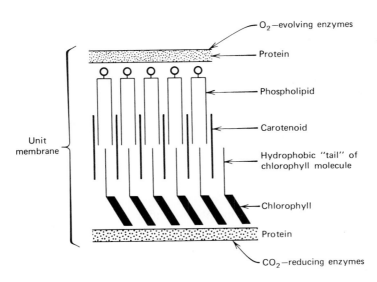

Figure 13-5 Molecular organization of chloroplast membranes. (Redrawn after J. H. Morrison, *Functional Organelles*, Reinhold Publishing Corp., 1966, after Hodge.)

THE MITOCHONDRION

The cytoplasmic organelles known as mitochondria occur in practically all aerobic plant and animal cells and were discovered by Kölliker in 1857 and isolated by him in 1888. They were called "bioblasts" by Altmann in 1894, and first received the name mitochondria when Benda designated them as such in 1897. In 1900, Michaelis first demonstrated the Janus Green supravital staining of mitochondria.

Mitochondria constitute most of the so-called "large granule fraction" (Warburg) when the components of homogenized cells are separated by centrifugation. Warburg found that the majority of the enzymes that catalyze oxidations in the Krebs cycle were confined to this fraction. In 1912 B. F. Kingsbury suggested that mitochondria were the organelles in which cellular respiration occurred, but it was not until 1948 that mitochondria were shown to be the sites of oxidative energy release and ATP synthesis of cellular respiration. This earned them the nickname of "powerhouses of the cell."

(i) Morphology and ultrastructure

In the light microscope, mitochondria are seen to be pleomorphic and usually appear as roughly spherical, elongated, or filamentous cytoplasmic bodies of relatively constant width (0.5 μ), which may vary from 0.2 to 2.0 μ, according to cell type. The length may be up to 10 μ or, rarely, 40 μ. The morphology of mitochondria is more or less constant in cells of a similar type or in cells that carry out similar functions.

When ultrathin, osmium-stained sections of mitochondria are examined in the electron microscope, complex membranous ultrastructure is evident, as in chloroplasts, but its architecture is somewhat simpler and easier to understand (Fig. 13-6). The mitochondrion is a double membranous organelle consisting of an outer membrane about 60 Å thick, much like a typical plasma or endoplasmic reticulum membrane in structure and composition, and an inner membrane 60 Å thick just inside the outer and separated from it by a space about 80 Å wide. The inner membrane is folded in a complex manner into tubular (plant cells) or platelike (animal cells) invaginations called mitochondrial crests or cristae that project into the inner mitochondrial space (Figs. 13-7 and 13-8). Compositionally and structurally, the inner membrane differs strikingly from the outer. It contains considerably less lipid, more protein, and an array of enzymes.

If a mitochondrion is caused to swell and break by immersion in a hypotonic solution, and the membranes are then negatively stained with phosphotungstate and examined in the electron microscope, the inner membrane is seen to be covered on its inner surface with knoblike particles 80 to 100 Å in diameter, each attached to the inner membrane by a stalk about 50 Å long (Fig. 13-7). These have been designated F_1 particles, or elementary particles, and they are

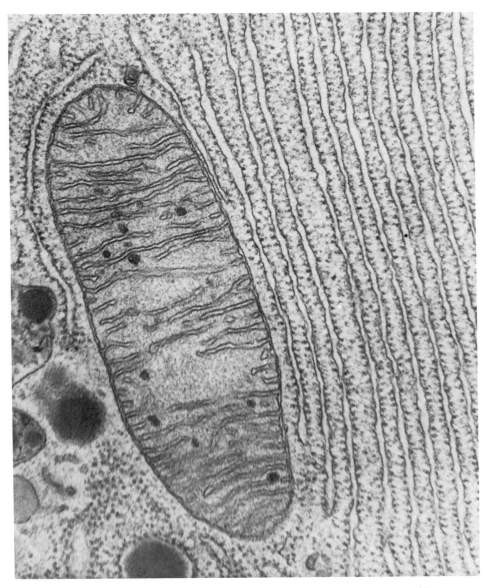

Figure 13-6 Electron micrograph of thin-sectioned mitochondrion. (Courtesy of Dr K. R. Porter. From *An Atlas of Fine Structure* by D. W. Fawcett. W. B. Saunders Company, 1966. Reprinted by permission.)

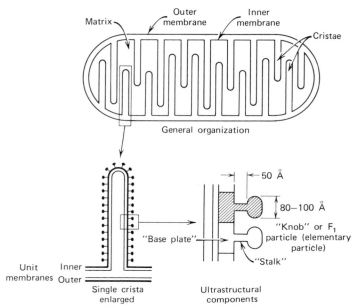

Figure 13-7 Mitochondrial organization. (*Note.* The term "elementary particle" is used by some authors to designate the entire base-plate-stalk-knob complex.)

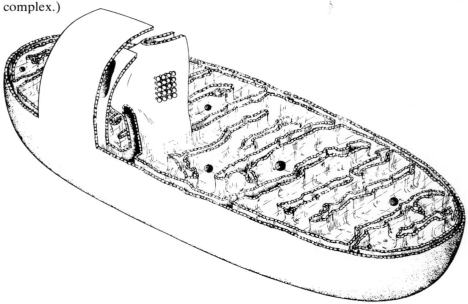

Figure 13-8 Three-dimensional reconstruction of mitochrondrial structure. (From *Cell and Molecular Biology* by E. J. DuPraw. Academic Press, 1968. Reprinted by permission.)

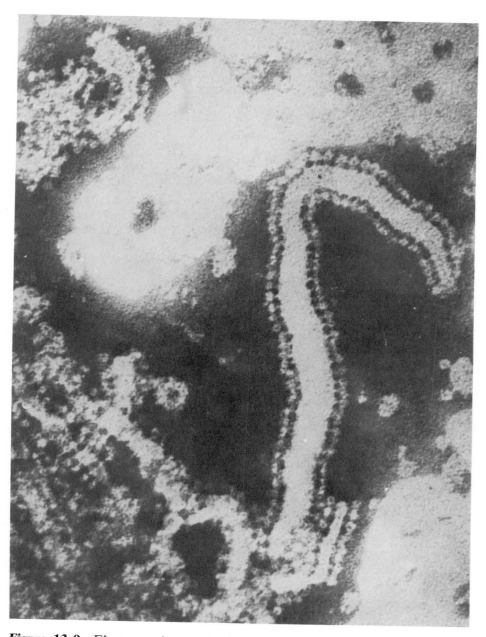

Figure 13-9 Electron micrograph showing elementary particles attached to mitochondrial membranes. (Courtesy of Dr B. Chance. Science *142*: 1176–1180, November 29, 1963. Copyright 1963 by the American Association for the Advancement of Science.)

spaced at regular intervals about 100 Å apart. They are normally contained within the inner membrane and hence are not seen in sections, but are revealed by hypotonic disruption of the mitochondrion followed by negative staining (Fig. 13-9). For this reason they were once considered to be artifactual, but similar negative staining of other membrane systems does not produce this effect.

(ii) Mitochondrial compartmentalization and spatial organization of molecular components

The presence of two membranes divides the mitochondrion into two compartments, an outer space (continuous with the intracristal spaces) and an inner space. The inner space is gellike and is called the matrix. It contains the soluble enzymes of fatty acid oxidation and the soluble enzymes and cofactors of the Krebs tricarboxylic acid cycle. Glucose, fatty acids, and amino acids can all be converted to two-carbon fragments that enter the Krebs cycle as acetyl coenzyme A. As seen earlier, after a complex series of reactions, the two-carbon fragment is ultimately degraded to carbon dioxide and water. At several points in this reaction sequence, pairs of electrons or, to be more precise, hydrogen atoms are detached by dehydrogenases and transferred to the electron carrier chain, where their energy is tapped off and utilized to synthesize ATP. The components of the electron transport system (flavoproteins, cytochromes, cytochrome oxidase) are part of the "insoluble" mitochondrial fraction and are actually keyed into the inner membrane in the form of highly organized and spatially arranged "solid-state" assemblies. At three points in the electron transport chain, the oxidation energy released in electron transport is coupled to ATP synthesis by "coupling factors." Since the ATP synthesizing mechanism is localized to the inner membrane, it is not surprising that the most metabolically active cells possess the greatest number of cristae and the most intricate folding of the inner membrane; there is a correlation between the number of cristae per unit volume and the rate of respiration.

Evidently, there must be a high degree of spatial organization within the inner membrane at the molecular level if the electrons are to be passed in correct sequence over the electron carriers in increasing order of redox potential, and if the oxidative phosphorylation of ADP is to be coupled to this and at the right places. In fact, the flavoproteins and cytochromes are known to occur in simple whole number ratios to each other, and the molecules of the enzymes of the respiratory chain are closely and precisely arranged in the order in which they interact. A complete organized molecular unit consisting of one molecule each of the different respiratory enzymes is called a respiratory assembly, and they are arranged in a regular manner (like bricks in a wall, according to Lehninger) on the outer surface of the crista membranes. In a liver mitochondrion there may be 15,000 assemblies, and a mitochondrion of an insect's flight muscles may have

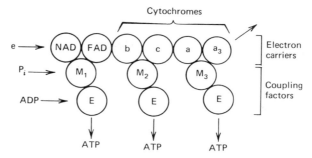

Figure 13-10 A mitochondrial respiratory assembly (according to A. Lehninger). M_1, M_2, and M_3 are coupling factors that may be specific for each of the three energy-conserving sites. E is the terminal phosphate-transferring enzyme. (Redrawn from A. Lehninger.)

100,000 assemblies. Within these assemblies the molecular components are spatially arranged so that the reactions proceed sequentially. The possible organization of a single respiratory assembly, as postulated by Lehninger, is shown in Fig. 13-10.

When mitochondrial membranes are disintegrated, treated with detergents, and centrifuged at high speed, very small particles are found in the sediment that are capable of carrying out the reactions of electron transport from NADH and succinate to oxygen but without coupled phosphorylation. These "electron transport particles" (ETP) were first isolated by Crane and others from beef heart sarcosomes in 1956. Each ETP can be further "dissected" into four subunits or complexes, each of which represents the smallest functional units so far separated from mitochondria. If they are further subdivided their activity ceases.

The four complexes that carry out different parts of the electron transport chain are as follows.

Complex I.   Catalyzes oxidation of NADH coupled to reduction of ubiquinone (coenzyme Q).
Complex II.   Catalyzes oxidation of succinate coupled to reduction of ubiquinone.
Complex III. Catalyzes oxidation of ubiquinone coupled to reduction of cytochrome C.
Complex IV.   Catalyzes oxidation of cytochrome C by molecular oxygen.

The four complexes can be recombined to form a complete electron transport chain. One model, by D. Green, is shown in Fig. 13-11. Green put forward the model of the inner mitochondrial membrane of repeating similar structural units, each consisting of a "base plate" and a "stalk" to which a "knob" was attached.

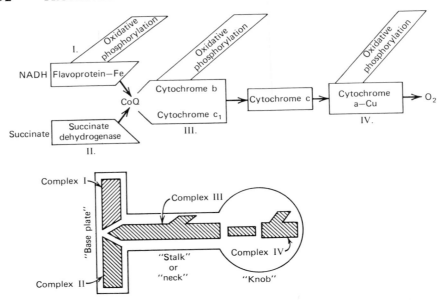

Figure 13-11 Organization of electron transport chain in an elementary particle (according to D. E. Green). (Redrawn after J. Paul, *Cell Biology*, Stanford University Press, 1964.)

Complexes I and II occur in the base plate, complex III is in the stalk, and complex IV is found in the knob. This concept was criticized in the early 1960s by E. Racker and others whose experiments indicated that the knobs of the crista membranes do not contain an essential part of the electron transport chain, but that their presence is necessary for coupled phosphorylation and that they contain the coupling factors. One of these, the F_1 factor isolated by Racker, is apparently the ATPase responsible for linking electron transport to ATP synthesis, because mitochondrial membrane fragments that lack this protein cannot couple electron transport to ATP synthesis and lack the inner membrane knobs; addition of the purified F_1 substances leads to ATP synthesis again, and the knobs reappear on the inner membrane. It is interesting to note that the isolated F_1 factor acts as an ATPase, but that it functions as an ATP synthetase when it forms part of the complete and intact respiratory assembly. Its integration with the other membrane components thus has a marked effect on its function. The association of ATPase activity with F_1 particles recalls the presence of ATPase in the quantasome particles of chloroplast membranes and suggests a possible analogy between quantasomes and the F_1 particles of the inner mitochondrial membrane.

(iii) Proposed mechanisms of mitochondrial ATP synthesis

One postulated mechanism of mitochondrial ATP synthesis depends on the existence of localized accumulations ("sinks") of protons and hydroxyl ions that result from electron flow in the respiratory chain toward molecular oxygen. These H^+ and OH' sinks provide the energy necessary for a dehydration reaction in which ATP is synthesized by removal of the elements of water from ADP and P_i.

$$ADP + P_i \longrightarrow ATP + H_2O$$

H^+ and OH' sinks arise because upon oxidation of the respiratory flavoprotein enzymes, two electrons are passed to the cytochrome system and two protons go out into the surrounding medium. At the other end of the respiratory chain, hydroxyl ions are formed when the terminal cytochrome reacts with oxygen. Because of the spatial orientations of the respiratory enzymes and the terminal cytochrome, respectively, the protons build up on one side of the mitochondrial membrane and the hydroxyl groups build up on the other. Because of the presumed impermeability of mitochondrial membranes to H^+ and OH', a pH gradient is thus formed which, under the influence of an ATPase in the mitochondrial membrane, causes the elements of water (H^+, OH') to be withdrawn from ADP and P_i to yield ATP.

Another mechanism, proposed by Lehninger, is similar to the energy-conserving steps in glycolysis, especially the oxidation of 3-phosphoglyceraldehyde to 1,3-diphosphoglycerate, where oxidation energy is conserved in the phosphate groups that are finally transferred to ADP to give ATP. The energy released on oxidation of $NADH_2$ by FAD in the respiratory chain is considered to be similarly conserved by formation of high-energy intermediate compounds of NAD with "coupling factors" (designated X). The high-energy intermediate of NAD with coupling factor reacts with inorganic phosphate to give a high-energy phosphate intermediate compound with the coupling factor, and this reacts with ADP to give ATP.

(i) $NADH_2 + X \longrightarrow NADH_2 - X$
(ii) $NADH_2 - X + FAD \longrightarrow NAD \sim X + FADH_2$
(iii) $NAD \sim X + P_i \longrightarrow NAD + P_i \sim X$
(iv) $P_i \sim X + ADP \longrightarrow ATP + X$

As is well known, mitochondria undergo cycles of swelling and contraction. Proteins in the mitochondrial membrane may contract and lose water, which results in the anhydrous conditions required by a third postulated mechanism of ATP synthesis. Here, ATP synthesis is pictured as resulting from removal of hydroxyl ions from inorganic phosphate.

$$ADP^{3-} + HPO_4^{2-} \longrightarrow ATP + OH'$$

In this model, the cytochrome chain components are considered to alternate with insoluble phospholipids; as electrons traverse this system, they are thought to draw protons along with them. When the electrons enter a phospholipid region of the chain, the protons become bound to the phospholipid because of the electron's negative charge. When electrons transfer to a cytochrome, however, the protons become free in the medium which, being anhydrous, allows displacement of hydroxyl ions from inorganic phosphate by the protons. The proton is now drawn away again by the electrons passing on to the next electron carrier, and the hydroxyl ion, instead of reacting with the proton to form water, moves out of the mitochondrion and is replaced by inorganic phosphate from outside; Krebs cycle substrates also enter. When the mitochondrion swells, water moves in and ATP synthesis ceases.

VERTEBRATE PHOTORECEPTOR CELLS

(i) Development and ultrastructure

Comparison of the submicroscopic organization of vertebrate photoreceptor cells with that of plant chloroplasts reveals striking similarities between them. We again find a complex internal ultrastructure of interweaving lipoprotein membranes. The photosensitive tips (outer segments) of vertebrate retinal rod cells will be described as an example.

The outer segment of the rod cell has a curious developmental origin, because it starts as a nonmotile cilium growing out from the rod cell surface. The photoreceptive outer segment is derived from this by a process of growth and invagination. First, the ciliumlike bulge increases in size and becomes filled with membranous vesicles that are arranged to one side. The vesicles gradually flatten and rearrange themselves in a stack, like a pile of coins, and fill the major part of the outer segment. Finally, they lose continuity with the plasma membrane and behave as intracellular organelles. These are the thylakoid sacs of the fully developed outer segment of the rod cell (Figs. 13-12 and 13-13). One wonders whether the striking similarity to chloroplast thylakoids and grana is fortuitous or whether this type of arrangement is a fundamental feature of light-transducing organs. The disc membranes are 30 to 40 Å thick, and the space between the sacs is 50 to 120 Å. The outer segments of cone cells have a similar structure, with only minor differences.

The fully developed outer segment stays connected to the rod cell by a stalk whose ultrastructure betrays its ciliary origin in that an outer ring of nine double fibers is seen in a transverse section; the two central fibers are missing, however. Other sensory structures such as the sensitive hairs within the gravity receptors and auditory organs of vertebrates also develop from nonmotile cilia. Other animal phyla in which photoreceptor cells develop from nonmotile cilia are the

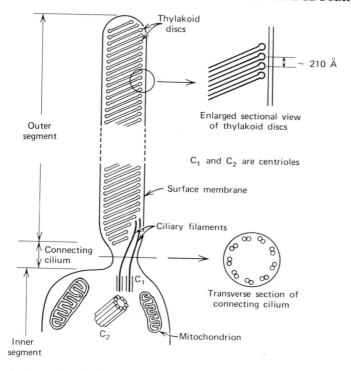

Figure 13-12 Longitudinal section of the outer segment of a rod cell. (Redrawn from E. D. P. De Robertis, W. W. Nowinski, and F. A. Saez, *Cell Biology*, 5th ed., W. B. Saunders Co., 1970.)

coelenterates, echinoderms, molluscs, platyhelminthes, and annelids, and this may possibly have phylogenetic significance, according to R. M. Eakin.

(ii) The visual pigment

A freshly removed frog's retina has a bright red color that, when exposed to light, becomes yellow and finally white; the same is observed in the retina of a live frog if it is exposed to a strong light. The color is restored when the frog is removed to a dark chamber. These observations were made by Boll in 1877. Soon after, the red, light-labile pigment was found in the retinas of other animals and was named rhodopsin by Kühne. In 1933 Wald discovered that vitamin A is formed when rhodopsin is whitened by light and disappears again when the rhodopsin color returns in darkness. Rhodopsin was found to be composed of a protein (opsin) combined with a carotenoid substance that Wald called retinene,

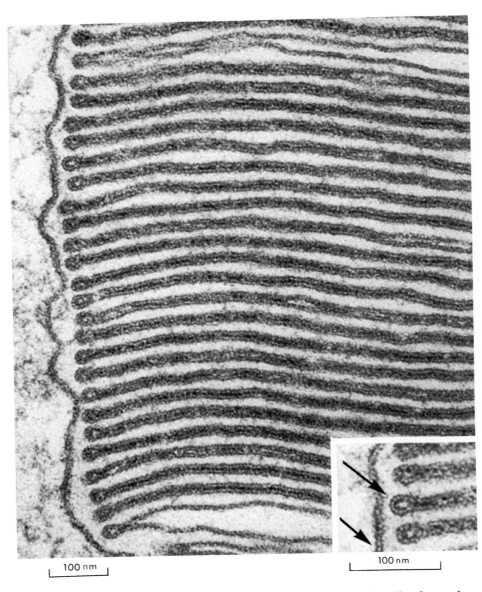

100 nm

100 nm

Figure 13-13 Electron micrograph of thin-sectioned rod cell of monkey (glutaraldehyde-fixed ×165,000; inset ×257,000). (Courtesy of Dr J. E. Dowling. From *Molecular Organisation and Biological Function* (Ed. J. M. Allen). Copyright 1967 by J. E. Allen. Reprinted by permission of Harper and Row, Publishers Inc.)

and this appeared to be the color-bearing (chromophoric) component of rhodopsin. Wald also showed that the color change from red to yellow when rhodopsin is exposed to light is due to dissociation of retinene from opsin, and that the change from yellow to white corresponds to the conversion of retinene to vitamin A. Subsequently retinene's name was changed to retinal when it was found in 1944 that retinene was an aldehyde polyene derived from vitamin A (an alcohol) by removal of two hydrogen atoms from the terminal carbon atom of the molecule. The nature of the linkage between retinal and opsin was eluci-dated by Akhtar et al. with the aid of tritiated rhodopsin prepared from tritiated 11-*cis* retinal. Irradiation of the labeled rhodopsin in the presence of sodium borohydride resulted in irreversible binding of the retinal to the protein. Further studies revealed that the retinal in this reduced rhodopsin derivative was attached to the epsilon-amino group of a lysine residue. These investigators suggest that the spectroscopic characteristics of rhodopsin and other visual pigments receive their most acceptable explanation in terms of a charge-transfer interaction between retinal and some other group (—X or —X·H) in the opsin.

(iii) The molecular basis of vision

Retinal, like chlorophyll, possesses a conjugated system of double bonds (Fig. 13-14). It is derived from dietary vitamin A, and its molecule exists in several isomeric forms, only two of which are important in the molecular events of

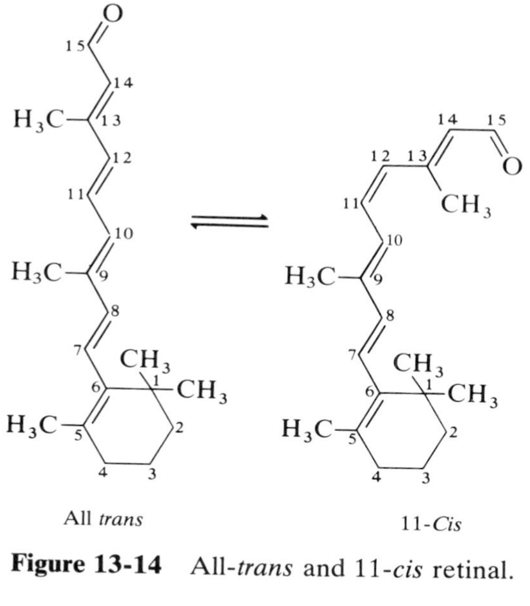

Figure 13-14   All-*trans* and 11-*cis* retinal.

vision, the "straight" (all-*trans*) form and one of the "bent" (11-*cis*) forms (Fig. 13-14). The essential molecular event in vision is photoisomerization of these two forms of the retinal molecule. The evidence indicates that 11-*cis* retinal, which has a higher energy content than the *trans* isomer, is favored by darkness and is the form that combines with opsin to give rhodopsin. Absorption of a light photon somehow causes the 11-*cis* retinal to "straighten out" into the *trans* form which, for some reason, does not fit the opsin molecule and so cannot stay combined with the protein, with the result that the rhodopsin complex dissociates (Fig. 13-15). This has the effect of causing a configurational change in

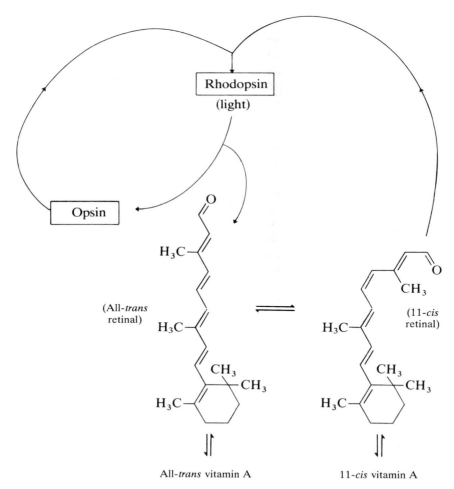

Figure 13-15   The "chemistry of vision."

the opsin molecule that decreases the permeability of the rod membranes to sodium ions and brings about a hyperpolarized state. This process presumably is related to the initiation of action potentials in the associated bipolar and ganglion cells of the retina. The all-*trans* retinal then adds on two hydrogen atoms, thus becoming all-*trans* vitamin A. This isomer is then converted to 11-*cis* vitamin A and then, by loss of two hydrogen atoms, to 11-*cis* retinal, which recombines with opsin to reform rhodopsin.

(iv) Chemical composition of the outer segment membranes

Vitamin A deficiency causes blindness in rats that may be cured when the vitamin is administered again. Electron microscope examination of sections of the rod cells, of deficient animals reveals degeneration of the intricate membranous ultrastructure of the outer segments. If the membranes are largely composed of rhodopsin and if the combination of retinal with opsin somehow stabilizes the latter, then absence of retinal would result in an unstable protein, and degeneration of the membranes would be the expected result. The idea that retinal may stabilize opsin is pure supposition, but it may be analogous to the case of the myoglobin molecule which, as mentioned in a previous chapter, is stabilized by its combination with heme. The effect of vitamin A deficiency on the integrity of the thylakoids at least suggests that rhodopsin is a constituent of the membranes. The plasma membrane of the outer segment appears not to contain rhodopsin, since it is apparently unaffected by vitamin A deficiency.

Definite evidence showing that rhodopsin is a constituent of the thylakoid disc membranes was obtained by Dewey et al. using the immunofluorescent technique. They prepared a specific immune serum from rabbits against frog rhodopsin. Sections of frog retina were treated with the rabbit antirhodopsin gamma globulin and then with fluorescein-labeled sheep antirabbit gamma globulin. Fluorescence microscopy revealed uniform fluorescent staining in isolated whole rod outer segments and in isolated disc membranes. These observations showed that the photopigment is an integral part of the disc membranes and also of the external membrane of the outer segment; the latter finding is apparently at variance with the vitamin A deficiency study described above.

Localization of photopigment at the molecular level was accomplished by Blasie et al. using electron microscopy and low-angle X-ray diffraction in studies of disc membranes treated with antirhodopsin serum. Wet photoreceptor disc membranes were treated separately with either antirhodopsin serum, other nonimmune sera, or were not treated at all. The X-ray diffraction patterns obtained of these preparations indicated that genuine conjugation of the antirhodopsin molecules with antigen in the disc membranes had occurred, and not nonspecific adsorption of protein. The planar arrangement and ordering of the adsorbed antirhodopsin molecules on the disc membrane surface practically coincided with 40 to 50 Å particles that are seen with the electron microscope inside the untreated membranes. From these and other studies it was concluded that the 40 to 50 Å particles are most likely the nonpolar cores of photopigment molecules and that they exist in a planar, liquidlike arrangement.

CONCLUSIONS

Cellular energy transducers are remarkably similar in their ultrastructural organization, consisting as they do of complex stacked arrays of phospholipid membranes, into which are keyed the reactive molecular components. In the case of chloroplast and mitochondrion membranes, and probably also in photoreceptor cells, these are spatially arranged in precise geometric arrays so that they resemble solid-state systems; there is, in fact, evidence indicating that perhaps some of the phenomena of solid-state physics may occur in chloroplasts. Although described in an earlier chapter under fibrous proteins, the vertebrate sarcomere is, of course, also an energy transducer and converts the chemical energy liberated in ATP hydrolysis to the mechanical energy of contraction. In the sarcomere we also see the development of an intricately organized membranous ultrastructure—the sarcoplasmic reticulum— which is really the highly elaborated endoplasmic reticulum of the muscle fibers.

Both mitochondria and chloroplasts contain similar enzymes and electron carrier molecules, and these are arranged into assemblies that carry out electron transport. Both of these organelles effect synthesis of ATP by coupling electron transport to phosphorylation of ADP.

An energy cycle exists within eucaryotic cells that sustains their activities and is mediated by the chloroplast and the mitochondrion, both of which carry out an electron flow maintained by sunlight energy. Starting with the low-energy electrons of water, the chloroplast "captures" solar energy and transfers it to electrons that ultimately become incorporated into the high-energy molecules of carbohydrates, fats, and proteins. Biologically useful energy is released when these materials or derivatives of them are oxidatively degraded in aerobic respiration and the energy of the electron is coupled to ATP synthesis in the mitochondrion. Eventually, when all their energy is released, the electrons and associated protons unite with oxygen to form water again.

Thus, as Albert Szent Györgi expresses it, light-absorbing molecules at the surface of the earth interact with solar photons, and electrons are thereby promoted to higher energy levels. As they return to their ground state, "life," in effect, interposes itself in the path of the "falling" electrons and utilizes the energy released for its maintenance.

PART FOUR

PROTEIN-NUCLEIC ACID ASSEMBLIES. BIOLOGICAL REPLICATION. THE STORAGE AND TRANSMISSION OF GENETIC INFORMATION

CHAPTER FOURTEEN
The transmission of genetic information

Scientists and philosophers have often attempted—unsuccessfully—to define what is meant by "life" or "being alive." When all is said, we appear to be no more successful even today. Perhaps, after all, we are only playing with words; one author has suggested that it is incorrect to define biology as the "science of life" because "life" is really an abstract noun and biology is not the study of abstract nouns. Biology, he says, is more properly defined as the study of "living things," so we need to define what is meant by a living thing. Maybe we are now on firmer ground than when trying to define life; as a first step, we could attempt to determine what are the characteristics of living things that are not shared by nonliving entities, an exercise that also leads to difficulties.

Two characteristics can illustrate what most people would regard as living things. One is that living things have the power of spontaneous movement—but so have solid microscopic particles suspended in liquid when observed under the microscope, a phenomenon called "Brownian motion." In Brownian motion the liquid molecules bombard the solid particles, which causes their ceaseless movements. The phenomenon was first observed by the botanist Robert Brown in pollen grains suspended in water.

Growth has frequently been cited as characteristic of living things. This is usually criticized by pointing out that crystals will also "grow" when placed in a saturated solution of the same material, but here there is a difference from true organic growth. A crystal merely becomes bigger by surface deposition from the solution of material similar to itself. True organic growth, on the other hand, involves taking in food materials from the environment and chemically altering them, a process called metabolism. The food is first broken down (catabolism) and then built up again (anabolism or biosynthesis) into the compounds characteristic of that living thing. Alas, even this does not seem to be truly unique to living things. The clever experiments of A. I. Oparin with artificial cell-like structures (coacervates) have shown that even metabolism can be strikingly mimicked in a nonliving system. Other characteristics of living things may be selected in an effort to separate them from nonliving things, but it seems that almost always, we run up against a brick wall.

There appears to be only one characteristic of living things that has no parallel in the nonliving world. This is the ability to replicate or reproduce with simultaneous transfer of "genetic information" to the progeny. Genetic information may be defined as the "coded instructions" contained within living entities that induce and direct the elaboration of the myriad variety of proteins that are necessary for the structural integrity and normal function of the progeny.

Biological replication and information transfer are associated with and effected by protein-nucleic acid assemblies (i.e., nucleoproteins). As will be seen later, there are at least three biological entities that structurally are nucleoprotein assemblies. Two of them—chromosomes and ribosomes—are components of entities that are easily recognized as being alive—cells. The third—viruses— seem to stand on what is often regarded as the border between the living and nonliving. The viruses are self-replicating nucleoprotein particles and thus possess the unique criterion of living things, yet in other ways they do not seem to be alive; they cannot replicate outside of living cells and they can be crystallized. Perhaps it is best not to attempt any further to try to define living things. Whatever the pros and cons, one thing is clear—the study of protein-nucleic acid assemblies seems to bring us very close to the heart of life, or being alive, however one defines these terms.

Long before chromosomes were discovered, an Austrian monk named Gregor Mendel investigated and carefully formulated the fundamental patterns and mechanism by which hereditary characteristics are transmitted to successive generations of organisms. Subsequently, these became known as Mendel's laws of inheritance; in effect, they are a set of conclusions arising from observations of the patterns of transmission of "hereditary factors" from parent to offspring in extensive breeding experiments that he carried out in the common garden pea, *Pisum sativum.*

MENDELIAN INHERITANCE

The pea plant exhibits several well-defined, outwardly visible characteristics; among them are stature (tall or short), flower color (red or white), seed color (yellow or green), and seed coat texture (smooth or wrinkled). Mendel possessed "pure strains" of plants showing these characteristics. Here, "pure" means that the plants had always "bred true" when self-pollinated for several generations (i.e., no deviation from the parental characteristics was ever observed in the offspring).

Mendel decided to try crossing two pure strains showing different forms of the same type of characteristic (e.g., a white-flowered strain with a red-flowered strain). A lay person might intuitively feel that the two flower colors would blend and give rise to offspring all with pink flowers, like mixing different colored paints. Mendel found—surprisingly—that the offspring of this cross all bore *red* flowers. This generation was called the first filial (F_1) generation, and the parents of this were called the P generation. He allowed the F_1 plants to self-pollinate, planted the seeds, and awaited results. The offspring (the F_2 generation) were both red and white flowered, the red-flowered plants outnumbering the white-flowered plants by about three to one. Evidently, the white-flowered character was somehow masked in the F_1 generation and reappeared again in the F_2. Precisely similar results were obtained when Mendel crossed pure strains of other contrasting characteristics such as tall with short, yellow-seeded plants with green-seeded plants, smooth-seeded with wrinkled. In each case, one only of the characters appeared in plants of the F_1 generation and, when these were self-pollinated, this same character appeared again in the F_2 generation along with the other characteristic previously "masked" or hidden in the F_1. The plants showing the previously masked character were always outnumbered by the plants showing the other character by about three to one. Mendel designated the characteristic that seemed to be masked in the F_1 generation as "recessive" and the characteristic whose effects seemed to hide it as "dominant." Thus, the red flower character is said to be dominant to white, and the tall plant character is dominant to short. The results of a crossing experiment and Mendel's interpretation are illustrated in Fig. 14-1. Mendel also tried the effect of crossing two strains breeding true for two different characteristics. He crossed a strain that bore smooth yellow seeds with one bearing wrinkled green seeds. Previously, he had found that smoothness and yellow color in pea seeds were dominant to wrinkled and green, and so, not surprisingly, all the F_1 generation of such a cross yielded plants bearing smooth yellow seeds (i.e., both dominant characters were exhibited). When these F_1 plants were self-pollinated, the resulting F_2 generation consisted of plants bearing different proportions of four different types of seeds, some of which resembled the original parental (P) types and others with combinations of characteristics not found in the original parent generation, in the ratios of $9:3:3:1$ (Fig. 14-2). Evidently, the two traits had assorted independently of each other; the experiment illustrates Mendel's law of independent assortment, meaning that the inheritance of any one characteristic is not influenced by the presence of other characteristics.

Therefore, contrary to the "paint mixing" idea of inheritance in which hereditary characteristics are supposed to blend, like mixing different colored paints,

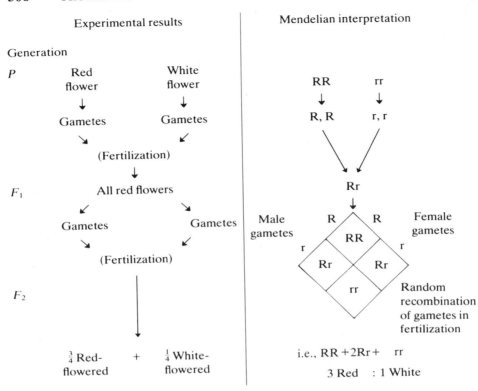

Figure 14-1 Inheritance of flower color in garden peas.

the actual mechanism is more like the mixing and sorting of colored marbles (i.e., heredity appears to operate by a particulate mechanism). The particlelike hereditary factors for a given characteristic segregate independently of each other in gametogenesis (Mendel's law of independent segregation) and retain their individuality from generation to generation, although the outwardly visible effects of one form of a factor may be temporarily concealed by its other form in some of the generations.

Mendel summarized these results and explained them in what are now known as Mendel's laws of inheritance, which are as follows.

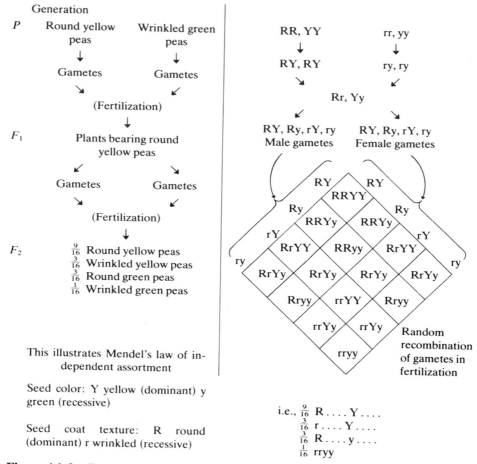

Figure 14-2   Result of crossing pea plants differing in two characters: seed coat texture and seed color.

1. Each hereditary characteristic is determined by a *pair* of hereditary factors.
2. During gamete formation, hereditary factors separate; each one of a pair passes independently of the other into a gamete.
3. At fertilization the hereditary factors are randomly brought together again in pairs in the zygote.
4. One hereditary factor of a pair can dominate over the other and temporarily conceal its effects.

(The phenomenon of dominance and recessiveness is not universal; the differing forms of a single hereditary factor are often expressed equally when present together in the same individual.) Mendel's work was published in 1865, but its significance was not realized until more than 30 years later. Mendel's hereditary factors were renamed "genes" by W. Johanssen in 1911, and the name has been retained. The different forms of a gene are known as alleles. An organism's gene makeup is called its genotype, and the outward visible expression of the genes is called the phenotype. An organism possessing two dissimilar alleles for a given phenotypic trait is said to be heterozygous ("impure"), whereas one possessing two similar alleles is said to be homozygous ("pure"). Examples of heterozygotes are the F_1 hybrids produced in Mendel's experiments when he crossed the two true-breeding strains showing different forms of the same trait (e.g., tall and short pea plants); homozygous organisms are exemplified by the individuals of the parent (P) generation, the one quarter of the F_2 generations showing the recessive phenotype and one third of the proportion of the F_2 generation showing the dominant phenotype.

CHROMOSOME BEHAVIOR IN CELL REPLICATION

The nucleus of a living cell plays a key role in regulating the metabolism of the cell, and it is indispensable for cell reproduction (replication) and also differentiation in the tissues of multicellular organisms. Cells that normally do not contain a nucleus (e.g., mature mammalian red blood cells) are never observed to reproduce. If the nucleus is removed from an amoeba, the enucleated cell never divides again and does not survive for long. As will be shown in Chapter Fifteen, the cell nucleus and also the chromosomes consist almost entirely of nucleoprotein; one might therefore reasonably suspect that nucleoprotein has something to do with the storage of hereditary characteristics (i.e., genetic information).

Cell reproduction involves division of the cytoplasm (cytokinesis) and division of the nucleus. In tissues undergoing normal growth and repair, the type of nuclear division involved is called mitosis. In the specialized tissues destined to give rise to reproductive cells (gametes), a different type of nuclear division called meiosis is observed. Both of these types will be briefly described and the significance of each will be apparent later.

(i) Mitotic cell division (mitosis)

Mitotic cell division may be conveniently studied in thin, stained sections of rapidly dividing tissues such as the cells in the region just behind the root cap in young hyacinth roots or in an embryonic tissue such as the blastula of the whitefish.

The process of nuclear division is a continuous process; since different cells in a rapidly dividing tissue may be at different stages of the process, it is possible to reconstruct the mitotic "cycle" by studying the different stages in several cells that have been "frozen" at one particular stage by the killing and fixation procedures. Traditionally, mitotic cell division is usually viewed as four sequential stages for purposes of description, but one must remember that the entire process is continuous; each "stage" blends imperceptibly into the next.

When viewed with a sufficiently powerful microscope, the nucleus of a fixed and stained cell in the so-called "resting" or nondividing (interphase) stage appears to have a rather amorphous granular appearance (Fig. 14-3). The four stages of *mitosis* (nuclear division) are as follows (for simplicity, the role of accessory structures involved in the mechanics of cell division such as centrosomes and spindle fibers are omitted from the following description).

Prophase.   The first sign of the onset of mitosis is a change in the appearance of the nucleus. The previously granular chromatin begins to look like a tangled skein of delicate threads—the "spireme"—which become progressively shorter and thicker. Ultimately, the nuclear chromatin assumes the form of a number of

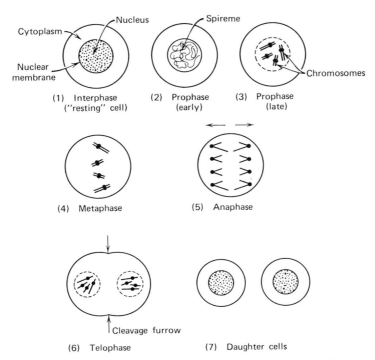

Figure 14-3 Mitotic cell division (schematic). For clarity, only two pairs of chromosomes are shown.

relatively short, thick, deeply staining rods called chromosomes ("colored bodies"). Careful studies have shown that the number of the chromosomes in the normal nongametic cells of a given species is constant and characteristic of that species, and that each chromosome has a characteristic size and morphology.

A single chromosome consists of two similar threads called chromatids, and these appear to be held together at the centromere (Fig. 14-4). The centromere occupies a characteristic position in a given chromosome and, on the basis of this, chromosomes have been classified into four morphological types: telocentric, acrocentric, submetacentric, and metacentric, in which the centromere is, respectively, at the end of the chromosome, near the end, near the center, or almost exactly at the center. The first two chromosome types are rodlike in shape and the last two are V- or L-shaped, the centromere being situated at the angle of the L or V.

An important point is that each chromosome has a morphologically similar partner (i.e., the chromosomes occur in homologous pairs). Therefore, if there are n different morphological types of chromosomes in the nongametic cells of a given species, then the total number of chromosomes per cell is $2n$. This is known as the diploid chromosome number and is constant and characteristic in and of that species.

The morphological features of the chromosomes of each homologous pair, such as their size, shape, and centromere position, and the diploid number are

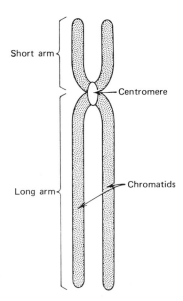

Figure 14-4   Diagram of a chromosome.

characteristic features of the chromosome makeup of an organism and are known as the karyotype.

The appearance of the fully condensed chromosomes marks the end of the prophase stage . Three phases follow prophase.

Metaphase.   The chromosomes move themselves into a discoidal arrangement (the "metaphase plate") at the center or "equator" of the cell.

Anaphase.   The centromeres of each chromosome divide, and each half moves in opposite directions toward opposite ends of the cell, at the same time pulling its attached chromatid along with it.

Telophase.   By this time the division of the cytoplasm is proceeding and each of the two separated sets of chromatids cluster together near opposite ends of the now dividing cell that will soon be two new cells. Gradually, the individual chromatids become thinner, and the tangled thread structure is again manifest. Finally, the amorphous granular appearance of the two nuclei is seen in the new two separate "daughter cells," as they are called, and the two new cells enter another "resting stage" or interphase period.

Sometime during the mitotic cycle, most likely immediately before or during prophase, the chromatids replicate themselves, because the mitotic chromosomes consist of two chromatids each, it will be remembered, whereas one chromatid of each chromosome passes into each of the daughter cells.

Thus, in mitotic cell division, the chromosome number of each daughter cell is the same as that of the parent cell; the biological significance of mitosis is therefore the exact division and equal distribution of nuclear material from the parent cell to the daughter (progeny) cells.

(ii) Meiosis

During the life cycle, many organisms undergo what is called a sexual process, and this is more or less closely associated with reproduction, depending on the organism. Essentially, the sexual process is the fusion of two cell nuclei; the single body thus formed is called a zygote. The zygote then undergoes mitotic cell division to give the adult organism. Since sexual processes involve the fusion of the chromosomes of two cell nuclei to make one nucleus, it follows that the chromosome number of a given cell or species would increase every time a sexual process occurred. This contradicts the statement made earlier that the diploid chromosome number of a given species is constant. Obviously, there must be some means of preventing the increasing chromosome number in sexual processes. The way in which this is accomplished is that in organisms exhibiting sexual processes, special cells are produced that possess only one of each type of chromosome; it is these cells that engage in the sexual process. Such cells containing half the usual number of chromosomes are said to be haploid [i.e., these "sex cells" (gametes) contain the n number of chromosomes]. When two gametes fuse (fertilization), a diploid zygote is the result. Hence there is an

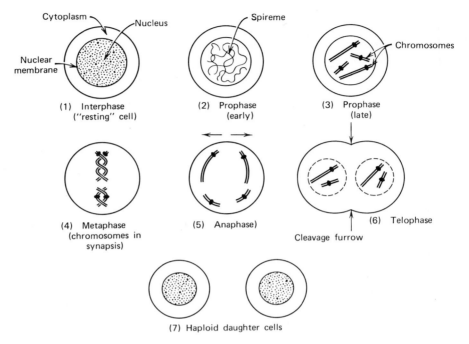

Figure 14-5 Meiotic cell division (schematic). For clarity, only two pairs of chromosomes are shown. After a brief pause (interkinesis), the two haploid cells resulting from the first (reduction) division undergo a second division resembling mitosis, the final result being the formation of four haploid cells.

alternation of haploid and diploid phases in the life cycle. The haploid gametes are produced in a form of cell division called meiosis, in which there are two cycles of division; the first cycle achieves the halving of the chromosome number (reduction division), and the second cycle resembles mitosis. The essential stages of meiosis follow (see Fig. 14-5).

DIVISION I (REDUCTION DIVISION)

Prophase. As in mitosis, the chromosomes gradually make their appearance as well-defined, distinct, rodlike structures, but now a difference is observed. In mitosis, there is not normally any observable interaction between the chromosomes of a homologous pair. At this stage, in meiosis, homologous chromosomes become closely apposed along their whole length, a process called synapsis. Apparently, there now occurs an exchange of material between chromatids of opposite homologous chromosomes; this phenomenon is usually

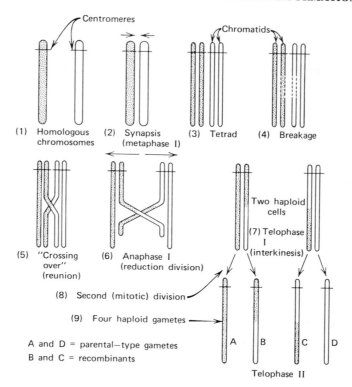

Figure 14-6 "Crossing over." A purely schematic representation of the "breakage-reunion" model.

called "crossing over." What happens appears to be that breaks occur at various places on the chromatids, followed by reunion with segments of the chromatid on the opposite side. The process is schematically represented in Fig. 14-6 in simplified form. This has the effect of introducing genetic variability into the gametes, as will be appreciated later.

Metaphase, anaphase and telophase. During these stages the halving of chromosome number and the formation of haploid cells is apparent. *Instead of centromeres dividing and chromatids separating, each member of a homologous pair of chromosomes migrates separately to opposite ends of the cell.* The cytoplasm divides during telophase and the nuclei of the two resulting daughter cells therefore contain the equivalent of one only of each type of chromosome.

Interkinesis. This is a brief pause between the two cycles of meiotic cell division.

DIVISION II (MITOTIC)

The chromosomes of these two haploid cells reappear during the prophase of this second division. At anaphase, the centromeres divide and chromatids separate, as in mitosis, and the final result is the production of *four* haploid cells.

Further "maturation" of these daughter cells occurs in higher plants and animals to give the distinctive male and female gametes that unite in fertilization, and the diploid chromosome number is thus restored in the zygote.

LINKAGE GROUPS, MUTATIONS, AND GENETIC MAPPING

The perceptive reader will no doubt have noticed that the behavior of genes during gametogenesis and their random recombination in the offspring parallels the behavior of chromosomes during meiosis to give haploid cells and the restoration of the diploid chromosome number at fertilization. This naturally led biologists to suspect a connection between genes and chromosomes when chromosomes were first observed and their complicated movements in cell division and gametogenesis were elucidated. That genes are located on or in chromosomes was first put forward independently by W. S. Sutton and by T. Boveri in 1903. This has now been well established, and chromosomes behave as though they were linear sequences of genes. The independent segregation of two different genes situated on two different, separate chromosomes thus receives an explanation in the observed independent segregation of the separate members of two homologous pairs of chromosomes during gametogenesis and their random recombination in fertilization. However, what would happen if the two genes were situated on the *same* chromosome? Obviously, the genes would not now segregate independently, so that the traits they determine would tend to be inherited together. Such genes are said to be linked or to exhibit linkage. Linked genes do, however, sometimes segregate independently as a result of the phenomenon of "crossing over" of chromatid material during early meiosis. Greater variability of gene combinations is thereby introduced into gametes and offspring that would not otherwise occur; this is significant in the evolutionary processes of organisms.

Mendel was indeed fortunate in hitting on characteristics of pea plants in which the genes responsible segregate and recombine independently during gametogenesis and fertilization, because it greatly simplified the interpretation of his experimental results.

Although the precise mechanism of crossing over does not seem to have been established, it appears to occur by a process involving the breakage of chromatids at various points, followed by reunion with fragments of the opposite chromatid, as shown in Fig. 14-6. Crossing over appears to take place during synapsis of homologous chromosomes in the reduction division of meiosis.

During synapsis, homologous chromosomes coil tightly around one another, and maybe this involves mechanical stress that causes breaks in the chromatid.

On the assumption that chromatid breaks occur randomly and that all parts of a given chromatid are equally likely to break, it follows that the further apart two genes are on a chromosome, the more likely it is that a break will occur between them (i.e., the more likely it is that the two genes will exhibit crossing over). The frequency of crossing over between two genes and their alleles situated on the same chromosome can therefore be deduced from the phenotypic ratios exhibited by offspring in breeding experiments. This has been done extensively in the fruit fly, *Drosophila melanogaster*. *Drosophila* was employed as an experimental animal in genetic studies by T. H. Morgan and is well suited for this. First, it has many easily recognized phenotypic characteristics such as variations of wing shape, wing size, body color, and eye color. Like other organisms, *Drosophila* exhibits *mutations*; a mutation is a change that suddenly occurs in the genetic apparatus of an organism that may involve whole chromosome segments or only single genes. Gene mutations are usually permanent and irreversible; they are transmitted to offspring and generally behave as recessives to the corresponding normal allele. Such permanent heritable changes in genes occur rarely (about one in every 10^6 cell divisions) and appear to arise spontaneously. Gene mutation rates can, however, be greatly accelerated by the action of certain chemicals (nitrous acid, mustard gas, urethane) and physical agents (x-radiation, ultraviolet light), which appear to interact with and cause alterations in the genetic material. Many mutations have now been recognized in *Drosophila*; these show up as well-defined phenotypic characteristics such as heritable variations in wing size and shape, eye color, and so forth. Second, the generation time of *Drosophila* is only two weeks, so that no long waiting periods are needed to see the results of the genetic experiments. Third, the chromosomal apparatus of *Drosophila* is quite simple, since the haploid number of chromosomes is only four. The salivary gland chromosomes are unusually large ("giant chromosomes") and are visible even during interphase; they are therefore said to be permanently "condensed."

In *Drosophila*, the gene determining normal wing shape and its recessive allele determining vestigial wings is situated on the same chromosome as a gene determining body color, the two allelic forms of which determine gray (dominant) or black body color (recessive). Figure 14-7 illustrates two possible results of crossing an F_1 hybrid (produced by a mating between parents, one of which is homozygous for both dominant traits and the other homozygous for both recessive traits) with the recessive parent. Note that characteristic phenotypes and phenotypic ratios are exhibited in the offspring according to whether the genes are supposed to be completely linked or not linked at all. The actual result of the F_1 backcross with the recessive parent does not give either of the above patterns, but something in between—80% of the offspring exhibit the parental phenotypic combinations, and 20% show phenotypes unlike that of the

Two Possibilities:

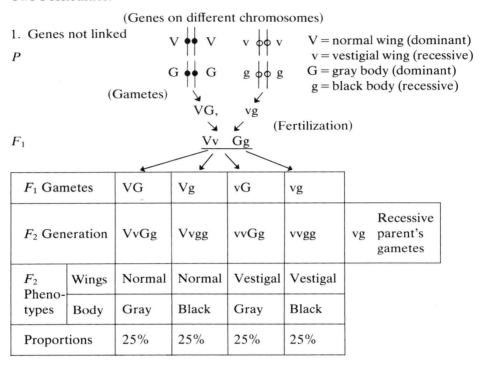

Figure 14-7 Inheritance of body color and wing shape in *Drosophila*.

parents. The reason for this is that crossing over has occurred during gameto-genesis. The 20% of offspring that are unlike the parents are therefore known as recombinants and have resulted from crossing over, which occurred in about 20% of the gametes in this example. The two genes therefore exhibit a crossover rate of 20%.

The "recombination rate" for unlinked genes is 50%, and anything less than this therefore indicates linkage. On the basis of extensive studies of inheritance patterns and recombination of genes in *Drosophila*, it has been possible to assign the genes to four linkage groups, which correspond, of course, to the four gametic chromosomes visible in the microscope. Morgan and other workers obtained complete confirmation of this in studies of the giant salivary gland chromosomes when it was found that specific mutations were correlated with visible morphological changes at the corresponding specific sites on the individual chromosomes. The frequency of crossing over has enabled actual gene

2. Genes linked

P

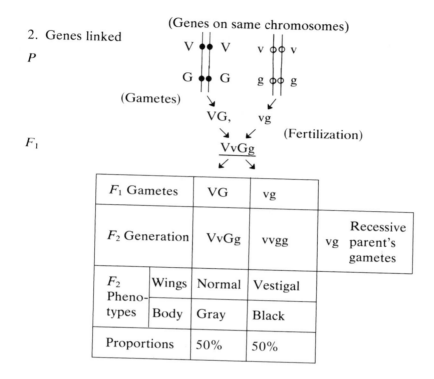

F_1

F_1 Gametes	VG	vg	
F_2 Generation	VvGg	vvgg	vg Recessive parent's gametes
F_2 Pheno-types	Wings	Normal	Vestigal
	Body	Gray	Black
Proportions	50%	50%	

Figure 14-7—continued

positions to be plotted on the chromosomes, thus permitting genetic "maps" to be constructed. Briefly, the procedure is as follows. If two linked genes A and B exhibit a crossover rate of 25%, they are said to be separated by a distance of 25 map units. Suppose also that gene B exhibits a crossover rate with a third gene C of 10%. Then gene C is 10 map units from B—which is ambiguous because it could be *between* A and B or on the side of B away from A. If, however, it is found that, say, the crossover rate of C with respect to A is 15%, then this places C unambiguously 10 units from B *between* A and B. Likewise, a crossover rate of 35% with A would place C 10 units from B on the side away from A.

The sequence of genes can be assumed to be correct, but the relative distance between some of them may be inaccurate. This is because the frequency of chromatid breakage and crossing over, on which estimates of gene distances are based, varies from one part of a chromosome to another and is therefore not completely random.

WHAT DO GENES DO?

The manner in which genes express their phenotypic effects is suggested by the existence of certain rare inherited biochemical human disorders, the so-called "inborn errors of metabolism," a term coined by the English physician Sir Archibald Garrod. Three such abnormalities arise from defective metabolism of the amino acids phenylalanine and tyrosine.

In people afflicted with the rare condition called imbecillitas (oligophrenia) phenylpyruvica, a form of insanity, there is always a small amount of phenylpyruvic acid excreted in the urine; this is not present in the urine of normal people. For this reason the condition is commonly called phenylketonuria or PKU disease. There appears to be impaired formation of the myelin sheaths of nerve fibers in this disease, and some of the damage at least seems to be due to the presence of the phenylpyruvic acid. The cause of this abnormality is either lack of or inactivity of the enzyme phenylalanine hydroxylase, which is present in the tissues of normal people; this enzyme catalyzes the conversion of phenylalanine to tyrosine. In its absence, phenylalanine undergoes transamination and loses its amino agroup to give phenylpyruvic acid. The condition is determined by a rare recessive gene, and only homozygous individuals are affected.

A second inherited disease involving tyrosine metabolism is the condition known as alcaptonuria. The urine of affected people turns black on exposure to air. (Mothers of affected babies are first made aware of the abnormality by the alarming dark staining of diapers.) In 1859 the German biochemist C. Bödeker showed that the condition was due to homogentisic acid (alcapton) excreted in the urine of alcaptonurics; this is oxidized on exposure to air, causing the urine to become black. Alcaptonurics cannot metabolize homogentisic acid, which is an intermediate breakdown product of tyrosine; the deamination of tyrosine is the first step. Normal individuals possess an enzyme (homogentisic acid oxidase) that converts homogentisic acid to fumaric and acetoacetic acids. In alcaptonurics, the enzyme is either missing or inactive. Again, the condition is caused by a rare recessive gene, and only homozygotes are affected.

A third condition also arising from faulty tyrosine metabolism is albinism, in which there is a total lack of melanin pigment in affected individuals. Yet again, this has been traced to absence of or inactivity of a specific enzyme (tyrosinase) that is present in normal people and catalyzes the conversion of tyrosine to dihydroxyphenylalanine ("DOPA"), which is the precursor of melanin. If tyrosine is not converted to DOPA, melanin cannot be produced. Albinism is transmitted by a recessive gene, and only homozygotes are affected.

The reactions and enzymes involved in these three metabolic disorders are shown in Fig. 14-8. These and other inherited human metabolic disorders all point to a relation between genes and enzymes. In normal homozygous individuals or in heterozygotes, the given enzyme is present and active, whereas ccasionally some individuals inherit two mutant alleles, one from each of two

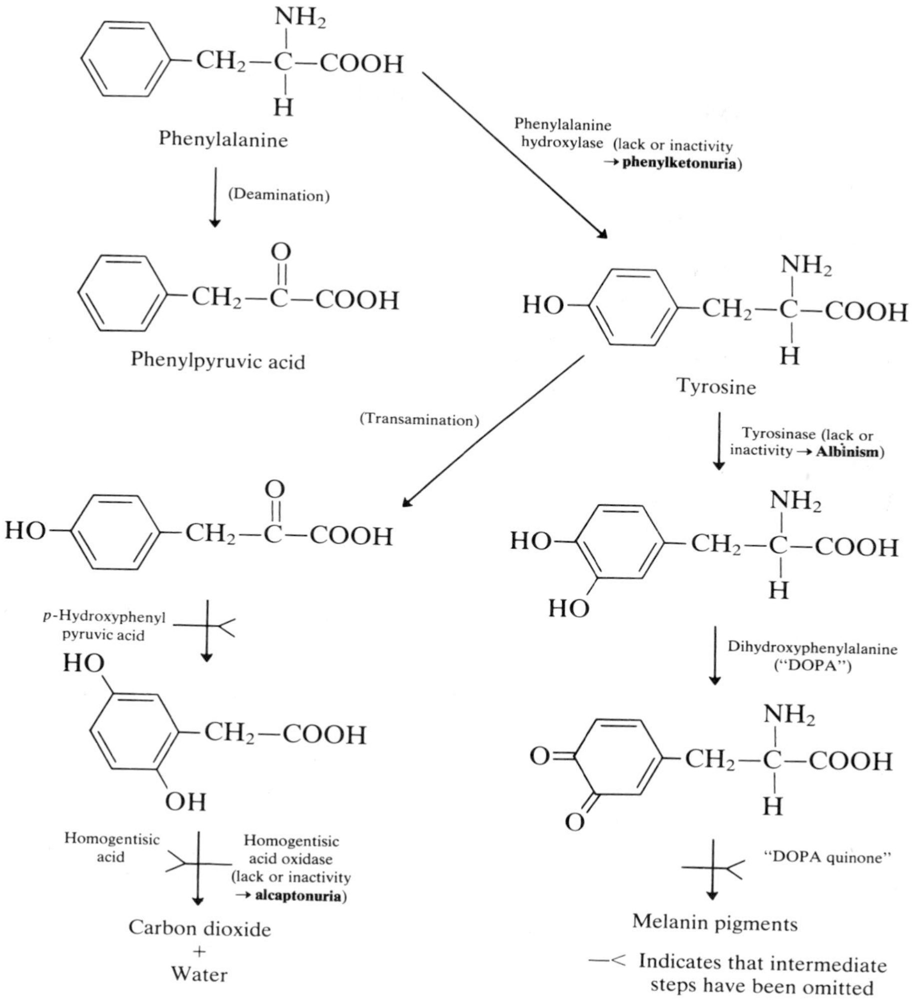

Phenylalanine

Phenylalanine hydroxylase (lack or inactivity → **phenylketonuria**)

(Deamination)

Phenylpyruvic acid

Tyrosine

(Transamination)

Tyrosinase (lack or inactivity → **Albinism**)

p-Hydroxyphenyl pyruvic acid

Dihydroxyphenylalanine ("DOPA")

Homogentisic acid

Homogentisic acid oxidase (lack or inactivity → **alcaptonuria**)

"DOPA quinone"

Carbon dioxide + Water

Melanin pigments

—< Indicates that intermediate steps have been omitted

Figure 14-8 Metabolism of phenylalanine and tyrosine in humans (abbreviated), showing the origin of three inborn errors of metabolism.

heterozygous parents and, as a result, inactive enzyme is produced or possibly there is no enzyme at all.

The precise relationship between genes and enzymes was elucidated by the researches of G. H. Beadle and E. L. Tatum in 1941; they studied various mutations and associated metabolic deficiencies in the common bread mold *Neurospora crassa*. Beadle and Tatum's selection of *Neurospora* as an organism for studying the effects of genes and mutations is another example of wise selection of material for studying a biological problem. Mendel's experiments were performed in pea plants, which are diploid organisms; this is a complicating factor, because the effects of recessive genes were masked by their dominant alleles in heterozygotes. Also, he had to wait a whole year each time to observe the results of crossing two strains (maybe only a monk would have such patience!) Among other advantages, *Neurospora*, by contrast, is a haploid organism, so that the effects of recessive mutant genes are immediately expressed in progeny and its life cycle lasts only 10 days between successive sexual spore generations.

GENES AND ENZYMES IN *NEUROSPORA*

The "wild-type" strain of *Neurospora* grows well on a "minimal medium" containing only sugar, salts, and biotin (a vitamin), and it is able to synthesize all of the amino acids required for its own protein biosynthesis. One of the amino acids, arginine, is synthesized in a biosynthetic pathway involving intermediate compounds, and each step is catalyzed by a specific enzyme. The sequence is as follows.

$$\text{Ornithine precursors} \xrightarrow{E_1} \text{Ornithine} \xrightarrow{E_2} \text{Citrulline} \xrightarrow{E_3} \text{Arginine}$$
$$(E_1, E_2, E_3 \text{ are enzymes})$$

By exposing the wild-type *Neurospora* to ultraviolet light or ionizing radiation, Beadle and Tatum succeeded in producing and isolating several stable mutant forms of *Neurospora*, none of which grew on the minimal medium unless arginine or one of its precursors were added. Hence, they were called arginine-requiring mutants. One type of mutant, for example, which may be designated M_3 for purposes of discussion, would grow only if supplied with ready-made arginine. Another mutant (M_2) could grow if either citrulline or arginine were singly added, but would not grow if ornithine was added, or its precursors. A third type (M_1) would grow if ornithine or citrulline or arginine were added singly, but not if ornithine precursors were supplied. Since the sequential order of the intermediates in arginine biosynthesis in *Neurospora* were known, it was possible to pinpoint the step at which biosynthesis was "blocked" in each of the above mutants. Because the biosynthetic steps are each catalyzed by a specific enzyme (E_1, E_2, E_3), the conclusion was that the metabolic blocks were due to absence of a particular enzyme or at least to production of a defective (inactive) enzyme. These results are summarized in Table 14-1.

That these three mutations are nonallelic was shown by taking advantage of the fact that in one part of the *Neurospora* life cycle, a stage called the heterokaryon occurs. This is produced when cells of two different strains of the same mating type fuse and produce a mixture of two genetically different types

TABLE 14-1 **Growth of Arginine-Requiring Neurospora Mutants on Minimal Medium Supplemented with Arginine and Intermediates in Arginine Biosynthesis**

		Grows on:			
		Minimal Medium Supplemented with			Inactive
Arginine-Requiring Mutant	**Minimal Medium**	**Ornithine**	**Citrulline**	**Arginine**	**or Absent Enzyme**
M_1	No	Yes	Yes	Yes	E_1
M_2	No	No	Yes	Yes	E_2
M_3	No	No	No	Yes	E_3

of nucleus in the same cytoplasm. It was found that a heterokaryon formed from two different types of the above mutants (e.g., M_1 and M_2) would grow on minimal medium, whereas there was no growth of a heterokaryon formed from mutants of the same type. The explanation of this is that in the heterokaryon formed from M_1 and M_2, the M_1 nuclei are producing enzymes E_3 and E_2 and the M_2 nuclei are producing enzymes E_1 and E_3. Hence, all three of the enzymes necessary for arginine biosynthesis are present, so that this heterokaryon can therefore grow on the minimal medium. The two different types of mutant are therefore said to complement each other, and this phenomenon of complementation, which is an important tool in genetic analysis, enables one to conclude that the two mutants involve two different genes. The heterokaryons produced from two similar mutants are therefore noncomplementary, and there is no growth on minimal medium.

Each of the above *Neurospora* mutations is therefore characterized by a loss of or inactivity of a specific enzyme that catalyzes one of the steps in the biosynthesis of arginine. This strongly suggests that a primary manifestation of gene function is the production of a specific enzyme. Because of this, Beadle and Tatum's results are often summed up in the phrase "One gene, one enzyme." The pigments of flower petals, for example, are produced in biosynthetic reactions catalyzed by specific enzymes. A mutated pigment gene may produce inactive enzyme or no enzyme at all, and so no pigment will be produced. In a heterozygote, the normal allele will still produce active enzyme and hence pigment, so that the normal allele will behave as "dominant" to the mutant form, which is "recessive." The same applies to many other phenotypic characters that are determined ultimately by enzyme-catalyzed biochemical reactions.

The Beadle and Tatum concept has since been expanded in the light of further discoveries that show that nonenzymic proteins are also determined by genes.

The hemoglobin polypeptides of a human are an example and are as much phenotypic characters as flower color, although they are not outwardly obvious and require special tests to reveal their presence.

The modern view is that a single gene determines the structure of a single specific polypeptide chain, although even now it is also conceded that not all genes specify the structure of a polypeptide. In fact, it is rather difficult to define precisely what is meant by a "gene."

WHAT IS A "GENE?"

From the foregoing one may picture the gene as the smallest unit of genetic function, a discrete particle that determines phenotypic characteristics, usually but not necessarily always through the activity of a specific polypeptide, the structure of which is determined by the gene. It may also be viewed as the smallest conceivable segment of a chromosome that can undergo mutation or as the shortest chromosomal segment between two crossover points. All three views apparently are in agreement with the idea that the gene is an indivisible ultimate genetic particle, a view that is similar to the older concept of the atom as the ultimate particle of matter. However, one must remember that the concept of the gene as the smallest chromosome segment capable of exhibiting, for example, crossing over, depends entirely on and is limited by the ability of classical genetic tests, involving the counting of sufficient numbers of offspring and enumerating phenotypes to detect rare recombinational events. In other words, when we define the gene as the unit of recombination, we are limiting the concept to whether or not recombination *between two genes* has ever been detected. The gene may possibly possess ultrastructure and consist of smaller subunits, so that crossing-over phenomena could conceivably occur between adjacent chromatids *within* the limits of the length occupied by a given gene on a chromosome. However, such processes may be so rare that they would not be detected by classical methods of analysis unless almost unmanageably large numbers of organisms were used in the genetic experiments.

Again, classical methods of genetic analysis would appear to be insufficiently sensitive to detect whether the whole gene or only a part of it changes in the process of mutation. If the gene possesses a complex ultrastructure, different parts may mutate, giving rise to morphologically different mutations but similar phenotypic effects, so that the mutations would be indistinguishable by ordinary methods.

The application of classical genetic techniques to the study of the genetics of certain viruses has led to the development of methods of genetic analysis with greatly increased "resolving power" because of the ease of handling very large numbers of rapidly multiplying microorganisms and of detection of very rare recombinational events. Such studies have established that—like the atom—the gene possesses a complex "fine structure." In particular it has been demonstrated that the genetic unit of function (i.e., the entity determining the structure of a functional polypeptide) is not the same as the units of recombination and mutation. The functional "genes" occupy considerably greater chromosomal lengths than the units of recombination and mutation and contain many structural subunits, all of which can exhibit mutation and recombination.

GENE FINE STRUCTURE

As previously shown, the positions and sequences of genes on individual chromosomes may be "mapped" because of the phenomenon of chromosomal crossing over and gene recombination; genes exhibit frequent or rare recombination according to whether they are far apart or close together on the chromosome.

In genetic experiments with *Drosophila* as the test organism, a few thousand individuals at most are employed in an experiment, and every one has to be laboriously examined for phenotypic characteristics. Each generation takes about 3 weeks. On the other hand, a microorganism such as a bacterial virus has a generation time of a matter of minutes, and the handling of literally billions of individuals is quite easy.

Techniques have been developed in which examination of the results of millions of crossings between microorganisms and isolation of organisms of special interest from a population of many millions is fairly simple, so that rare recombinational events can be easily detected. The greatly increased "resolving power" of such methods has made it possible to "dissect" genes and to "map" them in detail.

Two microorganisms that were often used in these researches were the colon bacillus, *Escherichia coli*, and a group of viruses that infect *E. coli*, the so-called T-even bacteriophages or "phages" for short. In the phages known as T_4 and T_2 several genes have been mapped. They all seem to belong to a single linkage group, and the phage behaves as a haploid organism. Of special interest has been the T_4 phage; its structure is shown in Fig. 14-9. When the tip of the tail of the T_4 phage makes contact with the surface of the *E. coli* cell, it sticks, the tail contracts, and the phage chromosome enters the bacterial cell in a process like hypodermic injection and rapidly replicates. As a result, the biosynthetic "machinery" of the *E. coli* cell is caused to synthesize phage protein components, which with the new phage DNA assemble to form progeny phage particles. Finally, about 20 minutes after infection, the bacterial cell disintegrates, a process called lysis, and the new phage particles are liberated to infect other *E. coli* cells.

When a suspension of T_4 phage particles is added to an even growth or "lawn" of either of two different strains of *E. coli* known as strain B and strain K on the

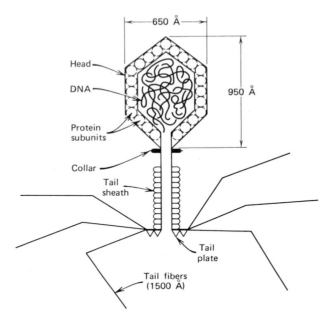

650 Å

Head

DNA

950 Å

Protein
subunits

Collar

Tail
sheath

Tail
plate

Tail fibers
(1500 Å)

Figure 14-9   Structure of T₄ bacteriophage. The bacteriophage attaches itself to an *E. coli* cell by its tail plate and fibers. The tail sheath contracts, its hollow core penetrates the bacterial cell wall, and the phage DNA is injected into the *E. coli* cell.

surface of an agar culture, "plaques" appear after a while; these are cirular "holes" or zones with sharp edges where *E. coli* cells have been attacked and lysed by phages. Plaques are formed within a few hours of infection by the phage. Each plaque may contain 10 million phage particles, and these are the progeny generated by a single particle. The size and appearance of plaques is a phenotypic characteristic of the phage.

Several mutant forms of the T₄ phage are known in which the life cycle appears to have been accelerated. These mutations have been shown to involve any one of at least three discrete regions on the phage chromosome and, because the life cycle is more *rapid* in these mutants, these chromosome segments have been assigned the symbols rI, rII, and rIII.

The accelerated life cycle is revealed by the formation of larger plaques with fuzzy edges in a B-strain culture of *E. coli* (r-type plaques) but, in K-strain cultures, normal, r-type plaques may be produced or no plaques at all (Table 14-2). When a liquid culture of B-strain *E. coli* is exposed to a mixed infection of rI and rII phage mutants, practically every bacterial cell becomes infected with both mutant-type phages. After the phage replication and bacterial lysis phase,

TABLE 14-2   **Four Genetic Types of Bacteriophage as Revealed by Plaque Characteristics in Strain B and Strain K *E. coli* Cultures**

	Plaque Type	
Phage type	**B strain**	**K strain**
Normal (wild)	Normal	Normal
rI	r-type	r-type
rII	r-type	None
rIII	r-type	Normal

the new progeny virus can be harvested and used to infect a "lawn" of strain B *E. coli* in agar culture. In addition to the expected r-type plaques, some normal ones are also formed. This result has been explained as arising from a process occurring during phage replication that is analogous to crossing over in the cells of higher organisms (Fig. 14-10). Exchange of segments between chromosomes derived from two different mutant phages gives rise to recombinant phages possessing either both mutations or an unmutated wild-type chromosome, the latter giving rise to the normal plaques. By similar mixed infection experiments,

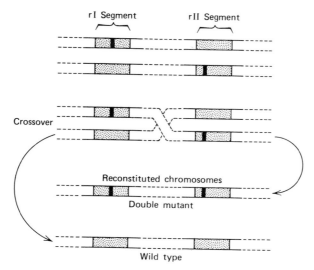

Figure 14-10   Recombination between chromosomes of two different T_4 phage mutants.

it has been possible to map the position of the three r regions in the phage chromosome; they have been found to lie in the order rII-rI-rIII.

From the point of view of genetic analysis an important property of the rII mutants is that they cannot grow on strain K *E coli* and hence can be easily recognized by this unambiguous phenotypic test. S. Benzer isolated several independently arising and stable rII mutants from mutant-type plaques on a culture of strain B *E. coli.* He inoculated a mixture of two of these rII mutants into liquid *E. coli* culture. When the culture was subsequently plated on agar cultures of strain K *E. coli* (on which neither rII mutant could grow), a few standard-type plaques appeared, thus showing that some standard-type phages were present. The explanation is that exchange of chromosomal segments had therefore occurred, this time *within* the rII region. In a few cases the result is either a chromosome carrying an rII region with both mutations or a standard type recombinant containing a normal rII region (Fig. 14-11). (That the standard type had appeared because of back mutation of either of the rII mutants to the wild type seems highly improbable.) This technique makes possible the detection of one recombinant even among many million progeny.

Other studies indicated that the rII region actually consists of two distinct functional segments, which Benzer called "cistrons," and he designated them rIIA and rIIB. Benzer coined the term cistron from the so-called *cis-trans* test, which

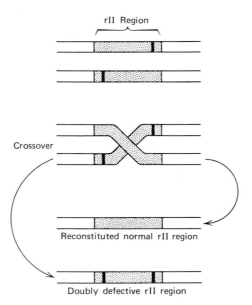

Figure 14-11 Recombination within the rII regions of T$_4$ bacteriophage rII mutants.

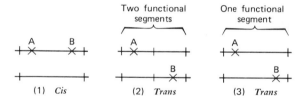

Figure 14-12   The *cis-trans* test. A and B are two mutational sites, either of which prevent phages from multiplying. If both are in the *cis* (Lat. = "on this side of") position, as in (1), the other chromosome is normal and supplies the function missing from the defective partner.

If in the *trans* (Lat. = "across") position, as in (2) and (3), the phage will grow and multiply only if the mutant sites are situated in different functional segments [as in (2)], because the intact segment of one will "complement" the defective segment of the other. If in the same functional segment [as in (3)], complementation is not possible.

is used to establish the presence of two functional segments; this is explained in Fig. 14-12. It is used to determine whether or not two mutants that produce a similar abnormal phenotype involve a single hereditary unit of function or whether they involve two interacting units, the products of which combine to produce the phenotypic change. The experiments that led Benzer to the conclusion that there were two such segments were as follows. Whatever position a mutation in the rII region occupies, the phenotypes will be similar (i.e., larger than normal plaques in B-strain *E. coli* are produced and the phage is unable to grow in K-strain *E. coli*). Imagine that there are three different rII mutants (M_1, M_2, M_3), meaning that in each, the mutated sites are in different parts of the rII region (Fig. 14-13). (The positions of the mutant sites have previously been determined.) If strain K *E. coli* is infected simultaneously with wild-type T_4 phage plus any *one* of the above mutants, it is found that both the wild-type and mutant phages flourish. Benzer's explanation of this extraordinary result was that the rII region of the wild-type phage controls production of a substance—possibly a polypeptide—that the mutant cannot synthesize and therefore makes enough of this product both for its own needs and to make good the deficiency in the mutant, which can grow in the K-strain *E. coli*, whereas it could not on its own. It is important to realize that whether or not the mutant grows is determined soon after infection and *before* the exchange of chromosomal segments (crossing over) or recombination have had a chance to occur. The phenomenon of complementation, which we encountered in Beadle and Tatum's work on *Neurospora* mutants, is again responsible and is quite different from recombination; two mutants carrying different mutations in the same cistron can recombine to give normal viable progeny. However, the two processes can be

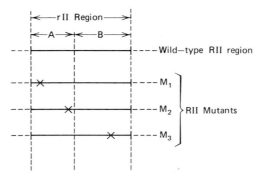

Results of mixed infections of strain K
E. coli cultures with wild-type and rII
mutant T_4 phages

Infection	Phage Growth			
	Wild Type	M_1	M_2	M_3
Wild type + M_1	Yes	Yes		
Wild type + M_2	Yes		Yes	
Wild type + M_3	Yes			Yes
M_1 + M_2		No	No	
M_1 + M_3		Yes		Yes
M_2 + M_3			Yes	Yes

Figure 14-13 Detection of two cistrons in the rII region of T_4 bacteriophage by the complementation test.

differentiated; *recombination* occurs only in very few cells of a population, but *complementation* affects all the cells.

If strain K *E. coli* is now infected with M_1 and M_2, no growth will be observed, but infection with either M_1 and M_3 or M_2 and M_3 results in the growth of both mutants in each case. Benzer explained this by suggesting that the rII region consists of two functional segments, each of which produces a substance, possibly a polypeptide, necessary for phage replication in strain K *E. coli* and that these, perhaps, combine to form an enzyme. M_1 and M_2 both contain defective rIIA segments and normal rIIB segments; therefore complementation cannot occur and the complete rII product cannot be produced. In M_3, the reverse is true; the rIIA and rIIB segments are normal and defective, respectively. Therefore mixed infection with either M_1 and M_3 or M_2 and M_3 results in synthesis of both A and B polypetides, because the normal segment of one complements the corresponding defective segment in the other mutant and hence plaques are formed. Since M_1 and M_2 cannot complement each other, no plaques are formed, and this indicates that the two mutant sites are in the same functional segment of both mutant phages. Detection of the smallest segment of a chromosome that controls the biosynthesis of a functional product was thus made possible by this complementation technique. Benzer designated the different complementation possibilities shown in Fig. 14-12 by the words *cis* and *trans*, which he borrowed from the terminology of organic chemistry. The technique of determining the size of functional segments is therefore also known as the *cis-trans* test and from these, as mentioned earlier, Benzer coined the term cistron, the name given to the shortest chromosomal segment that determines the structure and synthesis of a functional product. A mutation occurring within a cistron may have three possible effects.

1. Alteration of the polypeptide structure without loss of, or interference with, function.
2. Alteration of polypeptide structure with consequent loss of activity.
3. Inability of the cistron to bring about synthesis of the polypeptide.

"RUNNING THE MAP INTO THE GROUND"

Because genetic recombination occurs within the cistron, the methods used to prepare genetic maps of *Drosophila* chromosomes may also be used to map the cistron. Thanks to the increased "resolving power" made possible by the ease of handling millions of phage particles and of selective techniques for isolating certain recombinant classes, detection of recombination frequencies of hyndredths of 1% are possible, whereas frequencies of several percent only were possible using *Drosophila* as the test organism. That certain rare recombinants are not new back-mutations to the normal must always be checked, however.

Bateriophage has thus been a highly useful tool in the hands of geneticists for the mapping of finer and finer detail to the limit. This is called "running the map into the ground," a phrase Benzer attributed to Max Delbrück. Of all the important results that emerge from these refined analyses, the most significant is the correlation of the physical size of the ultimate units of recombination as determined in the genetic experiments, to the known dimensions of the DNA molecule (see Chapter Fifteen). Calculation has shown that the size of the smallest unit of recombination is about one to three times that of a DNA nucleotide pair. Considering the errors inherent in the assumptions and measurements involved in these calculations, it is remarkable that the estimate of the size of the smallest recombinational unit is so close to that which is theoretically the smallest conceivable. The smallest unit of mutation is also considered to be of the same magnitude, and there is evidence in support of this. What, then, is a "gene?"

T$_4$ phage genetics reveals that there is an rII *region* and that this consists of two *cistrons*, each of which contains very many mutable and recombinable *sites*. Which of these are we to call a "gene?" The fact is that the gene concept depends partly on the "resolving power" of the particular methods of genetic analysis employed and on the particular aspect of the "gene" that we are studying. Benzer introduced the terms *cistron, recon,* and *muton* to designate, respectively, the ultimate units of function, recombination, and mutation. The cistron would therefore correspond roughly with the Beadle and Tatum concept of the gene, but this is a much larger entity than the units of recombination and mutation. We should not be too eager to regard the term "gene" as obsolete merely because of its complexity or because of improved understanding of its organization and function. From the organism's point of view, function is the most important aspect of a gene, and however simple or complicated its structure, it is still useful and legitimate to view it as a genetic particle that functions in the whole organism, in much the same way as atoms, although possessing complex structure, also *function* as units of matter in ordinary chemical reactions.

CHAPTER FIFTEEN
The structure and replication of DNA

POLYNUCLEOTIDES

For a long time it has been known that the nuclei of cells stain intensely with basic dyes such as methylene blue or hematoxylin; for this reason the earlier microscopists called the nuclear material "chromatin" (Gk. *chroma* color). The cell nucleus is now known to consist almost entirely of protein conjugated with nucleic acids (i.e., it is nucleoprotein). Tissues in which the nucleus/cytoplasm ratio is high, such as thymus, pancreas, and sperm, will therefore be a good source of nucleoprotein.

The first extraction and purification of nucleic acid was achieved by F. Miescher in 1847; he obtained it from salmon sperm. In 1899, R. Altmann described the preparation of nucleoprotein from calf thymus by macerating it with large volumes of water and then saturating it with salt. This precipitates the protein component that was removed, and the addition of ethanol to the remaining solution causes precipitation of the nucleic acid in the form of fibers. Altmann also showed that nucleic acid can be extracted from yeast cells and that chemically it was somewhat different from that derived from calf thymus.

Two general types of nucleic acids are now known, the deoxyribonucleic acids (DNA) and the ribonucleic acids (RNA). Nucleic acid from calf thymus ("thymus nucleic acid") was shown to be DNA, and that from yeast ("yeast nucleic acid") was shown to be RNA. At one time, therefore, it was thought that DNA was confined to animal cells and RNA to plant cells; this was actually stated by

331

W. Jones in 1921. However, in 1924, R. Feulgen and H. Rossenbeck, using a specific histochemical method (the Feulgen reaction) demonstrated that "thymus nucleic acid" (DNA) occurs in plant cell nuclei as well as in animal cell nuclei. "Yeast nucleic acid" (RNA) was also found in animal cells at about the same time. Animal and plant cells therefore contain both forms, the DNA mostly confined to the cell nucleus and the RNA occurring most abundantly in the cytoplasm. DNA and RNA are collective terms like the word "protein." Many different specific DNAs are now known and characterized, and several different functional types of RNA have been recognized.

DNA and RNA are complex, linear, polymeric orthophosphoric acids, the "building blocks" of which are called nucleotides, so that structurally, DNA and RNA are polynucleotides. Nucleic acid structure is therefore analogous to protein structure insofar as proteins are also linear polymers of smaller structural units (amino acids).

Ribonucleoside
5'-Monophosphates

Names

Adenosine-5'-phosphoric acid
(adenylic acid)
Guanosine-5'-phosphoric acid
(guanylic acid)
Cytidine-5'-phosphoric acid
(cytidylic acid)
Uridine-5'-phosphoric acid
(uridylic acid)

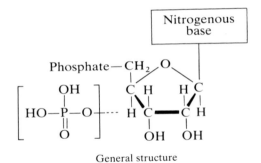

General structure

2'-Deoxyribonucleoside
5'-Monophosphates

Names

Deoxyadenosine-5'-phosphoric acid
(deoxyadenylic acid)
Deoxyguanosine-5'-phosphoric acid
(deoxyguanylic acid)
Deoxycytidine-5'-phosphoric acid
(deoxycytidylic acid)
Deoxythymidine-5'-phosphoric acid
(deoxythymidylic acid)

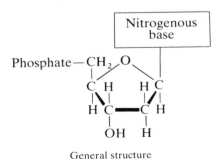

General structure

Figure 15-1   Structures of nucleotides.

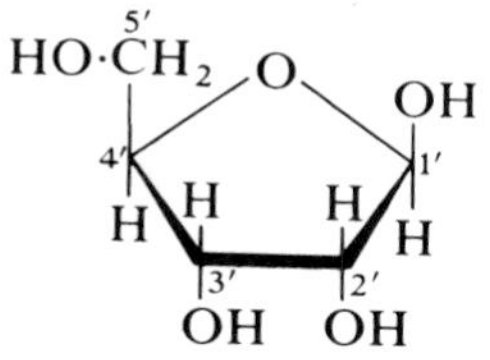

Ribose

(β-D-Ribofuranose)

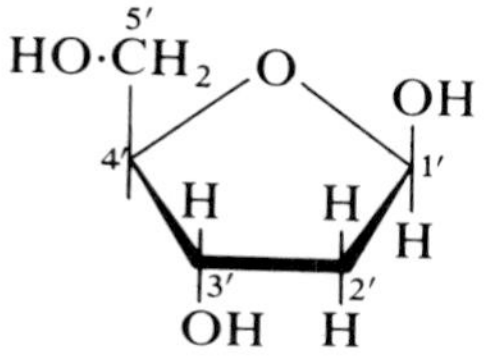

Deoxyribose

(β-D-2-Deoxyribofuranose)

Figure 15-2 Structures of ribose and deoxyribose.

Nucleotides themselves are built from three smaller units—a five-carbon sugar (pentose), a phosphate group, and a nitrogen-containing organic base. Nucleotides are, in fact, the 5′-phosphate esters of N-glycosides of the nitrogenous bases, or nucleosides as they are called. (Fig. 15-1).

Chemically, DNA and RNA differ in two respects. The first difference is in the nature of the pentose sugar component. In DNA this is β-D-2-deoxyribofuranose, and in RNA it is β-D-ribofuranose (Fig. 15-2). The second is in the nature of the nitrogenous base. Five different bases are principally involved in the structure of polynucleotides in general, and these fall into two categories. These are 6-keto and 6-amino purines and 6-keto and 6-amino pyrimidines. The purines are adenine and guanine, and the pyrimidines are thymine, cytosine and uracil (Fig. 15-3). Four of these enter into DNA structure, two purines (adenine, guanine) and two pyrimidines (cytosine, thymine). In RNA, the same two purines are found plus cytosine but, instead of thymine, there is uracil. Uracil is rarely found in DNA, and thymine rarely, if ever, in RNA. Both DNA and RNA contain phosphate. The complete hydrolysis of both types of polynucleotides therefore yields the following substances (in addition to inorganic phosphate):

	(Nucleus) DNA	**(Cytoplasm) RNA**
Pyrimidines	Cytosine, thymine	Cytosine, uracil
Purines	Adenine, guanine	Adenine, guanine
Pentose	D-2-deoxyribose	D-ribose

Purines

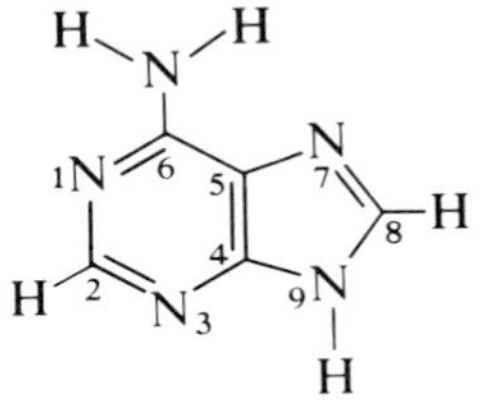

Adenine
(6-aminopurine)

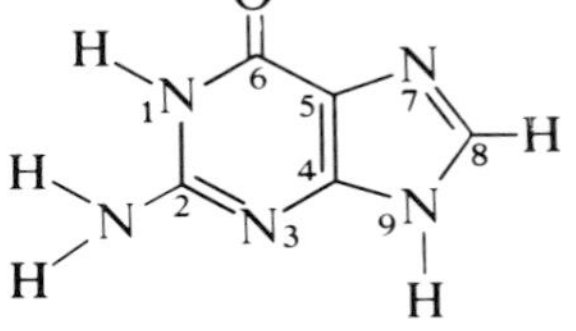

Guanine
(2-amino-6-ketopurine)

Pyrimidines

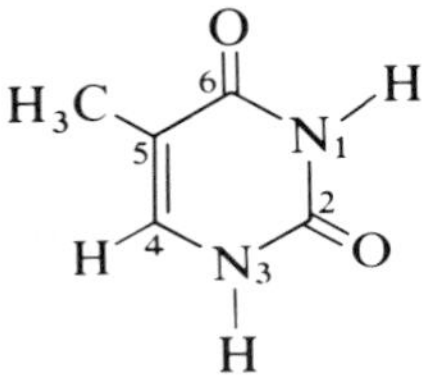

Thymine
(5-methyl-uracil)

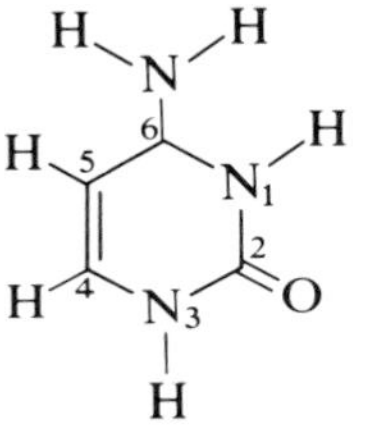

Cytosine
(2-keto-6-aminopyrimidine)

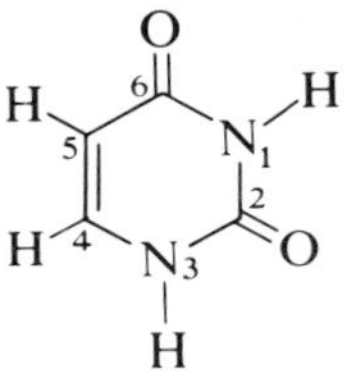

Uracil
(2:6-diketo pyrimidine)

Figure 15-3 Structures of purines and pyrimidines. (Shown here in the *keto* form where applicable.)

Careful studies of the products of the partial nonenzymatic and enzymatic hydrolysis of DNA and RNA indicate the manner in which the individual nucleotides are joined together in the polymeric structure. The existence of 3':5'-phosphodiester linkages between the phosphate and pentose components in DNA and RNA is now well established, and the structure of an RNA polynucleotide is shown in Fig. 15-4. DNA polynucleotides are identical except for the lack of the oxygen atom at the 2'-position in the deoxyribose.

STRUCTURE OF DNA

In 1950, E. Chargaff pointed out the curious fact that in most samples of DNA and in some samples of RNA, the molar ratios adenine:thymine, guanine:cytosine, and total purine:total pyrimidine are all approximately equal to 1. On the other hand, the ratio of adenine plus thymine to guanine plus cytosine is very variable. These empirical observations remained unexplained until the three-dimensional structure of DNA was elucidated.

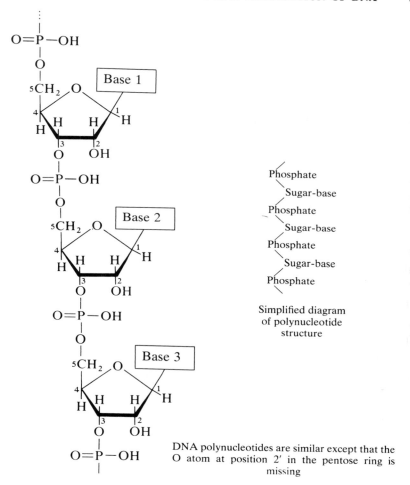

Phosphate
    Sugar-base
Phosphate
    Sugar-base
Phosphate
    Sugar-base
Phosphate

Simplified diagram
of polynucleotide
structure

DNA polynucleotides are similar except that the
O atom at position 2′ in the pentose ring is
missing

Figure 15-4 Structure of an RNA polynucleotide. DNA polynucleotides are similar except that the O atom at position 2′ in the pentose ring is missing.

X-ray diffraction studies of crystalline DNA and the physicochemical properties of DNA solutions indicate that the DNA molecule is a long, thin rod of almost uniform diameter (Approximately 20 Å). DNA rods are not fully flexible, but have a definite three-dimensional structure and, under certain conditions, undergo denaturation somewhat similar to that of proteins.

As far back as 1938, R. Signer et al. had inferred from the optical anisotropy of DNA that the planes of the purine and pyrimidine ring structures were oriented perpendicular to the long axis of the DNA molecule. Several attempts were made by investigators to construct models of the molecule that would be consistent with these features. In 1952, S. Furberg speculated that DNA consisted of a single polynucleotide chain in which the stacked based formed a column with the phosphate-sugar backbone, forming a spiral on the outside. In 1953, L. Pauling and R. B. Corey, also speculating, suggested that three twisted chains formed an elongated helix and that the bases were disposed on the outside of a central core of "backbone" phosphate groups.

At about the same time, more detailed X-ray diffraction pictures were secured by M. H. F. Wilkins, R. Franklin, and others working at King's College, London. They used concentrated DNA solutions from which threads were prepared containing DNA molecules in an almost crystalline orientation. Their results began to make clear that DNA, in fact, consisted of two intertwined helical molecules with the phosphate groups on the outside and with bases directed inward. This led J. D. Watson and F. H. C. Crick to experiment with scale molecular models, and they finally put forward their well-known double helical model of the DNA molecule in 1953; this model is universally accepted as being essentially correct. For this achievement, Watson, Crick, and Wilkins were jointly awarded the Nobel Prize in 1962. In this model, the two intertwined

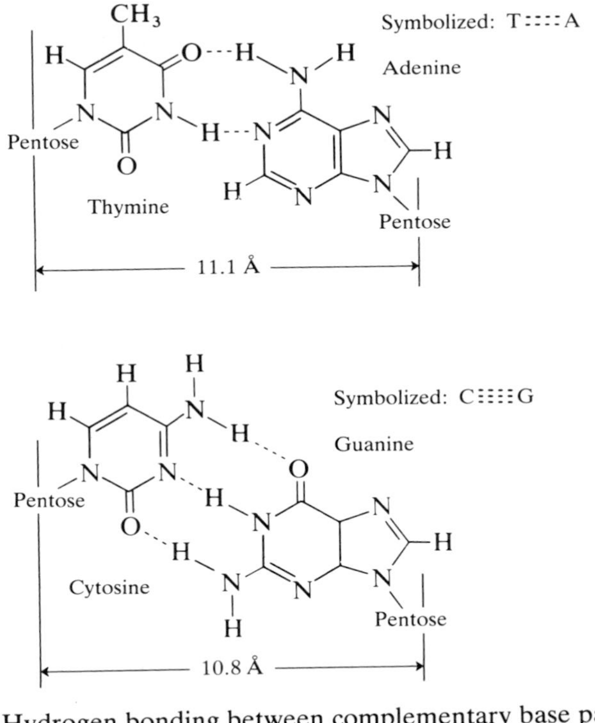

Figure 15-5 Hydrogen bonding between complementary base pairs in DNA.

helical polynucleotide chains have opposite polarity (i.e., they are antiparallel and "point" in opposite directions). They are held together by hydrogen bonds between pairs of bases; each base extends inward from the polynucleotide chain and perpendicular to it. The constant diameter of the double helix places restrictions on the types of possible base pairs, because the dimensions of the bases vary. Stereochemical constraints are such that a purine must face a pyrimidine; adenine always pairs with thymine, and guanine always pairs with cytosine (Fig. 15-5). This immediately explains Chargaff's empirical observation of the 1:1 molar ratios of adenine:thymine and guanine:cytosine. The two right-handed helical chains complete one turn every 34 Å, and there are about 10 base pairs per turn. The base pairs can be arranged in any sequence but the presence of, say, adenine on one DNA strand automatically determines that thymine must be present in the same position on the other strand; similarly, guanine on one strand determines cytosine on the other. Therefore, the two strands are complementary, each one being a template of the other. The sequence of the bases in the polynucleotide strands and the adenine:thymine and guanine:cytosine base pairing gives DNA its structural specificity. The double-stranded structure of DNA is shown in Fig. 15-6. If both the top and the

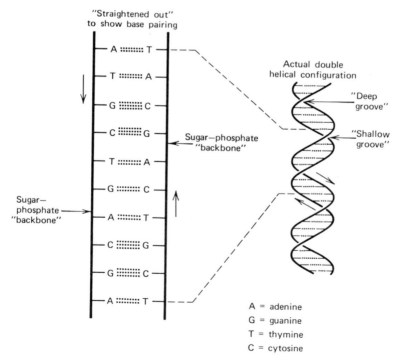

Figure 15-6   The double-stranded helical structure of the DNA molecule.

bottom of the ladderlike structure are twisted in an anticlockwise direction, a double helical formation will result, with the hydrogen-bonded base pairs forming the rungs of the spiral staircaselike structure. The molecular model shown in Fig. 15-7 conveys a more realistic picture of DNA.

The biological significance of the two-stranded model was not lost on Watson and Crick who, in their 1953 paper in *Nature*, modestly remark at the end: "It has not escaped our notice that the specific pairing (of the two DNA strands) we

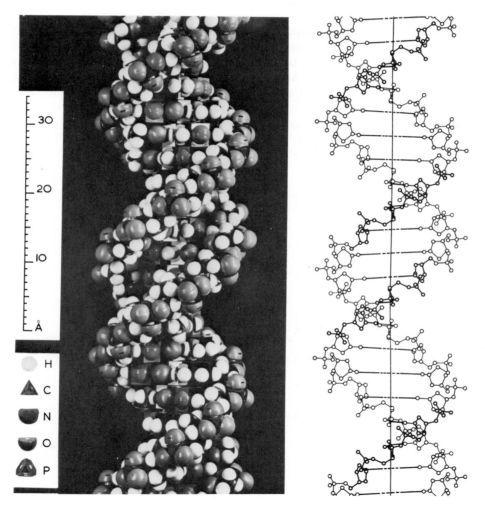

Figure 15-7   Space-filling molecular model of DNA (Courtesy of Professor M. H. F. Wilkins, The Biophysics Department, King's College, London, W.C.2).

have postulated immediately suggests a possible copying mechanism for the genetic material."

SYNTHESIS AND REPLICATION OF DNA

The fact that had not escaped Watson and Crick's notice was, of course, that either one of the two complementary strands of the DNA molecule could act as a template for the biosynthesis of the other. DNA thus has the potential to replicate itself, a property possessed by no other biological macromolecule. In its essentials, the replication mechanism of DNA is as follows. The first step is breaking of the hydrogen bonds between base pairs and separation of the two strands. Assuming that there is a supply of DNA nucleotides available, these precursors attach themselves to the single strands by hydrogen bonding with their complementary bases. The individual nucleotides are now joined into a polynucleotide chain under the influence of a specific enzyme that catalyzes formation of the phosphodiester bonds, and the result is two DNA molecules, both identical with the "parent."

That DNA can, in fact, be synthesized from a mixture of the component nucleotides (nucleoside monophosphates, see earlier) by an enzymatic mechanism was demonstrated in *in vitro* experiments by A. Kornberg, who was awarded the Nobel Prize for this feat in 1959. Kornberg succeeded in extracting an enzyme (DNA polymerase) from the bacterium *Escherichia coli* that was capable of catalyzing the *in vitro* synthesis of DNA. As DNA precursors, Kornberg employed deoxyribonucleoside triphosphates of adenine, thymine, guanine, and cytosine (i.e., nucleotides with two extra phosphate groups attached) instead of the monophosphates. This was because he anticipated that the nucleoside triphosphates would be more reactive by comparison with other biosynthetic reactions in cells, and hence would give a higher yield of DNA, with elimination of the extra phosphate groups as inorganic pyrophosphate.

Kornberg showed that DNA was synthesized from a mixture of nucleotide diphosphate precursors under the influence of DNA polymerase, provided that a little ready-made DNA was first added to act as a "template" or "primer." The primer did not necessarily have to be *E. coli* DNA. Subsequent analysis revealed that the new synthetic DNA formed in these experiments was identical in every way with the primer DNA. Both the base pairing and the phosphodiester bond formation occur under the influence of the DNA polymerase. There is more than one conceivable mechanism by which DNA replication might come about.

1. Watson and Crick's suggested mechanism in which the two separated halves act as templates for two more DNA molecules, each of these containing one "old" strand from the original molecule and one "new" strand. Since one half of the old DNA is conserved in the replicated DNA, they called this semiconservative replication (Fig. 15-8).

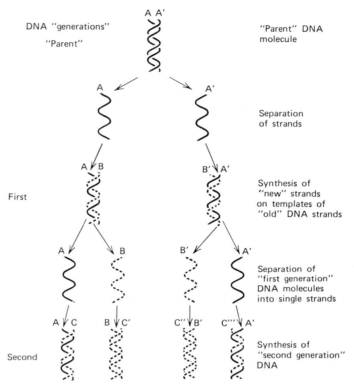

Figure 15-8 Semiconservative template mechanism of DNA replication as postulated by Watson and Crick.

2. Each double-stranded DNA molecule might direct the synthesis of a copy of itself that would then separate from it. In this case, the copy would contain two new strands so that new and old strands would not be found associated with each other. This is a conservative mechanism, because the old DNA is completely conserved.

3. The double-stranded molecule might fragment, the fragmented DNA acting as a template so that the new DNA molecules would contain portions of the old DNA and portions of newly synthesized DNA. This would be neither fully conservative nor fully semiconservative and can be called dispersive.

The template mechanism is strongly suggested by the complementary nature of the two polynucleotide strands of the double helix. As noted, one half of the DNA should be conserved after a replication cycle, and this could be verified if it were possible to follow the individual strands during replication. This can, in fact, be done, because individual DNA strands can be "labeled" with the heavy nitrogen isotope N^{15}. N^{15}-labeled DNA is denser than ordinary N^{14} DNA and can be separated from it by density gradient centrifugation. Watson and Crick's template mechanism was tested in the now well-known and classic experiment performed by M. Meselson and F. W. Stahl in 1957 at the California Institute of Technology. Meselson and Stahl grew *E. coli* in a medium containing the heavy isotope of nitrogen (N^{15}) as the only source of nitrogen. They allowed the bacteria to divide several times, so that the DNA of practically all of the bacteria in the culture contained N^{15}. A sample of these bacteria was then transferred to a medium containing the ordinary isotope of nitrogen (N^{14}). The cells were allowed to divide several times and, after each doubling of the cell population, the DNA was extracted from a sample of the resulting generation and the density determined in the ultracentrifuge. The DNA of the original N^{15}-containing bacteria gave a single band of density corresponding to pure N^{15} DNA. Cells from the first division after transferral to the N^{14} medium yielded DNA of density intermediate between that of pure N^{15} DNA and N^{14} DNA. Cells from the generation after this gave zones corresponding to intermediate and pure N^{14} DNA densities. Later bacterial generations yielded DNA in which the N^{14} zones became more pronounced, while the intermediate-density DNA zones became weaker (Fig. 15-9). The results of this experiment were exactly as would be predicted from the Watson–Crick semiconservative template mechanism, but they do not support conservative or dispersive mechanisms. If the conservative mechanism was operative, then the cells from the first division after transferral

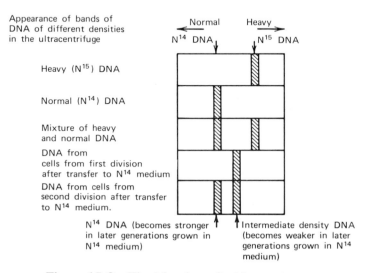

Figure 15-9   The Meselson–Stahl experiment.

to the N^{14} medium would have given two sharp bands of equal density corresponding to the N^{14} and N^{15} DNA, but no band corresponding to the N^{14}–N^{15} hybrid DNA. Subsequent generations would have yielded DNA in which the N^{15} band became progressively weaker and the N^{14} band progressively stronger.

The dispersive mechanism would have yielded a diffuse band intermediate in density between the N^{14} and N^{15} DNA bands which, over a series of generations, would become more disperse.

Of great interest in DNA replication is the stability of the original parent strands, as shown by their persisting intact through several cell divisions.

The mechanism of DNA replication in higher organisms was investigated by Taylor, who studied the process in chromosomes of the dividing cells of plant root tips. The methods he used are described in the later section regarding chromosomal organization. Both Taylor's results and similar experiments carried out on chromosomes of human cells in culture are again in excellent agreement with the results as predicted from the semiconservative replication mechanism.

BASIC PROTEINS OF THE CELL NUCLEUS

DNA does not exist as such in the cell, since it is a polyphosphoric acid and the cell could not tolerate such a strongly acid material. DNA is always found in association with basic proteins, and these appear to be mainly responsible for neutralizing the negatively charged phosphoric acid groups of DNA. There is a widespread belief that bacterial DNA is not complexed with basic proteins, but this idea may have to be changed when new data become available.

The basic proteins fall into two main groups, the protamines and the histones. Both are characterized by possession of a high proportion of arginine and lysine residues, and much of their aspartic and glutamic acid side chains are in the form of the corresponding amide derivatives. Protamine extracted from salmon sperm has a restricted amino acid sequence and contains only arginine, leucine, valine, alanine, glycine, proline, and serine, of which arginine comprises 68% as was found by M. J. Callanan et al. in 1957. The major nonpolar component is the "helix breaker" proline, which inhibits α-helix formation.

At physiological pH, the basic proteins therefore carry a large net positive charge, and the protein and nucleic acids in nucleoprotamines and nucleohistones are bound together mostly by electrostatic attraction. (Oddly, the only nucleoprotein in which the spatial organization of the nucleic acid and protein components are precisely known is the TMV, described in Chapter Seventeen.)

(i) Protamines

Protamines differ from histones in having a more restricted amino acid makeup. In 1874, F. Miescher discovered in salmon sperm what is now the best understood protamine-DNA complex. DNA-protamine (nucleoprotamine) is found only in spermatozoa, but not all sperm nuclei contain nucleoprotamine, as is commonly thought. X-ray diffraction shows that in the intact spermatozoon and when extracted therefrom, the DNA is present as a double helical molecule. In 1962, E. M. Bradbury et al. showed that hydrogen atoms of the protamine-DNA exchange rapidly with deuterium, which confirms that the protamine is, in fact, nonalpha-helical and is in the fully extended form. M. Feughelman et al. put forward a structure for the nucleoprotamine complex in which they showed that a fully extended polyarginine chain could be helically wound in the "shallow" groove of the DNA double helix in a manner that brings the positively charged basic side chains in close proximity to the negatively charged phosphate groups of the DNA. However, this makes it difficult to see how the nonbasic amino acids in an actual protamine can be accommodated; their presence would seemingly prevent such a regular 1:1 correspondence between the basic amino acid side chains and DNA phosphate groups. Those parts of the polypeptide chain containing nonbasic groups were therefore considered by Feughelman et al. to occur in outward-projecting "loops" of the backbone, provided that the neutral residues occur in pairs. The precise association of basic side chains and DNA-phosphate groups would therefore be maintained. That there are such "loops" in the protamine molecule was supported by the findings of K. Felix et al. in 1956

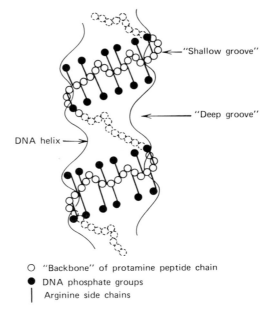

Figure 15-10 Structure of nucleoprotamines. (Redrawn and modified from Wilkins, 1956.)

that a frequently occurring sequence in herring sperm protamines is four consecutive arginines and that neutral residues tend to occur in pairs. These data would fit in with a protamine amino acid sequence in which four arginine residues and two neutral residues alternate in a repetitive and regular manner. Not all of the DNA phosphates are neutralized by basic amino acid side chains, however; about 5 to 10% of them are neutralized by sodium or potassium ions. A model of nucleoprotamine structure according to Wilkins is shown in Fig. 15-10.

(ii) Histones

Histones from the nucleus of a given species are more heterogeneous than protamines. Like protamines, histones are rich in arginine, but this amino acid is never present in amounts greater than 13% of the total, in contrast to protamines where the proportion of arginine is 68%. They also contain differing proportions of other ionized side chains, acidic as well as basic.

Using the method of hydrogen-deuterium exchange, Bradbury and other investigators in the early 1960s showed that, also in contrast to protamines, histones contain about 58% of the polypeptide chain in the α-helical form.

In the early 1940s A. E. Mirsky and others formulated criteria for defining and recognizing histones that were based on extraction conditions, solubility properties, and conditions favoring precipitation, among others. Around 1960, three major types of histones were recognized by E. W. Johns and others; they were characterized by being rich in lysine, less rich in lysine, or rich in arginine. The lysine-rich histones are designated H1. The others are further subdivided into two less lysine-rich types (H2a, H2b) and two arginine-rich types (H3, H4).

DNA in association with histones appears to be in the double helical form, and the histones are bound to the DNA principally by electrostatic linkages. As in the neucloprotamines, the ratio of DNA phosphate groups to basic protein groups is close to unity and, again, it seems that some of the DNA phosphates are neutralized by sodium ions. However, the fairly high proportion of α-helical secondary structure and the large proportion of nonbasic amino acids in histones makes it difficult to visualize how they are complexed to the DNA. The lysine-rich type has a high proline content, and so the polypeptide chain could exist in extended form, as in the protamines. X-ray diffraction and other studies of artificial DNA-lysine-rich histone complexes led Wilkins to believe that lysine-rich histones exist primarily in the fully extended form, whereas the nonlysine rich types possess α-helical secondary structure to a significant extent.

As noted in an earlier chapter, the alpha-helix is somewhat rigid and so α-helical sections of histones could not be considered to "wrap around" the DNA helix. However, according to G. Zubay and others, segments of α-helix could be accommodated in the "deep groove" of the DNA double helix, while the nonhelical segments would be able to bend, thus permitting the histone molecule, as a whole, to wind around the DNA molecule. However, H. Busch and also D. M. P. Phillips found that there is no regular arrangement or sequence of basic amino acid residues in the histones that they studied from calf thymus nuclei, so that there is difficulty in visualizing how the basic residues can be spatially oriented to neutralize the DNA phosphate groups. Any viable model of nucleohistone must be able to explain this and certain other general properties of histones.

Various investigators have concluded that apparently nucleoprotein has a regularly repeating structure, because its X-ray diffraction pattern consists of a series of regularly spaced diffraction rings. This regularity disappears if histones are removed. In 1970, J. Pardon, B. Richards, and M. H. F. Wilkins, working in England, suggested that if the DNA double helix was coated with histones and then supercoiled, the regularly spaced turns of the supercoil could be responsible for the X-ray diffraction pattern. However, the majority of investigators have now abandoned some of the elements of the supercoil model in favor of a different type of structure, to be discussed in the next chapter.

Before going on to the subject of genetic information storage, it will be appropriate to discuss next the organization of nucleoprotein in the interphase cell nucleus and in the chromosomes.

CHAPTER SIXTEEN

Chromatin fibers and chromosomal organization

ORGANIZATION OF THE INTERPHASE NUCLEUS IN EUCARYOTIC CELLS

The genetic function of nuclear components and the orderly behavior of chromosomes during cell division as well as the precision with which the nuclear chromatin replicates itself suggest that nuclear organization and ultrastructure would also be regular and intricate, perhaps with paracrystalline organization of its molecular and supramolecular components. Early electron microscope studies of thin sections of interphase nuclei were puzzling, because there was no apparent orderliness in the nuclear structure. All that was seen was a confused array of what looked like segments of fine threads. This unexpected and disorganized ultrastructure may have given rise to the idea that the threadlike appearance was a fixation artifact. Long, tangled threads were seen, however, in whole-mounted nuclei and, in many cases, single chromatin fibers were observed to be attached by their ends to raised, ringlike structures ("annuli") on the nuclear membrane surrounding the nuclear pores (Fig. 16-1). This and other evidence makes it highly probable therefore that the fibers are really present in the living state and are not artifactual. The nuclear pores are formed by fusion of the inner and outer membranes of the nuclear envelope and have a fairly uniform diameter of 700 to 750 Å. The annulus forming the perimeter of a nuclear pore has eightfold symmetry because of the eight, symmetrically ar-

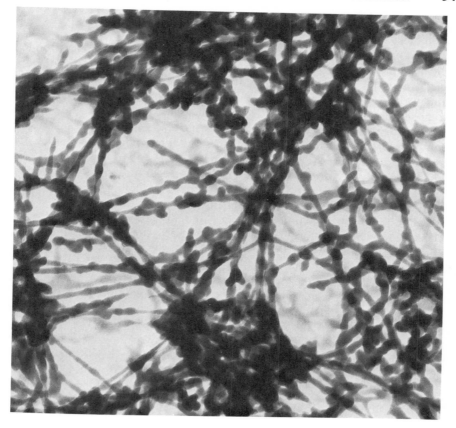

Figure 16-1 Electron micrograph of whole-mounted chromatin fibers of embryonic honeybee (×89,200) (From E. J. DuPraw. Proc. Natl. Acad. Sci. U.S.A., *53*: 161,1965).

ranged, roughly spherical subunits that form the annulus. The annular material, believed to be protein in nature, thus forms a kind of cylinder that penetrates right through the nuclear envelope alongside the pore and to either side of the margin of the pore. The nuclear pore, along with the annulus, are together referred to as the pore complex.

Different investigators have reported that the pore is plugged by what appears to be structureless, electron-dense material, a thin diaphragm or "annulus membrane," or ribonucleoprotein particles. These appearances plus the osmotic properties of nuclei indicate that the pores are specialized selective permeability structures, and not holes in the nuclear envelope. Fibrils extending into the lumen of the pore from the annulus subunits have been demonstrated and identified as ribonucleoprotein.

On the nuclear side, the pores are usually aligned with channels through the nuclear material that lie between clumped masses of denser chromatin. Certain nuclear materials and structures are intimately associated with the inner surface of the nuclear envelope such as the condensed heterochromatic human female X chromosome. The meiotic prophase chromosomes in plants and animals are often seen with attachments to the nuclear envelope when they pair, and the synaptinemal complex of each pair of homologous chromosomes has been shown to terminate on the nuclear envelope.

The previously mentioned connections of chromatin threads to the annuli may be related to the mechanism that controls the supercoiling of DNA and the condensation of heterochromatin and euchromatin (see later).

At various times, associations of one kind or another between structural nuclear components and the ribonucleoprotein fibrils in the lumen of the nuclear pores have been suggested, but the exact nature of these relationships is conjectural at present.

The fibers, which are the principal component of interphase nuclei, have a "nobbly" appearance; the diameters vary from less than 200 Å to over 300 Å. E. J. Dupraw showed that each consists of protein wrapped round a nonprotein central core. Trypsin treatment removes the protein, leaving a central trypsin-resistant thread that has the dimensions of a single Watson–Crick double DNA helix and that is specifically digested by DNAase. In other words, there appears to be a one-to-one correspondence between the single chromatin fibers and a long, solitary, Watson–Crick double DNA helix. Two distinct types of chromatin fibers have been detected in the interphase nucleus.

1. *Type "A" fibers*, which apparently consist of a single DNA double helix, which is itself coiled again (supercoiled) into a superhelix (i.e., the double helix and associated proteins is itself coiled).
2. *Type "B" fibers*, which are type "A" fibers further supercoiled (i.e., the superhelix of the type "A" fiber is coiled yet again and with additional proteins, the latter possibly playing a role in stabilizing the coiling).

The beaded appearance of chromatin fibers is probably due to alternating type A and type B regions.

In thin sections of interphase nuclei two states of the chromatin are usually recognizable. One is designated euchromatin, which consists of fine filaments; the other is heterochromatin, which has a more compact granular structure. The fibers composing euchromatin are smaller in diameter than those of heterochromatin, which suggests that the former may be in a less compact (less supercoiled?) state than the latter. This would be consistent with what is known about the relative metabolic activity of the two types of chromatin. Euchromatin incorporates DNA and RNA precursors quite actively, whereas heterochromatin is almost inactive in this respect. The relatively uncoiled state of euchromatin is what would be expected if its DNA is acting as a template for synthesis of fresh DNA, whereas the DNA template would be masked by the tighter coiling of the heterochromatin.

A strongly "beaded" appearance of the chromatin fibers prepared from various animal cell nuclei is apparent in electron micrographs published by A. Olins and D. Olins in 1974 (Fig. 16-2). The chromatin fibers have the appearance of linear sequences of spherical particles, often called nucleosomes or

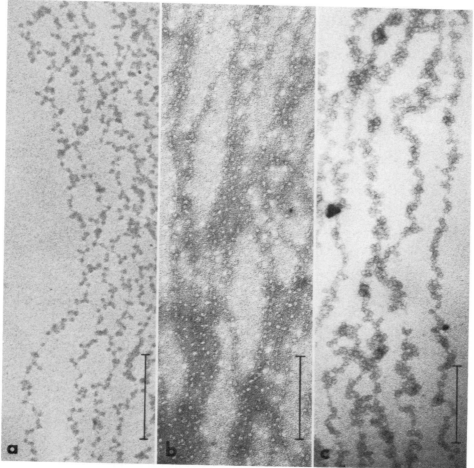

Figure 16-2 Electron micrograph of chromatin fibers showing the "string-of-beads" appearance. (Courtesy of Dr A. L. Olins. Science *183*:330–332, January 25, 1974. Copyright 1974 by the American Association for the Advancement of Science.)

"nu" particles, connected by thin filaments, an appearance that does not appear to be artifactual, since other independent evidence is consistent with the "string-of-beads" structure of the chromatin fibers (see later). This is the most direct of many kinds of evidence that, as mentioned in the previous chapter, has now led most investigators to turn away from certain aspects of the supercoil model in favor of the view that nuclear DNA fibers have a string-of-beads structure.

Five types of histones appear to be important to this structure, and four of these are needed for nucleosome formation. M. Senior showed that when nuclear chromatin is treated with a specific DNAase and then examined in the electron microscope, it has the appearance of separate particles, resembling the nucleosomes. The DNA in the nucleosomes seems to be shielded from enzymatic attack, presumably by the four associated histones, whereas the connecting threads of DNA and the associated nonspecific protein (the fifth histone?) are not so protected.

The four nucleosome histones do not appear to interact with or to "recognize" specific DNA sequences, but they do interact with themselves.

A first attempt to coordinate the X-ray data, DNAase degradation studies, and histone stoichiometry was made by R. Kornberg and J. Thomas. They found that in solution, the nucleosome histones designated F2A1 and F3 (H4 and H3 in the alternate nomenclature; see previous chapter) associate to form a pair, and so do the other two histones, F2A2 and F2B (H2a and H2b); they suggested that possibly the same pairs occur in the histones bound to DNA, although whether or not the behavior of histones in solution in this respect is the same as when associated with DNA should first be determined. Isenberg found that not only do the above histone pairs occur in solution, but also the F2A1–F2B (H4–H2b) pair. Hence, it would seem that the four histones could all be in contact in the chromatin nucleosomes. Kornberg and Thomas propose that nucleosomes consist of two each of the four histones H2a, H2b, H3, H4, plus 200 base pairs of DNA.

Probably one H1 chain is also present. If H1 is involved in the chromatin nucleosome structure, its role would not seem to be as critical as that of the other four histones. This is suggested by the fact that the X-ray diffraction pattern of chromatin fibers reconstituted *in vitro* and lacking H1 is the same as that of native chromatin fibers, whereas reconstituted chromatin from which any one of the other four histones is missing does not give the native chromatin pattern.

An alternative proposal is that H1 may be associated with the DNA in the threads connecting the nucleosomes.

An interesting point is that the histones, especially H3 and H4, have undergone hardly any primary structure variation during evolution. For example, the H4 histone from pea seedlings differs from that of calf thymus in only two of the 102 amino acid residues that comprise each of these two histones. Hence, there has been considerable selection pressure during evolution to conserve the histone structure. This suggests that these histones play an important part in the formation of the nucleosomes, because the form and dimensions of these exhibit practically no variation in eucaryotic chromatin, whatever its source.

The well-known variations in the degree of "condensation" of chromatin during the cell cycle, from the compact state in metaphase chromosomes to the more extended state in the interphase nucleus, are not accompanied by any change in the fundamental arrangement of the chromatin components. However, it seems that the function of the compact chromosomes is associated with chemical changes in the histones involving acetylation, methylation, and phosphorylation, and that these changes are cyclical in nature. The apparent heterogeneity that was earlier observed in histones was probably attributable to the existence of these chemical modifications. It is interesting to note that in the case of histone H1, maximum phosphorylation of this occurs in late interphase and reaches its lowest level at the end of mitosis, an observation that would again seem to argue in favor of the noninvolvement of H1 in the structure of the chromatin nucleosomes.

Neutron diffraction studies have shown that the diffraction pattern of chromatin is generated by regularly repeating protein instead of DNA units. Thus J. P. Baldwin, P. G. Bosely, and E. M. Bradbury were able to reconcile the string-of-beads model of chromatin with the X-ray diffraction pattern, which had previously been interpreted by Wilkins and others as being generated by the turns of a supercoil structure.

CHROMOSOMAL ORGANIZATION

Early models of chromosomal organization, some of which will be described here for their historical interest, show evidence of the intuitive feeling that there must be orderliness and precision in the arrangement of chromosomal components. This might reasonably be expected even more than in the case of the interphase nucleus, since the precise and orderly behavior of chromosomes in mitosis and in meiosis, and their genetic function seem to demand it.

The oldest model of chromosomal organization is that of A. E. Mirsky et al.; this model was put forward long before the Watson–Crick double helical structure of the DNA molecule was known. In this theoretical model, the chromosome is visualized as having a central axis consisting of nonhistone protein surrounded laterally by DNA-histone molecules. The ribose phosphate of the nucleic acids forms a layer around the central protein axis. Outside this and on the surface is a layer of nitrogenous bases. Plainly implied in this concept is that the principal material involved in genetic transmission is DNA (Fig. 16-3a).

There is some experimental support for a model of chromosome organization that was put forward shortly after Watson and Crick announced the double helical structure of DNA and its genetic implications. In this concept of

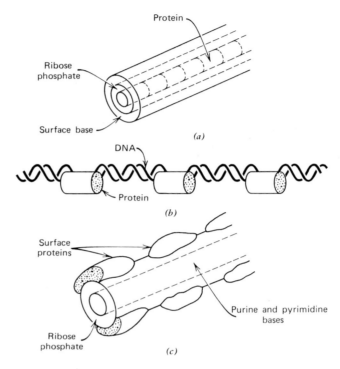

Figure 16-3 Models of chromosome organization. (Redrawn after various authors.)

chromosome structure, DNA double helices alternate with protein, the latter possibly acting as a cement material (Fig. 16-3*b*), The primary importance of DNA is again evident, but not so much as in the previous model.

The results of experimental studies by A. L. Dounce are interesting in connection with this model of chromosome structure. Dounce investigated the physical and chemical properties of nucleoprotein gels, and he feels that his data support the idea that in mammalian chromosomes, DNA segments alternate with segments of what he calls "residual protein." However, he does not go so far as to claim that he has proved the existence of protein *links* between chromosomal DNA segments. The nucleoprotein gels were prepared from isolated nuclei and chromosomes by subjecting them to pH 8 to 10 in distilled water or sucrose solution or by suspending them in 1 to 2 molar sodium chloride at pH 7.0 or in 0.5 to 2.0% sodium dodecyl sulfate at pH 5.5 to 10.0. The resulting gels were extraordinary in that they persisted even when diluted to the extent that the DNA concentration was reduced to about 0.01%. The residual protein is firmly bound to DNA and is thought to be an important structural component of chromosomes. It is distinct from histone, which is believed to be involved in regulating and inhibiting the action of genes. Each DNA segment is considered to be attached at each end to residual protein, except where a DNA segment is at the end of the structure. The linkages between DNA and protein might involve only one strand of DNA, so that the other is available for transcription. Dounce and his colleagues claim that they have found nothing inconsistent with this model, although they admit that their experimental results are capable of more than one interpretation.

G. Zubay and P. Doty visualized the core of the chromosome as consisting of DNA, and they suggest that the basic proteins are on the surface of the DNA chain. From this model it is apparent that proteins may play a role in genetic determination (Fig. 16-3c).

Two or more of the structures just described may possibly be present simultaneously in a single chromosome. For example, the second model could represent the protein linkages between DNA molecules, and the third could represent the structure of the segments in which DNA is concentrated.

A viable model of chromosomal organization must be able to account for the manner in which chromosomes duplicate themselves. As is well known, the two chromatids composing a whole chromosome separate during mitotic cell division, and each one passes into one of the two daughter cells. During interphase the chromatids apparently disappear but, when the next cell division is imminent, they reappear with new partners. Evidently, the chromatids have made copies of themselves during interphase.

The mechanism of chromatid duplication was investigated by J. H. Taylor and his collaborators who, with the aid of radioautographic techniques, studied the uptake of tritium-labeled thymidine by successive generations of chromosomes in cultured plant root cells. They found that in the cells that had taken up the radioactive thymidine prior to division, the two chromatids comprising each chromosome were uniformly radioactive. These cells were then transferred to a medium containing nonradioactive thymidine. When the chromosomes appeared prior to the next cell division, only one chromatid of each pair was found to be radioactive.

Taylor therefore envisioned the chromatid as a two-part structure, each half of which acts as a template during replication. Some of the later experiments of Taylor and his group suggested, in addition, that the two halves of a chromatid have complementary structures. This concept is therefore strongly reminiscent of the semiconservative replication of DNA and its structure of two complementary strands. One may therefore infer that perhaps chromatids are single DNA duplexes, but the great disparity in size between a DNA duplex and that of a chromatid makes this highly unlikely at first sight. Taylor points out that if fully extended, the DNA in a single chromatid would be more than a meter long and the two polynucleotide strands would intertwine in excess of 300 million times. Furthermore, we have to try to imagine how this long thread can be folded and packed—along with structural protein—into such a small space as is occupied by a chromatid. It is hard to imagine how such a long and highly coiled structure could become completely unraveled as it must do during the replication process.

Taylor proposed a model of the chromosome in which parallel rows of DNA double helices are attached by their ends to a bipartite protein backbone. The end of one member of each double helix is attached to one of the two halves of the backbone, and the end of the other helix is attached to the other half (Fig. 16-4). Prior to replication, the two halves of the backbone move away from one another, and the DNA helices separate from one another. Thus, each protein-DNA half structure acts as a template for synthesis of the other new complementary half. Electron micrographs were obtained of chromatids whose appearance bore a striking resemblance to the model just described, inasmuch as a backbone (protein?) is visible surrounded by a fuzz of some material (DNA?).

Models of chromosomal organization involving a protein backbone can no longer be considered seriously because of experimental findings and electron microscope studies that are not in accord with this idea. Several workers have shown, for example, that all parts of "lampbrush" chromosomes are fragmented when exposed to DNAase, but not when exposed to RNAase or trypsin. Both RNAase and trypsin do, however, strip away a large part of the coating from the lateral loops, but do not disrupt the central axis. DNA instead of protein would therefore appear to be the principal structural component of these

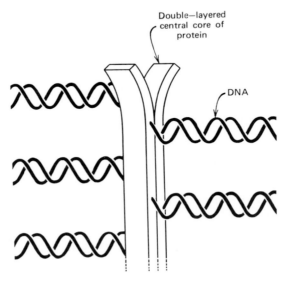

Figure 16-4 Chromosome structure and replication (according to J. H. Taylor). (Redrawn after J. H. Taylor.)

chromosomes. D. M. Prescott has shown that DNA and not protein is conserved during chromosomal replication in the Chinese hamster.

ELECTRON MICROSCOPY OF CHROMOSOMES

As with interphase nuclei, examination of thin sections of chromosomes with the electron microscope revealed a confused picture of randomly coiled threads of variable diameter, which seemed difficult to reconcile with the orderly and precise behavior of chromosomes in cell division. These fibers were seen to be more closely packed together in metaphase chromosomes than in thin-sectioned interphase nuclei.

H. Ris introduced methods for the electron microscopy of whole chromosomes and nuclei and obtained results that confirmed the fibrous ultrastructure of chromosomes. E. J. DuPraw examined whole-mounted metaphase human and embryonic honeybee chromosomes in the electron microscope. His method of isolating and preparing human metaphase chromosomes preserved the characteristic morphology, as observed in light microscopy. The high resolving power of the electron microscope revealed that, as in sectioned material, the chromatids are composed of an irregularly packed mass of chromatin fibers of diameter varying from 200 to 500 Å (Fig. 16-5). These results were confirmed in whole human chromosomes by other workers. Two further observations were informative. The first was that in whole-mounted chromosomes in a good state of preservation, hardly any "loose ends" of fibers were seen even at the ends of chromatids (telomeres); fiber loops were nearly always seen projecting from the surface. Second, in the centromere region, many fibers were seen linking together the two arms of each chromatid. These morphological features are what one might expect if the chromatid consisted of a single fiber of great length that was compactly folded in both transverse and longitudinal directions. Autoradiography does, in fact, lend support to this idea and shows that all of the DNA of a chromatid is in the form of a pair of dissimilar entities that show semiconservative segregation during chromosome replication. This recalls the complementary double strand structure and replication pattern of the DNA molecule. No sign of a protein backbone was seen in the chromatids. If it did exist, it would surely have been noticed, because the electron beam completely transilluminates the chromosomes. DuPraw produced evidence to show that chromatids consist entirely of chromatin fiber material.

THE FOLDED FIBER MODEL OF METAPHASE CHROMOSOMES

These experiments and observations led DuPraw to present his folded fiber model of chromosomal organization. In the interphase nucleus, prior to

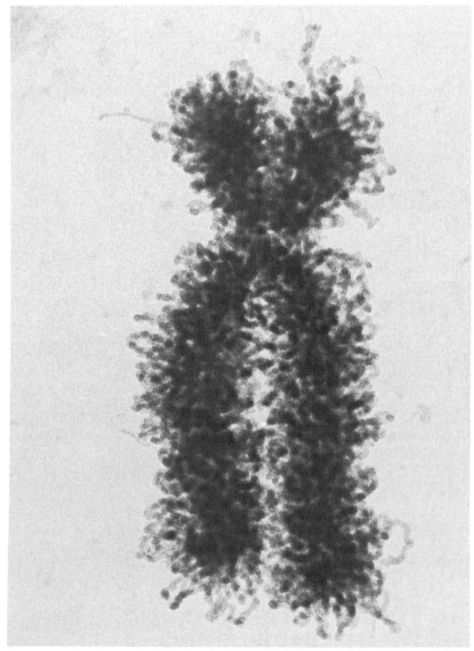

Figure 16-5 Electron micrograph of a whole-mounted human metaphase chromosome (×60,400) (From E. J. DuPraw. *DNA and Chromosomes*. Holt, Rinehart and Winston, New York, 1970).

replication, the chromosome is thought to be represented by a long DNA double helix invested with its associated proteins and RNA. This structure is called a unit chromatid. The unit chromatid is supercoiled to give rise to the 200 to 300 Å type B fiber. The supercoiling may be produced and stabilized by basic alpha-helical proteins such as histones. During replication, the type B fiber is pictured as unwinding at both ends to form two replication forks; each arm increases in length as the uncoiling at both ends approaches the center of the unit chromatid (Fig. 16-6). The two arms of each replication fork correspond to a pair of metaphase chromatids and are held together by unreplicated portions of the fiber, as occurs in the centromere region. The daughter chromatids become intricately and compactly folded during prophase and metaphase, giving rise to the chromosome morphology familiar from light microscopy. Chromosomes consisting of a single folded chromatin fiber are referred to as simple chromosomes. Chromosomes consisting of more than one fiber (see later) are designated compound chromosomes.

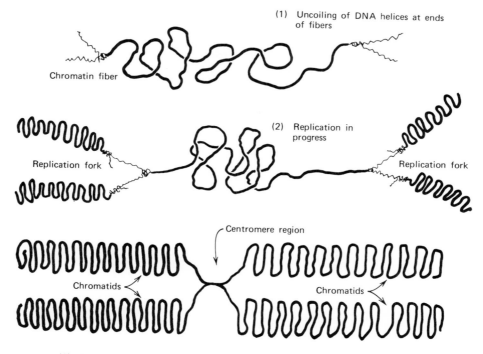

(1)   Uncoiling of DNA helices at ends of fibers

Chromatin fiber

(2)   Replication in progress

Replication fork

Replication fork

Centromere region

Chromatids

Chromatids

(3)   Metaphase chromosome — transversely folded chromatin fibers (chromatids)

Figure 16-6   Chromosome organization and replication. (After E. J. Dupraw.)

IMPLICATIONS OF AN ALTERNATIVE FOLDED FIBER MODEL

Another model of the organization of metaphase chromosomes, according to DuPraw, which involves longitudinal as well as transverse folding of chromatin fibers, is shown in Fig. 16-7. In this model the following features are apparent.

1. *The centromere region.* This is seen to consist of at least three components that are not visibly separate at the light microscope level of resolution. They are:

    a. *The kinetochores.* Spherical or discoidal structures to which spindle microtubules become attached.

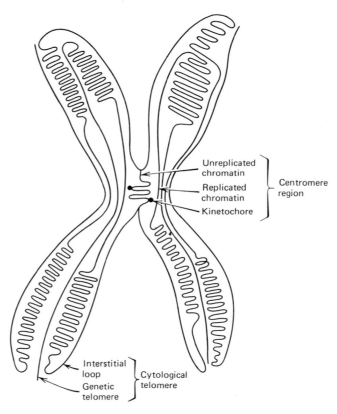

Figure 16-7 Alternative folded fiber model of a metaphase chromosome. (*Note.* In this model there is longitudinal as well as transverse folding of the chromatin fibers.) (After E. J. Dupraw.)

b. Chromatin fibers specifically associated with the kinetochores.

c. Chromatin fibers distinct from those associated with the kinetochores.

The well-known function of the centromere in holding chromatids together and attaching them to the spindle is therefore performed by different and separate entities, according to this model. (Note that in this model the centromere region consists of both replicated and unreplicated chromatin).

2. *Genetic and cytological telomeres.* A glance at Fig. 16-7 shows that cytological telomeres (morphological ends of the chromosomes) do not necessarily correspond to genetic telomeres (end of a genetic linkage group). As a result of the intricate folding pattern, cytological telomeres may contain fiber segments that are actually far apart on the extended chromatin fiber. Interstitial loops as well as true genetic telomeres may therefore lie close together in the cytological telomeres, bringing together genetic loci that actually are widely separated on the genetic map.

GIANT SALIVARY GLAND CHROMOSOMES

The interphase cell nuclei of certain tissues of insects such as *Drosophila* and *Chironomus* sp. contain permanently condensed chromosomes of relatively enormous size; this fact is well known to geneticists. These chromosomes are found in the larval salivary glands, the rectum, and the Malpighian tubules. When stained by the Feulgen method, these giant chromosomes exhibit a complex pattern, easily resolvable in the light microscope, of densely staining transverse bands with apparently clear spaces (interband regions) between them (Fig. 16-8). At one time it was thought that DNA occurred only in the bands and that the interband regions were empty of DNA. Later, the presence of DNA in the interband regions was demonstrated.

Synthesis of RNA in these chromosomes is accompanied by localized swellings or "puffs" at specific bands (Fig. 16-9).

The tendency of these chromosomes to produce frayed areas in which individual parallel fibers are visible gave a clue to their organization. DuPraw considers that giant chromosomes consist of many identical chromatin fibers (unit chromatids) in parallel array, a condition known as polyteny. Each of these fibers is analogous to an unfolded metaphase chromatid. Local areas of dense coiling (as in metaphase chromosomes) alternate with relatively straight fiber segments; these correspond to the band and interband regions, respectively (Fig. 16-10). The dense Feulgen staining of band regions is therefore due to the relatively high DNA concentration as compared with the interband regions. Chromosomal puffs are interpreted as local uncoiling of the fibers in the band regions, which exposes the DNA fibers and makes transcription (RNA synthesis) possible (Fig. 16-10).

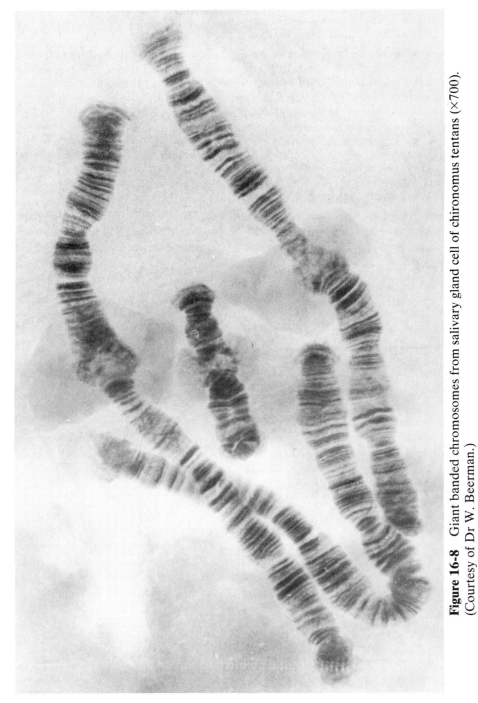

Figure 16-8 Giant banded chromosomes from salivary gland cell of chironomus tentans (×700). (Courtesy of Dr W. Beerman.)

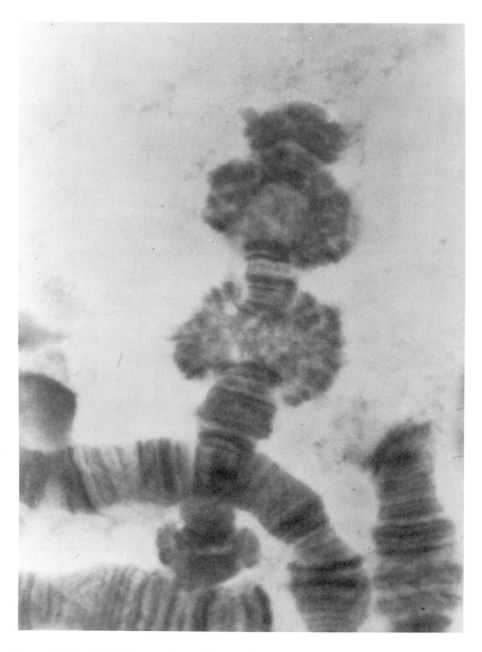

Figure 16-9 "Puffs" on giant salivary gland chromosomes. (Courtesy of Mrs Ingrid Clever.)

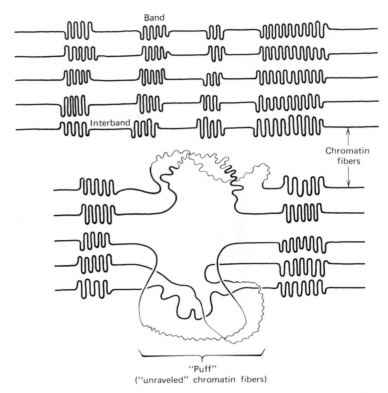

Figure 16-10 Giant salivary gland chromosome organization. (After E. J. Dupraw.)

SUPRACHROMOSOMAL ORGANIZATION

Light microscope observations of chromosomes during cell division have often revealed unmistakable regularities in the movement and arrangements of chromosomes that are quite unexpected in what are presumably independent, physically separate entities. Such phenomena have usually been attributed to "unknown forces." For example, multivalent or multipolar association of non-homologous chromosomes in the metaphase plate are seen in which the pairs of metaphase chromatids often lie close together and appear to be closely connected, although they would be expected to assume random positions in the metaphase plate. Chromosomes also will often be disposed in an orderly manner with, say, the short arms all facing each other.

E. J. DuPraw points out that the electron microscope has definitely revealed that separate nonhomologous chromosomes are often joined together by delicate threads that are invisible in the light microscope (Fig. 16-11); G. C. Hoskins

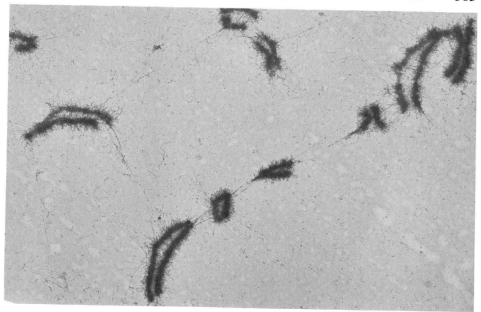

Figure 16-11 Electron micrograph of whole-mounted human metaphase chromosomes showing interchromosomal connecting fibers. (Courtesy of Dr F. Lampert. Lampert, F., Bahr, G. F. and DuPraw, E. J. Cancer *24*:367–376, 1969. Reprinted by permission.)

has shown that these contain DNA as an important structural component.

If chromosomes are physically connected, how can this be reconciled with the well-established, independent assortment behavior of nonhomologous chromosomes during cell division?

Although the number of genetic linkage groups in an organism usually corresponds to the haploid number of chromosomes, this is not necessarily inconsistent with the physical linkage of chromosomes. The interchromosomal connectives might be easily broken and rejoined and, moreover, in an individual chromosome, widely separated genes may appear to segregate independently if the recombination distance is more than 50 map units (Chapter Fourteen). Such genes are considered to belong to the same linkage groups only because linkage of both to a third gene in the same group is observed. Therefore, it follows that independent segregation of at least some of the chromosomes of a haploid set could occur even if they were joined by DNA threads, provided that the chromosomes were joined by noninformational DNA threads that were long enough to give 50% recombination. In fact, "nonsense DNA" consisting of monotonous -A-T-A-T-A-T-A- sequences has been found in considerable quantities in many organisms and could well be the DNA that joins chromosomes together. Whether or not this is so, what is interesting is the fact that such DNA is absent from bacteria, the haploid chromosome number of which is 1.

On the other hand, interchromosomal connections should cause genes located on separate chromosomes to exhibit a detectable but probably low level of linkage. There are, in fact, data indicating that this is the case.

DuPraw suggests, speculatively, that the haploid chromosome set of a eucaryotic cell represents local condensations of a very large single circular DNA double helix. If so, then the eucaryotic haploid chromosome complement in its basic structure is no different, except in size, from the single, circular chromosomes of bacteria (Chapter Nineteen).

CHAPTER SEVENTEEN
Storage of Genetic information. The genetic code

The primary manifestation of a gene is its direction of the synthesis of a specific protein. The protein itself may be a phenotypic character (e.g., a hemoglobin molecule), or it may be an enzyme functioning as a catalyst in a reaction or reaction sequence, resulting in the appearance of a phenotypic character such as the pigment in a flower petal. Genetic information may be regarded as "coded instructions," carried and transmitted by genetic material, which specify the structure and direct the synthesis of specific proteins.

IS DNA THE REPOSITORY FOR GENETIC INFORMATION? A DISCUSSION OF THE EVIDENCE

An enormous variety of proteins is found in living cells and, at one time, it was thought that genetic material itself must be protein. Because they are constructed from 20 or so different amino acids, proteins exhibit almost endless variations of structure and, they seemed to be the obvious candidates for the role of genetic material. Polysaccharides seemed highly unlikely because of their simple repetitive structure of identical monosaccharide units. Even nucleic acids, such as DNA, with variable structure made possible by four types of nucleotide "building blocks," did not seem equal to the task of storing and transmitting the complex specifications for every protein in a cell. However, much experimental evidence strongly indicates that DNA and not protein is the genetic material. The following critical review of the different types of evidence on which the case for the genetic function of DNA is based should give an idea of the validity of the almost universally held conviction among biologists that DNA is the primary genetic material.

(i) Constancy of DNA content of cell nuclei

Much cytological evidence indicates that hereditary determinants are located in the cell nucleus—or, to be precise, in the chromosomes—instead of in the cytoplasm. DNA is confined almost exclusively to the nucleus, which suggests that DNA may have a genetic function. Furthermore, the amount of DNA in the nuclei of somatic cells of a given species is constant. This is exactly what would be expected for genetic material. For example, in the rat the amount of DNA in a somatic cell nucleus, whatever the tissue, is 6.5–6.7×10^{-12} g. In the fowl, the figure is 2.3–2.6×10^{-12} g. This is consistent with the fact that the nuclei of the somatic cells of a given species all have a similar chromosome complement, whereas another species will have a different complement. Gametes, which are haploid, are found to have exactly half the amount of DNA than the diploid somatic cells, which correlates with their possession of half the chromosome number of the diploid tissues. However, the histone fraction of somatic cell nuclei also is usually found to have a similar quantitative distribution, so that constancy of DNA content of cell nuclei by itself is not conclusive evidence for the genetic role of DNA. One could speculate that DNA is merely a structural support for the protein genetic material.

(ii) Stability of DNA and its intact transmission in cell division

Genetic evidence indicates that the individuality and stability of genes is preserved from generation to generation. One may reasonably conclude, therefore, that genes are composed of material that is metabolically stable and that retains its identity during cell division. Furthermore, by using techniques in which DNA is labeled with radioactive elements, it has been shown that in cell division, nuclear DNA is transmitted intact to the daughter cells without first being broken down and then resynthesized. The same cannot be said of other biological macromolecules. The stability and intact transmission of DNA suggests that it may be the material from which genes are constructed. On the other hand, if it were possible to perform analogous experiments with labeled histone, a similar continuity and metabolic stability might also be demonstrated, so that the static nature of DNA does not by itself prove a unique genetic role.

(iii) The nature of physical and chemical mutagens

Generally, physical and chemical agents that increase the mutation rate of organisms act on and cause changes in the structure of DNA. Ultraviolet light, a mutagenic agent, is absorbed by DNA, and the wavelengths that are most effective in causing mutations are also those that are most strongly absorbed by DNA.

Thymine is one of the four nitrogenous bases entering into the structure of DNA. Its analog 5-bromo uracil is incorporated into DNA if it is added to a culture medium in which cells are growing and dividing. When this occurs, there is a great increase in the mutation rate because of production of abnormal DNA containing the thymine analog.

In light of this evidence, the conclusion that DNA is genetic material seems reasonable. However, one might also argue that the damage and structural changes in DNA induced by mutagens could cause changes or distortions in the protein, which is closely associated with DNA, and that it is these changes in the protein that are responsible for mutations.

(iv) Bacterial transformation

The phenomenon of bacterial transformation was demonstrated in England in 1928 by F. Griffith, who performed a series of experiments with two strains of pneumococci; one of them was virulent and killed mice when it was injected into them, and the other was nonvirulent. Heat-killed virulent pneumococci did not kill mice, but a mixture of the heat-killed virulent strain and living nonvirulent strain (neither of which alone were lethal to mice) did kill mice when it was injected into them. Moreover, *living* virulent pneumococci were found afterward in the tissues of the dead mice, a surprising and unexpected result.

In 1944, O. T. Avery, C. M. McLeod, and M. McCarty and later R. D. Hotchkiss, S. Zamenhoff and others fractionated dead cells of the virulent pneumococcus. They extracted a material that, when added to a suspension of living nonvirulent pneumococci, transformed them into the virulent strain. That this transforming agent was DNA was shown by treating it with an enzyme that specifically destroys DNA. The agent was inactivated by this treatment, but proteases did not affect it. The experiment therefore provided conclusive evidence that the transforming agent was DNA and not protein.

At first sight, such results seem to provide irrefutable evidence for the genetic role of DNA. An alternative possibility is that the DNA might have been acting as a specific mutagenic agent that caused the recipient bacterial strain to take on certain characteristics of the donor strain. The mutated genes would be transmitted to the offspring, but this does not mean that the genes themselves consist of DNA. However, there are other features of the transformation phenomenon that cannot be explained by a mutation hypothesis and that support the idea that actual transfer of genetic characters by isolated DNA takes place. For example, DNA has been separated from bacterial cells that have been transformed for two different traits. J. Marmur and R. D. Hotchkiss took a strain of streptomycin-resistant streptococci that had undergone mutation to a mannitol-fermenting form, and DNA prepared from these cells was added to a culture of wild-type streptococci. Most of the latter were transformed to either streptomycin-resistant forms or to mannitol-fermenting forms, but a few were found to have been transformed for both characters. Since the frequency of this double transformation greatly exceeded the product of the frequencies of the single transformations, the only genetic interpretation possible was that linked transfer of two characters by a single event had occurred. This strongly suggests the direct transfer and incorporation of a single fragment of DNA containing the information necessary for specifying two distinct physiological systems. Furthermore, it indicates that the DNA fragment is a portion of the normal genetic material of the donor bacterial cell.

Although DNA behaved in this way *in vitro* under specified conditions, it does not necessarily follow that this is the normal and universal role of DNA in all living cells.

(v) Virus replication, infectivity, and transformation of cells

Viruses are submicroscopic, self-replicating, nucleoprotein complexes that have some of the attributes of "life." They are responsible for many diseases in humans, in other animals, and in plants, and they are best regarded as infectious agents instead of as living entities. They cannot replicate outside of living cells. Viruses exhibit a wide diversity of size and morphology, and all consist basically of a nucleic acid "core" surrounded by a protein "coat."

Convincing evidence for the genetic role of DNA is provided by the phenomenon of virus infection of bacteria. The best-known example of this is the infection of *Escherichia coli* cells by the T_4 bacteriophage described in Chapter Fourteen. After the bacteriophage has injected its DNA into the *E. coli* cell, the empty head and tail (known as a "ghost") remains attached to the cell surface.

In 1952 A. D. Hershey and M. Chase performed experiments to determine the relative roles of DNA and protein in T_4 bacteriophage infection of *E. coli* cells. The fact that DNA contains phosphorus but no sulfur and that protein contains some sulfur but almost always no phosphorus was of critical importance in these studies. Hershey and Chase cultured bacteriophage in *E. coli* growing in a medium containing the radioactive isotopes of phosphorus and sulfur, P^{32} and S^{35}, respectively, so that bacteriophage particles containing P^{32} in their DNA and S^{35} in their proteins were produced. Nonradioactive *E. coli* were then infected with the radioactive bacteriophage particles and, after infection had occurred, the *E. coli* were agitated in a Waring blender to remove the empty adherent bacteriophage "ghosts." These virus remnants were found to contain only S^{35}, but the infected *E. coli* contained P^{32} and little or no S^{35}. This result therefore demonstrated that it was the bacteriophage DNA only that entered the bacterial cells and not the protein and that therefore the DNA alone was responsible for "reprogramming" the bacterial cells to produce viral instead of bacterial proteins and nucleic acids.

Further experiments with labeled atoms indicated that the atoms composing bacteriophage protein do not appear in the progeny virus, but more than half of the atoms in the DNA of the parent virus appear in the progeny.

A similar demonstration of the genetic role of nucleic acids is provided by the phenomena of infectivity and replication of tobacco mosaic virus. The nucleic acid in this virus is RNA, so it might be objected that the bearing of experiments with tobacco mosaic virus (TMV) on the genetic role of DNA in higher organisms can only be indirect. However, DNA and RNA have so many aspects of structure in common and the role of RNA in TMV is so similar to that of DNA in other organisms that the experiments and results with TMV RNA have strongly promoted acceptance of the information-carrying role of DNA. The tobacco mosaic virus (Fig. 17-1) is the agent responsible for causing mosaic disease of tobacco plants, in which necrotic blotches appear on the leaves. TMV is one of the larger viruses; it can be obtained in relatively large quantities and is easily purified. For these reasons it has been intensively studied, and its structure and composition are known in considerable detail.

The TMV particle has a cylindrical shape, is approximately 3000 Å long × 150 Å in diameter, and has a cylindrical hole about 40 Å in diameter running through its center. Physical and chemical studies show that the virus particle has a molecular weight of about 40×10^{6}, of which 94.5% is due to protein and 5.5% to RNA. Dissociation of the protein from the RNA can be brought about by treating the virus with weakly alkaline solutions giving protein fragments with a molecular weight of about 100,000. When the pH is lowered again, the pure protein will spontaneously reaggregate to give tubular structures, even in the absence of the RNA, that have the same internal and external diameters as the whole virus. These tubes are not infectious, however, and do not cause mosaic disease. X-ray diffraction studies have revealed that in both whole virus and in the reaggregated protein tubes, the protein is in the form of identical compact subunits; each one measures about 70 Å × 23 Å × 21 Å and is shaped somewhat like a shoe. The subunit molecular weight is about 17,000.

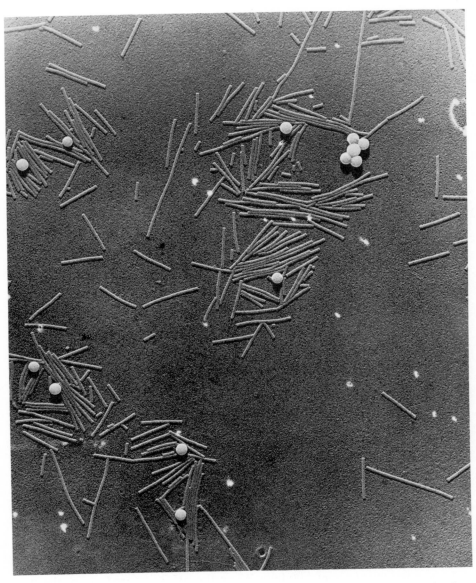

Figure 17-1 Electron micrograph of chromium-shadowed tobacco mosaic virus particles (in presence of 880 Å polystyrene calibration spheres). (Courtesy of Dr R. B. Park. From *Cell Ultrastructure* by William A. Jensen and Roderic B. Park. Copyright 1967 by Wadsworth Publishing Company Inc., Belmont, California, 94003. Reprinted by permission of the publisher.)

The RNA of TMV is a single-stranded polynucleotide in the form of a helix around which the protein subunits cluster in a helical arrangement. In the intact virus about 2200 helically arranged protein subunits form a tube 3000 Å long, and $16\frac{1}{3}$ subunits form one complete turn of the helix, or 49 form three turns. The structure of TMV is shown in Fig. 17-2.

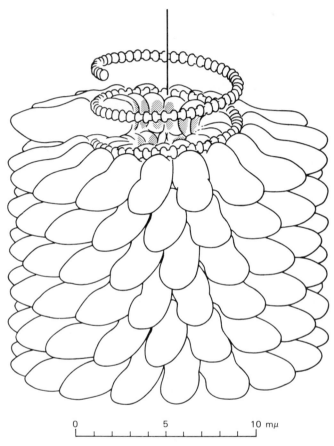

0   5   10 mμ

Figure 17-2 Structure of tobacco mosaic virus. The shoe-shaped objects are the protein subunits arranged around the central molecule of RNA, the helical structure of which is stabilized by the protein. Some of the protein subunits at the top of the drawing have been removed to show the helical RNA molecule which, in reality, would not retain its native configuration in the absence of the protein. From D. L. Caspar and A. Klug with permission. From "Physical Principles in the Construction of Regular Viruses," *Cold Spring Harbor Symp. Quant. Biol.* 27: 1–24 p. 4 (1962).

This arrangement of the protein subunits appears to stabilize the RNA in its helical configuration, because free RNA dissociated from the protein assumes a complex tertiary type of folding; this isolated RNA is infectious, but the infectivity is greatly attenuated as compared with intact virus.

The aggregation of the protein subunits is influenced by the RNA; the reaggregated protein tubes are not as stable as whole virus, and the lengths of the tubes vary.

When healthy tobacco plants are infected with TMV and the disease is allowed to develop, viral nucleoprotein can be extracted in quantities greatly exceeding those of the original inoculum. Hence, given the right cellular environment, the virus is able to provide for the replication of its RNA and to direct the biosynthesis of its own specific protein.

Various mutant forms or "strains" of TMV have different infectivity characteristics and induce various forms of the disease, each one having its own distinctive symptoms.

The protein and RNA components of TMV are easily separated and can be isolated in very pure form. Treatment of a virus suspension with phenol-water mixture denatures and precipitates the protein, leaving the RNA dissolved in the aqueous phase and essentially intact. Fraenkel-Conrat showed that when a solution of TMV RNA is mixed with isolated TMV protein, a spontaneous reconstitution process takes place with re-formation of infectious virus particles indistinguishable from the natural TMV. Reconstitution exhibits no great specificity, inasmuch as the RNA from one viral strain can combine with protein from a different strain to give a "hybrid" virus that is fully infectious. What is significant is that such hybrids produce disease symptoms in tobacco plants characteristic of the strain from which the RNA, but not the protein, was derived, and the protein of the progeny virus is identical with that from the strain from which the RNA was derived.

Isolated TMV RNA itself retains a small fraction of the infectivity of the intact virus from which it was derived and causes disease symptoms in tobacco plants similar to those caused by the donor virus. From plants infected in this way complete virus identical with the donor is extractable in quantity from the infected tissues. Evidently, the infectious RNA has directed the synthesis of the specific protein characteristic of the donor virus.

Transformation of cells by viruses is an alternative to death and lysis of the cell, brought about by "takeover" of the biochemical machinery of the cell and virus replication. Cell transformation involves incorporation of the viral nucleic acid into the cell's genome without replication to form new virus particles. The process does not result in cell death, but the characteristics of the cell are altered, and it is said to be transformed. The new characteristics are transmitted to the daughter cells when the transformed cell divides. Virus particles cannot be isolated from transformed cells, but virus-specific antigens are found at the cell surface. A transformed cell exhibits unrestrained reproduction and does not respond to the presence of nearby cells.

Virus infection and transformation of cells demonstrate that nucleic acids and not proteins encode and transmit the information necessary for specifying the structure and directing the synthesis of specific proteins. These phenomena provide one of the strongest pieces of evidence for the genetic function of nucleic acids.

CONCLUSION

Taken singly, each of the types of evidence for the genetic role of DNA is not conclusive, but the evidence taken together adds up to a strong circumstantial case for believing that genetic information is encoded in the molecular structure of DNA and that it is the material from which genes are made

THE SEQUENCE PROBLEM IN PROTEIN BIOSYNTHESIS

We have established with considerable certainty that DNA is the principal information-storing macromolecule necessary for specifying the structures of the myriad different types of protein found in living organisms. Such information is necessary because not only is it essential for cells to possess an enzymatic mechanism for linking together amino acid units into a polypeptide chain, but the correct order or sequence of amino acids and the right chain length for the polypeptide must be accurately reproduced. The biological properties of a given protein ultimately depend on these because, as we have seen earlier, the amino acid sequence determines the secondary and tertiary structures of protein molecules. Protein biosynthesis is therefore complicated by a "sequence problem" in addition to the biosynthetic process. In the case of other biopolymers, such as the starches and cellulose, the polymeric structure is a repetitive chain of identical monomers (glucose units), and a sequence problem does not arise.

CELLULAR TEMPLATES AND PROTEIN BIOSYNTHESIS

How does the DNA molecule direct the proper sequencing of amino acids in protein molecules? Its function in this respect may be analogous to a template and, to illustrate this idea, a mechanical analogy may help. Suppose we have a number of geometrical cardboard shapes and we are instructed to align these in one specific linear sequence. Apart from being instructed verbally on how to arrange them, the same information could be encoded in the form of a template, which is a cardboard shape along one edge of which are cut out shapes complementary to the shapes we need to arrange. Assuming that the cardboard shapes must be fitted only into their complementary shapes on the template, one

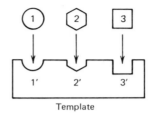

Template

Figure 17-3 Determination of a unique sequence of different geometrical shapes by a template.

specific sequence of them is thus automatically determined, and no other is permitted (Fig. 17-3).

Functionally, the DNA molecule may therefore be regarded as a template; the next thing to determine is what sequential features of the DNA molecule correspond to the sequence of geometrical notches in the cardboard analog. The only variable sequential feature of the DNA molecule is the order of the four nitrogenous bases along its length. Assuming that the sequence of individual bases or some combination of them in DNA is the cellular template or code, there must be *a priori* some simplest structure for each amino acid determinant corresponding to the notches in the cardboard model.

THE GENETIC CODE

In 1954 the physicist George Gamow speculated about the nature of the genetic code, as the cellular template directing protein synthesis is now called. Gamow reasoned that since there are only four different bases and about 20 amino acids for which to code, single bases cannot act as amino acid determinants, since only four amino acids could be thus encoded. Pairs of bases likewise would be inadequate because only 16 (4^2) possible combinations are possible of four different bases taken in pairs.

AA	AT	AC	AG
TA	TT	TC	TG
CA	CT	CC	CG
GA	GT	GC	GG

Bases in groups of three (triplets) would be more than adequate, because 64 (4^3) combinations are possible, and groups of four or five bases would provide even more possible combinations (256 and 1204, respectively). A code consisting of base triplets would be the simplest that is equal to the task of coding for the sequences of 20 different amino acids in protein molecules.

If, in fact, the genetic code is made up of a sequence of base triplets, the further question arises as to whether the code is overlapping or nonoverlapping. An overlapping triplet code would make for economy of DNA, as was pointed out by Gamow. If, for example, we have the base sequence TGCATAGCT, and there is no overlap in the triplet code, three code words are obtained: TGC   ATA   GCT. If there is overlap of one base in each triplet, then four code words can be derived: TGC   CAT   TAG   GCT. A two-base overlap would produce even more code words:

TGC   GCA   CAT   ATA   TAG   AGC   GCT

Although an overlapping code thus allows greater coding capacity than a non-overlapping one from a given amount of DNA, this advantage carries a snag; in a triplet code involving one-base overlaps, there are only 16 different amino acids that can follow any one particular amino acid, because the last base of a given code word automatically determines the first base in the next. Similarly, a two-base overlap determines the first two bases of the next triplet and reduces the number of different amino acids that can follow a given amino acid to four. Since sequence studies of polypeptides had not revealed any such restrictions in sequence pattern at the time that these speculations were being made, it therefore seemed certain that the triplet code—if it existed—could not be overlapping, despite the lack at that time of definite data bearing on this question. Another thing that would be expected from an overlapping code is that a mutation involving alteration of a single base would alter more than 1 amino acid residue in the corresponding protein, whereas in a nonoverlapping code only one amino acid would be changed. Studies of the effects of single base changes on protein structure, such as is found in hemoglobin variants, does, in fact, demonstrate that only single amino acid residues are changed, and so provide further evidence that the triplets are nonoverlapping.

Other questions that arise are whether some amino acids are coded for by base triplets and others by base pairs and whether anything analogous to "spacers" between the triplets exists. For answers to these questions, a priori reasoning can take us no further, and recourse must be had to experimental investigation.

That genetic information (i.e., the coded instructions for determining amino acid sequence in polypeptides) is carried by the DNA molecules in the form of a nonoverlapping triplet code of nitrogenous bases has been experimentally established, and the evidence for this was provided by the phenomenon of frame-shift mutation in the T_4 bacteriophage.

THE NONOVERLAPPING TRIPLET CODE

Toward the end of 1961, F. H. C. Crick and colleagues, working at the University of Cambridge in England, published results of experiments that show that, at least in the T_4 bacteriophage, the genetic code of DNA consists of a nonoverlapping sequence of bases taken in threes and that there is nothing analogous to gaps or "spacers" in the code. In their experiments they studied the effects of mutations in one of the cistrons of the rII region in the bacteriophage chromosome induced by the dye acridine orange. This has the effect of deleting or inserting a base pair in the DNA molecule, which is one type of mutational event. As described in Chapter Fourteen, mutations in the rII region of the bacteriophage chromosome are detected by variations in the form, size, or growth rate of the plaques formed in a "lawn" of *E. coli* cells (strain K) growing on a solid medium. Hence, insertions or deletions of DNA base pairs in the T_4 bacteriophage chromosome can be detected and analyzed in the same way as other mutations.

In recombination experiments with the mutant bacteriophage it was found that crossing over between two single insertion mutants (+ mutants) or two single deletion mutants (− mutants) always yielded mutant-type (nonfunctioning) recombinants, whereas a recombinant formed from single + and − mutants yielded a recombinant that was nearly normal. At times it was found that a recombinant carrying either three close insertions or three deletions was functional, but one carrying four close insertions or deletions was always nonfunctional.

Crick's essential findings were therefore that whereas one or two deletions or insertions in the cistron were revealed by characteristic changes in the plaques and represented malfunctioning, three deletions or insertions had no effect on normal functioning. In interpreting these results the reasoning was as follows. Let it be supposed that one of the DNA strands of the T_4 bacteriophage cistron has a base sequence as follows; furthermore, let us assume that the sequence "reads" from left to right in triplets starting at a marker (X). We therefore have:
X   AGCAGCAGCAGCAGCAGCAGC, which is read as:

AGC   AGC   AGC   AGC   AGC   AGC   AGC

If an extra base (C) is inserted, say after the A, which is the seventh letter from the left, we will now have the following sequence of triplets.

AGC   AGC   ACG   CAG   CAG   CAG   CAG   CAG

Notice that the effect of this insertion has been to alter all triplets to the right, but those to the left are unaffected. If a second base (C) is now inserted in this altered sequence, say after the twelfth letter (including the previously inserted C) from the left, we will now have this triplet sequence.

AGC   AGC   ACG   CAG   CCA   GCA   GCA   GCA

The effect of this second insertion is again to alter the meaning of triplets to the right of it. As in the first case, all of the triplets to the right of the first insertion are still altered, meaning that the protein for which this altered sequence codes would be defective; this manifests itself in malfunction and altered plaque morphology.

A similar result is found if deletions occur instead of insertions. Such mutations caused by single base deletions or insertions, which shift the reading frame one step to the left or right, respectively, are termed "frame-shift" mutations.

If a third insertion is now made between the first two (but not necessarily) say, after the tenth letter from the left, this has the following effect on the sequence:

AGC   AGC   ACG   CCA   GCC   AGC   AGC   AGC

Notice now that the triplets to the right of the localized region where the three insertions have occurred are now restored to normal. Three deletions would have a similar effect. One or two deletions or insertions thus alter all the triplet sequences to the right, which leads to a seriously defective protein and hence to malfunction; three close insertions would lead to a defect only within the relatively short stretch of the polypeptide chain delineated by the two insertions on either side of the middle one, and the result would be little or no malfunction. In the case of the recombinants formed from crossing over between a cistron with a deletion (which shifts the frame one base in one direction) and one with an insertion (which shifts the frame one base in the opposite direction), the frame shifting effects of the two mutations cancel each other out, and the triplet sequence reads correctly except for the segment between the insertion and the deletion. If the distance between them is quite short, the result is almost normal function.

Crick's results are therefore consistent with a genetic code made up of base triplets, and they also demonstrate that there is nothing analogous to gaps or spacers between them, since the reading of the code has to proceed uninterruptedly, beginning at one end of a cistron and going on to the other end. The experiment also shows that most of the triplets have a biological function. In the example where three insertions have occurred, abnormal triplets (ACG, CCA, GCC) are present in a consecutive sequence. If one or more of these did not code for any amino acid, then presumably there would be a gap in the assembly of the resultant polypeptide, which would break into two functionless fragments. Crick also obtained evidence for "markers" that ensure that "sentences" begin at the correct place.

Apparently, there is something in the nature of a molecular "full stop" because, as Crick's experiment also shows, the insertion or deletion of a single base pair resulted in inactivation of only one cistron at a time; something stops the "spreading" of the altered triplet effect to the right of the insertion or deletion into the next cistron. This was shown in a further experiment in which a different type of mutant discovered by Benzer was used. In this mutation the DNA in the region where the A and B cistrons of the T_4 bacteriophage join together has been lost, but the function of the B cistron is still normal. In the bacteriophage without this mutation, any mutation in the B cistron does not affect the A cistron, and vice versa. However, when this type of mutation is present, an acridine orange-induced base addition or deletion in one cistron spreads into the other and causes a malfunction. Hence it appears that the chromosomal segment deleted between the A and B cistrons contains some sort of marker whose function is to act as a full stop in the genetic code and to prevent spreading of the effects of acridine orange mutations into adjacent cistrons.

The Cambridge group therefore established that genetic information is contained in the polynucleotide structure of the DNA molecule in the form of linear sequences of nonoverlapping base triplets without spacers between them. However, experiments with mutations in the T_4 bacteriophage could not enable the code to be "broken," or deciphered (i.e., to determine what amino acids the triplets determine or code for in protein synthesis). In order to do this the process of protein biosynthesis must be understood. In order to appreciate how the genetic code was "cracked" we do not yet need to know all of the intimate details of the machinery of protein biosynthesis, but we do have to explain a seeming paradox, one that pertains to the "geography" of information storage and protein biosynthesis in both procaryotic and eucaryotic cells. Proteins are synthesized in the cell cytoplasm, and the small ribonucleoprotein assemblies called ribosomes are an essential component in this process. The information directing the proper sequence assemblage of amino acids in proteins is carried in the DNA molecules, which are virtually confined to the cell nucleus; the problem is how does the DNA exert its controlling influence on the distant protein "factories" in the cell cytoplasm?

TRANSCRIPTION OF THE GENETIC CODE. MESSENGER RNA SYNTHESIS

In some ways the cell's protein biosynthetic machinery and its control is organized like an automobile factory. Somewhere in the factory there will be a building containing the master blueprints and specifications of all the components manufactured in the factory (corresponding to the DNA in the cell nucleus). The components (amino acids) are assembled into finished cars (polypeptides) in other parts of the factory (cytoplasm) on special workbenches (ribosomes). The master plans and blueprints are too valuable to allow out into the factory areas, where they may be damaged or lost, but an unlimited number of working photocopies can be made and sent out to the workbenches; these copies are easily replaced if damaged or lost.

In the cell there is something analogous to working copies of the information stored in the DNA molecule that gets out into the cytoplasm and is the actual direct template on which amino acids are assembled into polypeptides. The working copies are the molecules of messenger RNA, so-called because they carry the genetic message from the DNA in the genes to the sites of protein biosynthesis in the cytoplasm.

As shown in Chapter Fifteen, RNA polynucleotides differ from DNA in the nature of the sugar component and in one of the nitrogenous bases; RNA contains ribose instead of deoxyribose and the base uracil in place of the closely similar thymine. Messenger RNA is a single polynucleotide strand synthesized on a single DNA strand that acts as a template. The double-stranded DNA molecule undergoes local unwinding, and the base sequences are exposed. One of these strands then acts as a template for the assembly of an RNA molecule.

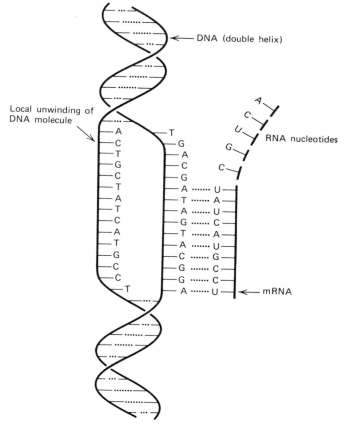

Figure 17-4 Messenger RNA synthesis (transcription). Local unwinding of the DNA double helix exposes one of the DNA strands that acts as a template for synthesis of mRNA. The RNA base sequence is determined by the base-pairing mechanism, and the RNA nucleotides are polymerized under the influence of RNA polymerase.

RNA nucleotides (ribonucleotides) "pair" with the complementary DNA bases and are joined together by an enzyme (RNA polymerase) into an RNA polynucleotide (Fig. 17-4). The sequence of base triplets on the messenger RNA (mRNA) molecule is the actual direct cellular template on which polypeptide chains are assembled. Notice that the base sequence is not strictly a copy of the DNA base sequence for a particular protein, but a transcription. The DNA base sequence is transcribed into a complementary base sequence in mRNA, in which uracil takes the place of thymine. The process of mRNA synthesis with base sequence complementary to that on the corresponding DNA is therefore called transcription. The synthesis of ribosomal RNA (rRNA) and transfer RNA (tRNA) (see later) are also processes of transcription.

Clearly, one of the two DNA polynucleotide strands is the genetic information store, because the other strand being a mirror image of it would not make biological sense. Somehow, it seems, the correct (coding) DNA strand is selected as the template for mRNA synthesis, but not the anticoding strand. The mRNA molecule finds its way through the nuclear pores into the cytoplasm, where it becomes associated with the ribosomes.

DECIPHERING THE GENETIC CODE

The protein-synthesizing system of living cells can be isolated and studied *in vitro*. M. W. Nirenberg and J. H. Matthaei showed that RNA fractions from various sources added to cell-free extracts of *E. coli* directed the synthesis of specific protein, admittedly in only minute amounts. In order to determine which mRNA triplets code for which amino acids, all that needs to be done is to add a mRNA of known base sequence to a cell-free extract capable of synthesizing proteins and then determine the amino acid sequence of the newly synthesized protein and correlate this with the base sequence of the mRNA. However, at the time that Nirenberg and Matthaei performed these experiments, the nucleotide sequences of naturally occurring mRNA molecules were not known. Instead, they used an artificially synthesized RNA consisting of one type of nucleotide only, the synthetic homopolymer of the uracil-containing nucleotide (uridylic acid), as an RNA template. This polymer (polyuridylic acid or poly-U) therefore had a repetitive sequence of UUU triplets. Around 1961, Nirenberg and Matthaei performed experiments in which poly-U and amino acids were incubated with an *in vitro*, protein-synthesizing system from *E. coli*, and the polypeptide recovered from the reaction mixture was polyphenylalanine (i.e., one consisting of phenylalanine residues only), thus clearly demonstrating that the mRNA triplet UUU coded for the amino acid phenylalanine.

Subsequently, Nirenberg and S. Ochoa showed that polyadenylic acid (poly-A) coded for lysine, and polycytidylic acid (poly-C) coded for proline. They next used synthetic random copolymers of different nucleotides (e.g., those containing the bases A and C). Such a poly-AC would contain the triplets AAA, AAC, CAA, ACC, CAC, CCA, CCC. The relative proportions of the different triplets would depend on the ratio of A and C in the copolymers. As well as the proline and lysine coded for respectively by CCC and AAA, the other amino acids incorporated in the resulting protein were asparagine, glutamine, threonine, and histidine. The relative proportions of the different amino acids in the proteins corresponded approximately with the predicted proportions of the different triplets in the synthetic RNA calculated on the basis of a random sequence of A and C. Hence, the base compositions of the corresponding triplets could be determined. However, this does not permit the base *sequence* of the triplet to be elucidated.

Experiments with other random copolymers enabled the base composition of further triplets coding for other amino acids to be worked out.

Synthetic mRNAs composed of two or more different nucleotides and with known repetitive sequences were then employed. Examples are repeating dinucleotides like UGUGUGUGUG. This contains two types of triplets with the sequences UGU and GUG, and so should code for two different amino acids. Similarly, the repeating trinucleotide AGC AGC AGC AGC contains the triplets AGC, GCA, and CAG. Depending on where the reading of the message begins, this should direct synthesis of three different homopolypeptides. This was found to be the case, and this provides further evidence that the code is read as a continuous sequence of nonoverlapping triplets. Using such methods, Khorana, Nirenberg, Ochoa, and others finally achieved the breaking of the genetic code, and the meanings of about 50 of the 64 possible triplets were discovered. The complete amino acid code word "dictionary" is shown in Table 17-1.

Synthetic RNAs of known base sequence have been likened to the Rosetta Stone in deciphering the genetic code.

FEATURES OF THE GENETIC CODE
(i) Degeneracy

Of the 64 possible triplets, 61 code for amino acids, and the remaining three are chain terminators. A glance at Table 17-1 shows that in most cases, more than one triplet codes for the same amino acid. The code is therefore said to be "degenerate." For example, two different triplets code for phenylalanine and as many as five code for serine. Only methionine and tryptophane are uniquely coded for by single triplets.

In many cases the degeneracy involves the third base (3' ends) of a mRNA coding triplet; for example, alanine is coded for by GCU, GCC, GCA, and GCG, and serine by UCU, UCC, UCA, and UCG. Apparently, specificity for these and some other amino acids lies in the first two bases. Degeneracy may

TABLE 17-1   **The Genetic Code (from J. D. Watson)**

First Position (5' end)	Second Position				Third Position (3' end)
	U	**C**	**A**	**G**	
U	Phe	Ser	Tyr	Cys	U
	Phe	Ser	Tyr	Cys	C
	Leu	Ser	(Term)[a]	(Term)[a]	A
	Leu	Ser	(Term)[a]	Trp	G
C	Leu	Pro	His	Arg	U
	Leu	Pro	His	Arg	C
	Leu	Pro	GluN	Arg	A
	Leu	Pro	GluN	Arg	G
A	Ileu	Thr	AspN	Ser	U
	Ileu	Thr	AspN	Ser	C
	Ileu	Thr	Lys	Arg	A
	Met	Thr	Lys	Arg	G
G	Val	Ala	Asp	Gly	U
	Val	Ala	Asp	Gly	C
	Val	Ala	Glu	Gly	A
	Val	Ala	Glu	Gly	G

[a] (Term) indicates a chain-terminating codon. These were formerly known as nonsense codons.

also reside in positions other than the third; for example, leucine is apparently coded for by UUG and UUA in addition to CUU, CUC, CUA, and CUG. In most cases, the triplets coding for a particular amino acid differ from each other in one base.

A hypothesis was advanced by F. H. C. Crick to explain the rather prevalent third base degeneracy in the genetic code as well as some other related phenomena, but this can be more appropriately discussed in Chapter Eighteen, which deals with protein biosynthesis.

Because the first two bases seem to be most important for determining a given amino acid, a suggestion has been made that the primitive genetic code may have coded for 15 "primordial" amino acids in doublets.

(ii) Properties of amino acids and their coding triplets

Close study of the triplets and the amino acids for which they code reveal interesting correlations (but probably *not* a cause-effect relationship) between the triplet base sequence and the physicochemical characteristics of the corresponding amino acid. It is the middle base in the triplet that seems to determine this. For example: (1) if the middle base is U, the amino acid coded for is nonpolar (i.e., having the type of side chain that is compatible with a hydrophobic environment), as is found "inside" globular protein molecules; (2) when the middle base is C, the corresponding amino acid is either nonpolar or has a nonionized polar side chain group that is compatible with a hydrophobic environment; (3) those triplets with A or G as the middle base code for amino acids that are found on the "outside" of globular proteins; (4) of the triplets coding for amino acids with ionized hydrophilic side chains, those with basic groups are coded for with triplets beginning with A or C followed by A or G; the third position is occupied by any of the four bases; and (5) amino acids with acidic hydrophilic side chain groups are coded for by triplets beginning with GA, the third position being any base.

Dickerson and Geis point out that these patterns have arisen as a result of natural selection and that the genetic code has the features of a fail-safe system, because the effects of a single mutation resulting in a single base change are minimized; for example, a change in the first or third base in any of the triplets coding for hydrophobic amino acids will result in a triplet coding for another hydrophobic amino acid that will be more or less an acceptable substitute, so there will be a good chance that the structure and function of the resulting protein will not be seriously affected. Again, if the third base of one of the six different triplets coding for arginine mutates, then the altered triplet will most likely still code for arginine; alterations in the first two bases of arginine-coding triplets will probably result in triplets coding either for arginine or for the other basic amino acids, lysine and histidine.

The genetic code therefore has the attributes of a self-correcting system whose design tends to minimize the possible disruptive effects of single base mutations on protein structure. In an organism with a random code the majority of mutations would result in altered triplets coding for amino acids physicochemically greatly different from the normal ones. The degeneracy of the genetic code therefore has survival value, since it stabilizes the genetic apparatus against mutation.

(iii) Presence of uncoded amino acids in proteins

In the chapter on structural proteins, the presence of amino acids uncoded for by mRNA triplets in certain structural proteins will be noted. These amino acids are hydroxyproline and hydrolysine in collagen, and desmosine and isodesmosine in elastin. In every case in which an uncoded amino acid has been found in proteins, it originates from a coded amino acid that is first incorporated in the polypeptide and then afterward is chemically changed. For example, hydroxyproline and hydroxylysine result from hydroxylation of proline and lysine, respectively. Desmosine also derives from lysine.

(iv) Colinearity of polypeptide chains and genes

If it were possible to demonstrate an exact correspondence between the linear sequence of coding triplets in a given mRNA molecule and the sequence of amino acids in the corresponding protein, this would go a long way toward proving the reality of the genetic code. The experiments of Crick and his colleagues on the frame-shift mutations in T_4 bacteriophage support the idea that there is a correspondence between the linear sequence of triplets in DNA and the amino acid sequence in a protein, but more direct evidence was provided by the experiments of C. Yanofsky and his colleagues. These workers investigated the effects of mutations on the structure of the enzyme tryptophane synthetase from E. coli. There are two identical "A" polypeptide chains and two identical "B" chains in this enzyme, and the mutational changes studied were in the A chain, whose wild-type primary structure has been determined. The changed A polypeptides from a number of E. coli mutants affecting the A chain were isolated, and the altered amino acid residues were identified with the help of the peptide mapping (fingerprinting) technique (see Chapter Nine). Genetic mapping enabled the linear sequence of mutations on the map of the A-chain gene to be compared with the positions occupied by the altered amino acids in the mutant A polypeptides. In a study of over 20 different mutants, the colinearity of the A-chain gene and the A polypeptide was definitely established.

The same was found by Brenner and his colleagues, who studied abnormal proteins of the T_4 bacteriophage "head" that resulted when the gene responsible underwent mutation to give rise to meaningless (nonsense) triplets that did not code for any amino acid. Such mutations lead to premature termination of polypeptide chain synthesis at various sites. The different lengths of the resulting polypeptides were perfectly correlated with the position at which the triplets were mutated.

All such findings including those that indicate the absence of spacers in the genetic code sentences and that the message is read in one direction only (see Chapter Eighteen) are all weighty evidence for the colinearity of genes and the polypeptides whose structures they determine.

(v) Universality

Although the genetic code was deciphered in experiments in cell-free, protein-synthesizing systems in E. coli, all evidence obtained so far strongly suggests that the code is universal. Experiments similar to those of Nirenberg have been carried out in protein-synthesizing systems from a wide variety of organisms, with results similar to those obtained by him; for example, the RNA from tobacco mosaic virus functions as a messenger for directing synthesis of the virus coat protein in cell-free extracts of *E. coli*. A striking demonstration of the universality of the code came from experiments performed by C. Merril, M. Geis, and J. Petricciani in 1971, in which they showed that a bacterial gene directed mRNA and protein synthesis in human cells. The human cells came from a patient with galactosemia, a genetically determined condition in which galactose cannot be metabolized. The cells used lacked a transferase enzyme that is necessary in a series of reactions in which galactose is chemically altered and isomerized to glucose. In these experiments a bacteriophage carrying an *E. coli* gene that accurately coded for the transferase was ingested by the cultured human cells; subsequently, the transferase gene directed the synthesis of transferase in the cells, which soon after were able to metabolize galactose in the normal way. This experiment therefore is strong evidence in favor of universality of the genetic code insofar as it shows that a bacterial gene can be correctly "interpreted" by the protein-synthesizing system of a human cell, thus showing that both types of cell possess a common genetic code.

Of the many forms that the evidence for universality takes, perhaps the most convincing comes from *in vitro* experiments in which it has been shown that bacterial protein-synthesizing systems will correctly translate mRNA from a variety of sources as diverse as viruses and mammals, into polypeptide structure characteristic of the species from which the mRNA was obtained.

In all living organisms so far investigated, the codes for the amino acids, polypeptide chain initiation and termination have all been found to be the same.

Apart from such experiments, perhaps the most compelling a priori reason for universality of the code is the difficulty of picturing how it could be altered, once it had evolved, without drastic consequences. If, say, the triplet UUU changed from coding for phenylalanine to coding for lysine, then the primary structure of virtually all the proteins in an organism's body would be changed, because these two amino acids are found in virtually all of the proteins in an organism.

CHAPTER EIGHTEEN

Translation of genetic information into protein structure

In the previous chapter the role of messenger RNA in directing protein synthesis in cell-free extracts and the elucidation of the genetic code were described. Apart from mRNA and various enzymes, two other cellular components are essential for the final translation, as the process is called, of the base triplet sequence of mRNA into the amino acid sequence of a polypeptide chain. These are the many different forms of transfer RNA molecules and the cellular nucleoprotein assemblies called ribosomes.

TRANSFER RNA

Amino acids differ from one another in the structure of their side chains, and it might be thought that these are in some way complementary in shape to the three-dimensional configurations of the base triplets on mRNA. Amino acids could be thought of as lining up directly on the mRNA template by the mechanism of complementary fit between their side chains and the mRNA triplets, prior to being enzymatically joined together by peptide bonds. However, although mRNA is certainly the cellular template, the amino acids do not polymerize directly on the mRNA; they do so through intermediary "adapter molecules," which are a second type of RNA called transfer RNA (tRNA).

The name transfer RNA pertains to its function of transferring amino acids to the mRNA template.

Molecules of tRNA are single-stranded, relatively small molecules (about 70 to 80 nucleotides) and, in places, "hairpin" bending occurs, so that parallel complementary segments of the single strand hydrogen bond with themselves and give local double helical regions that alternate with nonhelical "looped out" regions. Therefore, tRNA molecules have a cloverleaf configuration. There are three or more such loops in a tRNA molecule. The terminal ribose at the 3' end of the molecule is the attachment site for an amino acid and, in all tRNA molecules it has the base sequence CCA. Specificity for the amino acid appears to reside in the base sequence next to this group and also in the sequence at the other terminal of the single strand. The specificity of tRNAs is marked but not absolute; for example, fluorophenylalanine, if present in excess, can replace phenylalanine.

The middle loop of the tRNA molecule is formed from seven unpaired bases and carries an exposed base triplet that is designed to fit a complementary base triplet on a mRNA molecule; it is therefore called the anticodon. The anticodon is situated at the opposite end of the tRNA molecule from the amino acid attachment site. The 5' terminal base of all tRNA molecules is guanine.

For every amino acid there is more than one specific tRNA to which it becomes attached by an enzymatic mechanism, and each tRNA will possess the coding triplet specific to that amino acid. There are therefore more than 20 different tRNAs, each with specific binding affinity for an amino acid and with different coding triplets. The pattern of folding of the nucleotide chain assures that in all tRNAs so far studied, there is a constant distance between the coding triplet and the attached amino acid. The three-dimensional configuration of the different tRNA molecules appears to be similar, because mixtures of all of them give regularly formed crystals.

The general structure of a tRNA molecule and the functions of its parts are shown in Fig. 18-1.

The tRNA molecule functions to transfer amino acids to the mRNA template and to align them in correct sequence by fitting the coding triplet to the appropriate complementary triplet on the mRNA that codes for the amino acid attached to the tRNA. The nucleotide sequences of tRNAs are determined by transcription from a DNA template, but are not determined directly. A larger precursor RNA polynucleotide is first transcribed from DNA. The precursor is then successively degraded into smaller fragments by endo- and exonucleases; one of the final fragments becomes the mature tRNA molecule after chemical modification of a few bases. tRNAs contain unusual bases in addition to the usual A, U, G, and C, many of which are formed from A, U, G, or C by insertion of one or more methyl groups. Their function is still obscure.

The first determination of the nucleotide sequence of a tRNA was reported by R. W. Holley at the end of 1964. The source of this tRNA was yeast, and it binds

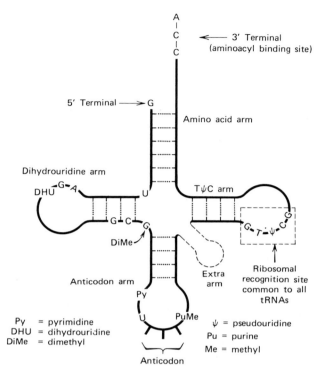

Figure 18-1   Structure of a transfer RNA molecule showing features common to all tRNA molecules.

specifically with alanine and so is called alanine-tRNA. It has 77 nucleotides and has an unusual number of the methylated bases. The primary nucleotide sequences of many tRNAs from a variety of organisms are now known.

At one time it was thought that specific tRNAs must exist for "recognition" of each one of the 61 mRNA coding triplets that determine the amino acids. Later studies of purified tRNAs in which the sequences were known showed that a given tRNA could recognize several different mRNA coding triplets. Also, many instances were found of tRNA molecules in which a base other than the four usual ones occurred in the anticodon triplet. This base was inosine, which is formed by deamination of the 6-carbon atom of adenine.

In order to explain these observations and also the third base degeneracy of the genetic code (described in Chapter Seventeen), F. H. C. Crick put forward what he called the "wobble" hypothesis, which is now supported by much evidence. This hypothesis states that whereas the hydrogen bonding between the first two bases of a mRNA coding triplet and the pair of complementary bases in the tRNA anticodon is rather rigid and specific, the third base (5′ end) of the tRNA anticodon is able to wobble and can form hydrogen bonds with bases on the mRNA coding triplet (3′ end) other than its usual partner. For example, the base U in the so-called wobble position in tRNA can form hydrogen bonds with A or G, while inosine (I) in the wobble position can pair with U, C, and A. The hydrogen bonding possibilities of bases in the wobble position of tRNA are summarized as follows.

Base in wobble position in anticodon	Base in mRNA coding triplet
I	A, U, C
U	A, G
A	U
C	G
G	U, C

One of the pieces of evidence that support the wobble concept is that a prediction made from it turned out to be correct. This was that there should be at least three tRNAs for the six mRNA triplets that code for serine (UCU, UCC, UCA, UCG, AGU, AGC).

THE RIBOSOME

Ribosomes are minute (100 to 250 Å diameter, depending on the source) cellular organelles that are essential components of the protein biosynthetic machinery of cells. They are complex protein–nucleic acid assemblies and contain virtually no lipid. The most studied are the ribosomes of E. coli. The ribosomes of E. coli are suspended in the protoplasm, whereas in eucarotic cells they may be free in the cytoplasmic matrix or they may be attached to the membranes of the endoplasmic reticulum. In cells actively synthesizing proteins, ribosomes are especially abundant. In cells manufacturing protein that is to be secreted (e.g., pancreatic enzymes), the ribosomes are mostly attached to the endoplasmic reticulum membranes. In a fast-growing embryonic cell, on the other hand, most of the ribosomes are free. Liver cells are intermediate in type; about three-quarters of the ribosomes are attached to endoplasmic reticulum membranes, and the rest are free.

Ribosomes consist of two subunits that can be reversibly dissociated. Lowering of the magnesium ion concentration causes dissociation, and restoring it to normal again brings about reassociation. A further increase of magnesium ion concentration causes intact ribosomes to associate into dimers (Figs. 18-2 and 18-3). The larger subunit is about double the size of the smaller one and is called

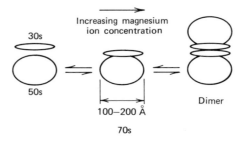

Figure 18-2   Association and dissociation of ribosome components.

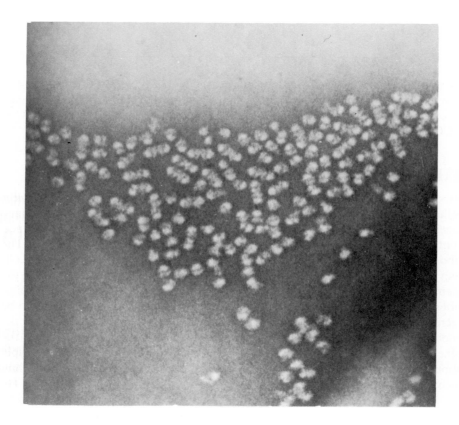

Figure 18-3   Electron micrograph of single 70S and dimeric *E. coli* ribosomes (negative contrast, ×200,000). (Courtesy of Dr H. E. Huxley. *J. Mol. Biol.*, *2*: 10–18, 1960.)

the 50S (50 Svedbergs) subunit; this designation indicates the rate (sedimentation constant) at which it sediments in the ultracentrifuge. The smaller is called the 30S subunit, and the complete ribosome is a 70S structure. Electron micrographs and models showing the detailed external appearance of *E. coli* ribosomes are shown in Fig. 18-4.

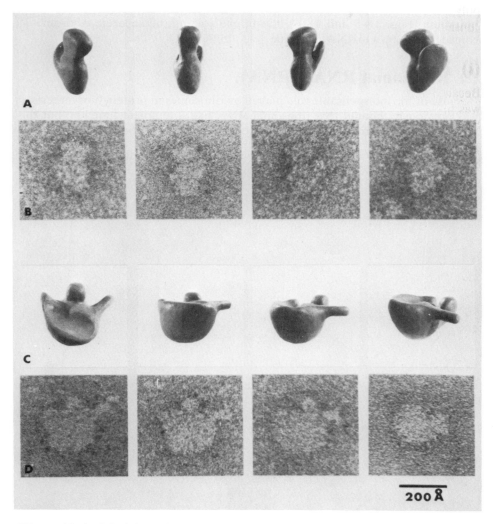

Figure 18-4   Models and electron micrographs of 30S and 50S *E. coli* ribosome subunits viewed in similar orientations from different angles. (Courtesy of Dr J. A. Lake. *J. Mol. Biol.*, *105*: 131–159, 1976.)

The ribosomes of *E. coli* and other procaryotic organisms are smaller than those of eucaryotes, whose sedimentation constants are larger (80S). Both subunits contain protein and RNA. The ratio of RNA to protein varies from about 2:1 in *E. coli* to 2:3 in other organisms.

The electron microscope reveals that ribosomes are usually found organized into structures consisting of many ribosomes joined together by a delicate thread attached to their smaller subunits. This structure is called a polyribosome or polysome (Figs. 18-5 and 18-6), the thread joining the separate ribosomes is considered to be a mRNA molecule.

(i) Ribosomal RNA (rRNA).

Because of the indispensable role played by ribosomes in protein biosynthesis, it was previously thought that ribosomal RNA had a template function, but this has since been disproved; the RNA in ribosomes is now believed to interact with

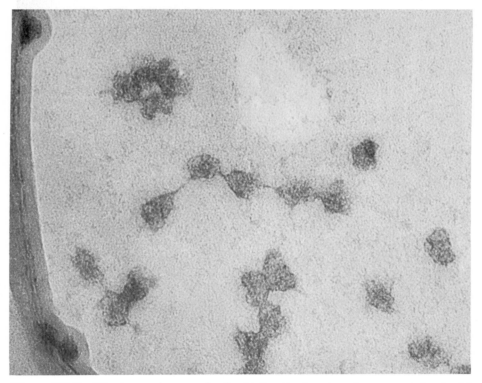

Figure 18-5   Electron micrograph of rabbit reticulocyte polysomes stained with uranyl acetate (×280,000). (Courtesy of Dr H. Slayter. *J. Mol. Biol.*, 7: 652–657, 1963.)

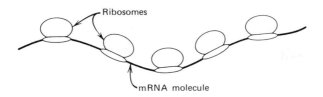

Figure 18-6 A polyribosome.

the ribosomal proteins in such a way as to maintain the three-dimensional structure of the ribosome.

rRNA, like mRNA, has been shown to be synthesized on a DNA template. This was proved by the formation of specific DNA and RNA hybrid molecules. When native DNA is heated, hydrogen bonds are broken and the two complementary strands separate. If the temperature is slowly lowered, the two strands spontaneously unite again by the complementary base-pairing mechanism to regenerate the original DNA. Certain DNA strands will thus specifically recombine with rRNA from the same organism to give hybrid rRNA-DNA helical molecules, but this will not happen with DNA from another species. The conclusion, therefore, is that the rRNA as well as mRNA is formed by transcription from a DNA template. (Note, therefore, that since DNA codes for both tRNA and rRNA, genes do not always determine specific amino acid sequences in proteins.)

Although many different proteins are found in ribosomes, there are only three kinds of RNA, and these are easily separable *in vitro*. Within the 30S subunit of bacterial ribosomes is one type of RNA (16S), and the 50S subunit contains a 23S and a 5S species of RNA. One copy of each RNA type is contained within the ribosome. Much of the nucleotide sequence of both the 5S and 16S RNA is known. The sequence data support the earlier belief that there are double-stranded regions within the single-stranded rRNA molecules that result from complementary base pairing. This arises from the molecule bending back on itself, which permits interaction of close parallel stretches of the molecule.

The 16S and 23S rRNA molecules are not separately transcribed, but are derived from nuclease-catalyzed degradation of a single, large, precursor 30S transcription ("pre-rRNA"). The initial breakdown products of the pre-rRNA are 17S and 25S fragments that are further degraded to the final mature 16S and 23S rRNAs.

(ii) Ribosomal proteins

As many as 50 or even 60 different proteins have been identified in the bacterial ribosome, of which 21 are associated with the 30S subunit and 34 with the 50S subunit.

The primary structures of several of these proteins have been determined. Each subunit has its own characteristic proteins; for example, C. G. Kurland and his colleagues, who purified 20 proteins from the 30S subunit, found that they differed with regard to molecular weight, amino acid composition, and the nature of the trypsin digestion products. The same investigators found about 30 proteins in the 50S subunit. Of the ribosomal proteins, most are basic, but several are neutral or acidic. A comparison of the peptides obtained by trypsin digestion of ribosomal proteins indicates that they are not homologous despite similarities of amino acid composition. Antibodies to single ribosomal proteins do not cross-react with antibodies to any other of the ribosomal proteins, and this further indicates that each ribosomal protein has a structure different from all the others.

Ribosomes derived from a single source have been found to be heterogeneous, and the evidence indicates that this is not artifactual. In the 30S subunit of *E. coli* ribosomes, for example, the proteins fall into three classes, according to the size of the "copies" of each of the individual proteins that are present: (1) *unit proteins*, which occur in amounts of between 0.8 and 1.2 copies per ribosome; (2) *fractional proteins*, which are present to the extent of less than 0.65 copies per ribosome, and (3) *marginal proteins*, which are present to the extent of 0.65 to 0.80 copies per ribosome.

All the available data suggest that no ribosomal protein occurs more than once, with the single exception of one of the 50S subunit proteins that occurs in acetylated and nonacetylated forms and of which three copies appear to exist in the 50S subunit. This indicates that both structurally and functionally, the ribosome is dissymmetric. The existence of many identical sites on the ribosome or the rearrangement of ribosomal subunits into many functionally similar configurations can therefore be excluded from speculations about the role of the ribosome in protein synthesis. This dissymmetry of the ribosome contrasts sharply with the great structural symmetry of the noncellular nucleoprotein assemblies called viruses.

Whether or not the surprising complexity of the 30S subunit is related to function has not yet been established.

(iii) Functions of ribosomal proteins

M. Meselson, M. Nomura, and S. Brenner found that centrifugation of ribosomes in a caesium chloride gradient causes them to separate into two fractions. One of these—"core particles"—was shown to be ribonucleoprotein particles left over after the stripping away from them of the other fraction, known as "split proteins." Incubation of core particles together with split proteins resulted in rapid spontaneous reassembly of functional 30S and 50S subunits. Two types of split protein were chromatographically isolated by P. Traub and M.

Nomura, an acidic fraction (A) and a basic fraction (B), and they were shown to be functionally different in *in vitro* experiments.

1. 30S cores plus 30S split B protein are active in protein synthesis.
2. 30S cores plus 30S split A protein are inactive.
3. 30S cores plus 30S split B protein plus 30S split A protein give greater activity than 30S split B protein alone.

(As indicated in the previous chapter, ribosome function can be studied in cell-free systems to which other necessary components, including mRNA amino acids and enzymes, have been added.)

Hence it was concluded that while the split A protein of the 30S subunit is not essential for protein synthesis, it performs a stimulating role. A similar situation was found in the case of the 50S subunit, except that the split B protein is the nonessential "stimulatory" protein and the split A protein is essential.

Nineteen of the 30S proteins were purified by Nomura and his co-workers. Mixtures of these proteins and ribosomal RNA lacking one of these individual proteins were prepared by them, and the effect of omitting each protein on both spontaneous reconstitution and ribosomal function was observed. Their results show that ribosomal proteins form a hierarchy of three functional types. The first type is necessary for both assembly of the 30S subunit and for recovery of function. The second type is not necessary for 30S assembly, but is required for function. The third group of proteins are not necessary for function or assembly, but have a stimulatory effect on the activity of reconstituted ribosomes; these are the so-called "dispensable" proteins.

Earlier in a similar investigation of the effects of omitting individual 30S and split B proteins from reconstitution mixtures, P. Traub and his co-workers found that several ribosomal functions were simultaneously and coordinately depressed when any one essential protein was missing. Thus, it was difficult to assign a specific function to any protein. The effects of omitting individual proteins are therefore quantitative instead of qualitative.

Such findings are valuable indications of ribosome structure and provide, for the first time, an insight into the process of assembly and constructional principles of a complex cell organelle.

(iv) Protein-rRNA interactions in ribosomes

Traub's and Nomura's experiments show that precise interactions between ribosomal proteins and rRNAs control the ribosomal assembly process and presumably also control ribosomal function. The results of certain experiments suggest that ribosomal proteins bind to helical regions of rRNA, and various investigators have been able to determine that each of the 30S and 50S rRNAs bind with specific ribosomal proteins. What nucleotide sequences are involved in protein binding and what are the complementary amino acid sequences on the bound protein are still obscure. Experiments with mutated ribosomal proteins with changed affinities for the binding sites on the ribosomes might help to answer these questions.

(v) Three-dimensional arrangement of ribosomal components

Ribosomes are too large for successful application of X-ray crystallographic techniques to structure determination. Experiments by G. Stöffler and his co-workers of the effect on ribosomal function of antibodies to specific ribosomal proteins have so far provided the most definite information regarding the three-dimensional organization of ribosomal components. Their observations show that some at least of the antigenic sites of every 30S protein are so situated that reaction with antibodies is possible. There are indications that this is true also of many, at least, of the 50S proteins. Other evidence indicates that all of the ribosome protein molecules appear to be partially embedded in the ribosome and partly exposed at the surface.

The RNA molecule of the 30S subunit appears not to be confined to the center of the subunit, and most likely it functions to hold together the proteins that are disposed as a monomolecular layer. The situation in the 50S subunit is probably more complicated.

The topographical distribution of the individual proteins within the ribosome have been explored by treating ribosomes with reagents that cross-link protein molecules. Complexes that consist of two different ribosomal proteins with bifunctional reagents have been isolated and characterized in such experiments and show that the two cross-linked proteins must be spatially close together within the ribosome. Such methods should eventually enable topographical regions of different "neighbor" proteins within the ribosome to be identified.

The technique of neutron scattering has been employed by D. M. Engelman and P. B. Moore in studying the spatial arrangement of the proteins in the 30S subunit of *E. coli* ribosomes. At the time of this writing, they have succeeded in determining the relative positions and distances apart of five of these proteins.

The method depends on the fact that neutrons are scattered by atomic nuclei, but not by electrons. The different isotopes of an element differ only negligibly in their chemical properties, but the neutron-scattering properties of the nuclei differ considerably. They are said to have different neutron-scattering lengths. One of the biggest differences in neutron-scattering length is that between hydrogen and its heavy isotope deuterium.

Traub and Nomura's discovery in 1968 of the spontaneous reconstitution of the 30S subunit components suggested to Engelman and Moore that they should be able to reassemble a 30S subunit in which some of the proteins were deuterium labeled and hence that their spatial disposition within the subunit might be determined by neutron scattering. *E. coli* cells were grown in a medium in which nondeuterated nutrients were dissolved in deuterium oxide instead of water. The ribosomal proteins of the progeny bacteria therefore contained deuterium instead of hydrogen. The deuterated ribosomes were harvested, and the 30S proteins were isolated and separated.

In one of their experiments, Engelman and Moore first removed two of the 30S proteins (S5 and S8) which were separated from the remaining 19 30S proteins of ordinary hydrogenated ribosomes prepared in a separate experiment. To the mixture of the 19 normal proteins were added the deuterated S5 and S8 proteins plus the normal purified RNA from the normal 30S subunit. The resultant reconstituted ribosomes were found to be morphologically and functionally indistinguishable from normal ribosomes.

Next, they prepared two ribosome solutions of equal concentration. In one, half of the ribosomes contained the two deuterated proteins; the other ribosomes were normal. In the other solution, half of the ribosomes contained one only of the deuterated proteins, and the other ribosomes contained the other deuterated protein. Using beams of slow neutrons of relatively low energy, the neutron-scattering properties of the two solutions were studied and compared and found to be different. The difference was due to the geometric arrangement of the two deuterated proteins in the solution of doubly deuterated ribosomes. Application of appropriate computer-aided mathematical analysis enabled them to determine that the distance between the centers of the S5 and S8 proteins in the 30S subunit is 35 Å.

In another experiment the distance between the S4 and S8 proteins was found to be 63 Å. Measuring the distance between pairs of these three proteins in different combinations thus defines a triangle, of which the lengths of the sides are known and at the apices of which are situated the centers of the protein molecules.

Distance measurements between these three proteins and of a fourth protein define a tetrahedron, but the position of the fourth with respect to the other three is ambiguous. It could lie on one or other of the sides of the triangle formed by the other three, so that the two possible tetrahedra would therefore be mirror images of each other. Engelman and Moore consider that this uncertainty regarding the "handedness" of the structure will probably remain, even when the entire spatial arrangement of the proteins in the 30S ribosome of *E. coli* has been elucidated.

(vi) Biogenesis of ribosomes

Early cytochemical studies by Caspersson and by Brachet revealed that in eucaryotic cells actively synthesizing protein, the nucleoli were large and rich in

RNA. Later, when the important role played by ribosomes in protein biosynthesis was realized, a connection between them and the cell nucleolus was suspected. Then the key role played by the nucleolus in the biogenesis of ribosomes was demonstrated. That nucleolar DNA is responsible for coding for rRNA synthesis was shown in DNA-RNA hybridization experiments. There is much duplication of the rRNA DNA (gene duplication or redundancy) in the nucleolar organizer. Initially, a large precursor rRNA molecule is transcribed from the ribosomal DNA, and this becomes enzymatically cleaved sequentially into smaller fragments. The definitive "mature" rRNAs are the end products, except for the 5S rRNA, which is synthesized outside the nucleus and has no relation to the nucleolar organizer. rRNA synthesis is coordinated with extranuclear biosynthesis of ribosomal proteins. Processing of ribosomal subunits occurs mostly in the nucleolus, and they are then released into the cytoplasm.

In procaryotic cells, ribosome biogenesis is simpler. In the bacterial chromosome there is a tight cluster of genes that code for the 5S, 23S, and 16S rRNA, and there are only a few copies of them. These genes are transcribed as a unit, and the RNA is detached immediately from the DNA, probably as the mature rRNA.

CELLULAR PROTEIN BIOSYNTHESIS (Translation)

The hydrolysis of polypeptides is a spontaneous exergonic reaction; the reverse process—amino acid polymerization with peptide bond formation—is not spontaneous and requires energy input.

(i) Amino acid activation

Before amino acids will react, they must first be "activated." This is accomplished by the chemical energy supplied by the high-energy bonds in ATP. Amino acids react with ATP under the catalytic influence of an amino acyl synthetase enzyme. For every amino acid there is a specific amino acyl synthetase. The amino acid thereby becomes attached to AMP by an energy-rich bond, which is a mixed anhydride bond formed from the phosphoric acid group joined to the 5'-C of AMP and the carboxyl group of the amino acid.

$$\text{Amino acid} + \text{ATP} \underset{\longrightarrow}{\overset{E}{\rightleftharpoons}} \text{Amino acid-AMP} + \text{P-P}$$

(E = amino acyl synthetase)

Amino acids thus activated will now react spontaneously to form peptide bonds. The same enzyme then catalyzes transfer of the amino acid from the AMP group to its specific tRNA molecule.

$$\text{Amino acid-AMP} + \text{tRNA} \xrightleftharpoons{\text{E}} \text{Amino acid-tRNA} + \text{AMP}$$

(E = amino acyl synthetase)

The amino acid becomes attached to the tRNA molecule by the 2′- or 3′-hydroxyl group of the free terminal ribose at the 3′-end of the molecule by esterification of the amino acid carboxyl group.

The tRNA molecule with its attached amino acid now becomes attached to the ribosomes.

(ii) The role of the ribosomes

The function of the ribosomes is to hold the mRNA template and the tRNA-amino acid in proper mutual orientation with the growing polypeptide chain so that peptide bond formation and correct "reading" of the genetic code of mRNA can occur. Each 70S ribosome possesses two binding sites, or cavities; each one is partly bounded by the 30S and partly by the 50S subunit. One of these is called the amino-acyl or acceptor (A) site and the other is the peptidyl or donor (P) site. The mRNA template is bound to the 30S subunit. The growing polypeptide chain is attached to the ribosomal site by an ester linkage formed from the carboxyl group of the C-terminal amino acid and the 2′- or 3′-OH of the 3-terminal ribose of the tRNA molecule specific for the C-terminal amino acid occupying the P site. The anticodon triplet of this tRNA molecule is hydrogen-bonded to the complementary triplet on the mRNA.

There are three main steps in the attachment of successive amino acids to the growing polypeptide chain (see Fig. 18-7).

1. A tRNA molecule attached to its specific amino acid "recognizes" its complementary mRNA triplet (or codon, as it is also called), this codon being situated alongside the A site and "fitting" onto it by the hydrogen bonding, base pairing mechanism (Fig. 18-7a). The amino acid is now correctly oriented for the next step.

2. The ester bond formed from the carboxyl group of the C-terminal amino acid and the ⁻OH of its tRNA's 3′-terminal ribose is broken, and a peptide bond is now formed from this C-terminal carboxyl and the amino group of the amino acid to be added next, and this is catalyzed by peptidyl transferase. Simultaneously, the "freed" tRNA occupying the P site leaves the ribosome (Fig. 18-7b).

3. The newly attached amino acid and its transfer RNA move into the P site, a process called translocation. At the same time, the mRNA molecule moves in the same direction, and the next codon is positioned alongside the A site, ready for addition of the next amino acid (Fig. 18-7c).

This process is repeated over and over again until the polypeptide chain is complete. The sequence of specific codons is thus "translated" into a cor-

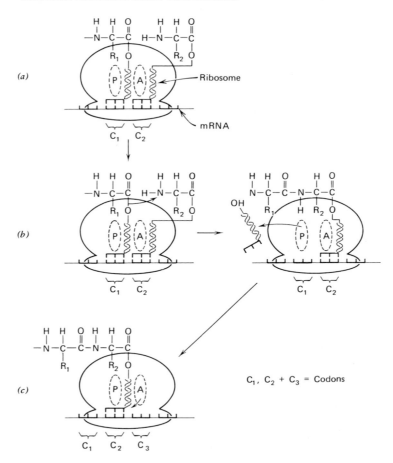

Figure 18-7   Polypeptide chain synthesis on ribosomes.

responding sequence of amino acid residues in a polypeptide. Note that amino acid residues are added to the C-terminal end of the growing chain so that the direction of polypeptide synthesis is from the N-terminal end to the C-terminal end. The polypeptide chain folds into its characteristic pattern of secondary and tertiary coiling as it is being synthesized on the ribosome.

(iii) Direction of reading of the genetic message

In describing the nature of the genetic code of mRNA in the previous chapter, we arbitrarily assumed for purposes of discussion that the genetic "message" begins at a starting point at the left end of the mRNA molecule and is "read" from left to right. Terms such as "left" and "right," of course, really have no meaning when applied to a mRNA molecule; however, the ends of an mRNA molecule differ and are distinguishable, so the idea of direction along the molecule does make sense. In Chapter Fifteen we noted that in both RNA and DNA, phosphate groups link ribose units through the 3' carbon atom of one sugar group to the 5' carbon of another, forming 3',5'-phosphodiester linkages. At the ends of a polynucleotide molecule, therefore, there will be a terminal ribose unit (in which the hydroxyl group of the 3' carbon atom is not involved in ester linkage) and at the other end the terminal ribose will similarly have a "free" 5' carbon atom. The polynucleotide molecule is therefore said to have a "3'-end" and a "5'-end." How, then, is the genetic message of mRNA read— from the 3' to the 5' end, or vice versa?

This question was answered by R. E. Thach and his co-workers using a synthetic hexanucleotide having the structure:

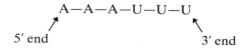

as a template in a cell-free, protein-synthesizing system. The resultant dipeptide had the structure:

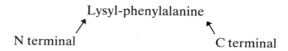

The dipeptide:

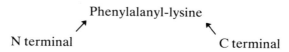

was not synthesized in any experiment. As already mentioned, synthesis of polypeptides begins with the N-terminal amino acid. Hence, this experiment clearly demonstrates that reading of the codons of mRNA begins at the 5' terminus and ends at the 3', which is the same direction in which mRNA synthesis occurs on the DNA template.

(iv) Polypeptide chain initiation and termination

During active cellular protein synthesis, free ribosomes in the cytoplasm are dissociated into their 30S and 50S subunits and reassociate to 70S ribosomes only when complexed with mRNA and actually involved in the act of translation. They dissociate again when released from the mRNA upon completion of a polypeptide.

At the commencement of synthesis of a polypeptide chain, the first event is the attachment of a 30S subunit to the first codon of the mRNA molecule to form the initiation complex. Binding of the first aminoacyl tRNA to the 30S subunit is followed by attachment of a 50S subunit to the 30S to give the functional 70S ribosome. The first (N-terminal) amino acid of a polypeptide chain, of course, has no C-terminal amino acid with which to form a peptide bond when it becomes attached with its tRNA to the P site on the ribosome. In bacteria, whatever protein is being synthesized, the first amino acid attached to the 30S subunit is methionine, in which the amino group is formylated (i.e., a —CHO group is attached, giving formyl methionine). The formyl group is added enzymatically after the methionine attaches to its tRNA. Formyl methionine tRNA is coded by the triplet AUG, which should therefore be at the beginning of every genetic message.

The formation of the initiation complex, the binding of the formyl-methionine tRNA and 50S subunit, and the dissociation of the 70S ribosome all require the presence of three proteins, called "initiation factors," plus GTP. They are released upon dissociation of the 70S ribosome and can be reused repeatedly. The blocked —NH$_2$ group of methionine makes it possible for the formyl-methionine tRNA to be bound at the P site; free methionine tRNA or any other free aminoacyl tRNA molecules cannot be so bound. The 30S subunit probably has a site that specifically recognizes formyl-methionine tRNA. The next aminoacyl tRNA attaches to the A site, and its free amino group forms a peptide bond with the —COOH group of the formyl methionine, which is left free upon release of the formyl Met-tRNA. The newly added aminoacyl tRNA is translocated to the now vacant P site and is ready to bond with the next incoming aminoacyl tRNA, which binds to the vacated A site.

The blocked —NH$_2$ group of the formyl methionine therefore compels peptide bond formation from the —NH$_2$ group of the next aminoacyl tRNA and the free —COOH group of the formyl methionine.

Soon after the beginning of polypeptide chain synthesis, the formyl group of the N-terminal formyl methionine is enzymatically removed. In a few polypeptides, the resulting N-terminal methionine residue is also subsequently removed enzymatically by an aminopeptidase, but the majority of bacterial proteins commence with methionine.

Polypeptide chain termination is brought about by the presence of termination codons (UAA, UGA, UAG) at the end of the message (cistron). When the final (C-terminal) aminoacyl tRNA has attached itself to the penultimate amino acid residue to give the C-terminal peptide bond, it is still attached by ester linkage to its tRNA. A protein "release factor" or factors to which GTP is bound and its complex with a termination codon causes the splitting of this ester bond, possibly by inducing peptidyl transferase to catalyze the reaction with water instead of an —NH$_2$ group. The tRNA is thus released with simultaneous detachment of the completed polypeptide chain.

Most probably, the polypeptide chain assumes its approximate three-dimensional configuration as it is being synthesized. Many native proteins contain disulfide bridges that cross-link the separate polypeptide chains of the complete protein molecule or parts of a single polypeptide chain. The approximately correct three-dimensional folding of the nascent protein probably brings the "right" side chain —SH groups into close enough contact for them to be enzymatically oxidized to give the —S—S— links. Some proteins may be synthesized as inactive precursors (e.g., certain enzymes, insulin, fibrinogen). In the case of insulin, proinsulin is first synthesized and is a single polypeptide chain with three disulfide bridges. Enzymatic cleavage at two sites generates the two chains of active insulin that are joined by two interchain —S—S— bonds. Clearly, determination of the C- and N-terminal amino acids in such proteins is no indication of the actual number and types of C- and N-terminal residues as coded in the genome.

(v) Polyribosomes (Polysomes)

At one time it was thought that one ribosome at a time passed along a mRNA molecule. This idea was difficult to reconcile with the high rate of protein biosynthesis in some cells and the relatively low concentration of mRNA in them. Later, it was discovered that mRNA molecules could be associated with many ribosomes (depending on the length of the mRNA molecule), so that several polypeptide chains, all in different stages of completion, could be synthesized simultaneously from the one mRNA molecule (Fig. 18-8). Rapid rates of protein synthesis could therefore be explained. Subsequently, polysomes were actually visualized in the electron microscope, and the delicate thread of mRNA joining them could be seen. However, each individual ribosome acts independently of the others and does not depend on the presence of the others for its proper functioning.

The length of a mRNA molecule depends, of course, on the length of the polypeptide chain for which it codes. In the case of hemoglobin, which consists of polypeptide chains of approximately 150 residues, the corresponding mRNA will have about 450 nucleotides, and this corresponds to a length of about 1500 Å. Ribosomes are about 220 Å in diameter. We may anticipate that several ribosomes could be accommodated on a mRNA strand in this case, with plenty of room to spare.

In the case of polyribosomes isolated from hemoglobin-synthesizing cells (e.g., rabbit reticulocytes), polysomes with up to six ribosomes to each mRNA molecule have been seen with gaps of 50 to 100 Å between the ribosomes; synthesis of the much larger muscle protein myosin is associated with polysomes with up to 100 ribosomes. Interestingly, polysomes exhibit a three-dimensional structure dependent on their size. Polysomes with at least six ribosomes form whorls, with the smaller subunit facing inward. When more than six ribosomes are present,

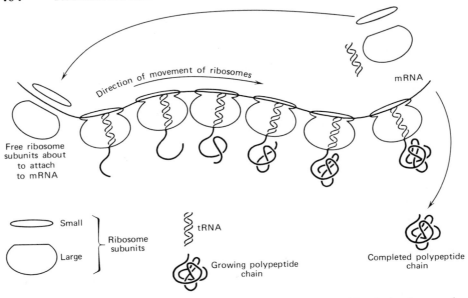

Figure 18-8   Simultaneous production of several polypeptide chains by a polyribosome.

coils are formed. If certain tissues of higher animals (e.g., cultured chick embryo cells) are kept chilled at 0°C and then ultrathin sections studied in the electron microscope, ribosomes are seen to be arranged into regular crystallike lattices, as shown by B. Byers. Possibly, ribosomes function in some sort of three-dimensional array within eucaryotic cells instead of in random disarray.

(vi) Monocistronic and polycistronic messenger RNA

Some mRNA molecules code for one polypeptide chain only (e.g., the mRNA coding for the separate α and β polypeptides of hemoglobin). Such mRNA is said to be monocistronic. Other single mRNA molecules code for several different polypeptides and therefore are said to be polycistronic. The existence of several cistrons in one mRNA molecule raises the question of how the cistron is translated. In Fig. 18-9 three possible modes of translation are indicated. In Fig. 18-9a, the ribosome attaches at one end of the mRNA, passes over the cistrons (1, 2, 3, 4, 5) in sequence, and detaches at the other end. This mechanism, in which there is only one ribosomal attachment and detachment site, therefore results in equal amounts of each polypeptide being synthesized. This would seem to be not very appropriate in some cases (e.g., polio virus contains at least five different proteins in varying amounts; there are two coat proteins, an RNA polymerase, and two proteins of uncertain identity, and these

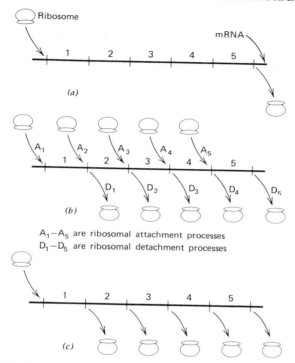

Figure 18-9   Possible modes of translation of polycistronic messenger RNA.

are coded for by a single strand of polycistronic mRNA). The coat proteins are needed in much larger numbers than the RNA polymerase molecules, and the translational mechanism of Fig. 18-9a would result in an unnecessary excess of RNA polymerase if sufficient coat protein is to be produced. A second mechanism of translation would be for each cistron to have its own attachment and detachment site, so that each cistron can be translated independently of the others and at different rates (Fig. 18-9b). Hence, proteins can be synthesized in nonequivalent amounts, and this would appear, therefore, to be more flexible and much less wasteful than the mechanism in Fig. 18-9a. There is some evidence available that the mechanism in Fig. 18.9b can occur, at least in *E. coli*. As will be described later in this chapter, *E. coli* simultaneously synthesizes three metabolic enzymes (β-galactosidase, galactoside permease, and galactoside acetylase) in response to the presence of the disaccharide lactose if this is the sole carbon source. These lactose-induced enzymes are almost certainly coded for by a polycistronic messenger RNA. They are produced in unequal amounts; for example, 10 times as much β-galactosidase as acetylase is usually produced. This would seem to be in favor of the mechanism in which the cistrons were separately and independently translated. Electron microscope studies have, in fact, shown that polyribosomes isolated from *E. coli* actively synthesizing these enzymes appear to have gaps alternating with closely packed groups of ribosomes, which would be consistent with the cistrons having separate ribosomal attachment and detachment sites. The rates at which the three *E. coli* enzyme cistrons are translated are temperature-dependent, but how the modulation of the translation rate is achieved is not known.

A third type of mechanism is for the mRNA to have an attachment site and several detachment sites (Fig. 18-9*c*), in which case the production of the polypeptides would be in the ratios $1 > 2 > 3 > 4 > 5$. However, there is presently insufficient evidence to indicate which of these mechanisms is operative in a given situation.

Puromycin

Aminoacyl-tRNA

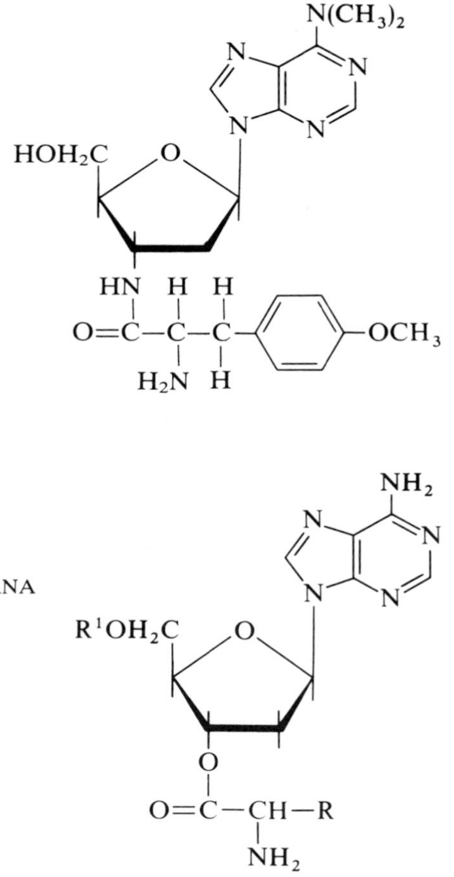

Figure 18-10 Structural resemblance between aminoacyl transfer RNA and puromycin.

(vii) Antibiotics and their effects on protein biosynthesis

Antibiotics have been employed for many years in the treatment of bacterial infections in humans and at least some of them exert their bacteriostatic effects by interfering with protein synthesis in various ways. *Puromycin* for example, results in the formation of incomplete polypeptide chains of various lengths and acts, by interfering with chain elongation. It does this by occupying the A site of ribosomes, owing to its striking structural resemblance to amino-acyl tRNAs (Fig. 18-10); it competitively inhibits the attachment of amino-acyl tRNAs to the A site. The puromycin becomes incorporated as the terminal residue, chain growth is thereby stopped, and incomplete polypeptides detach from the ribosome. The inhibition of protein synthesis and production of incomplete polypeptides by the action of puromycin is important evidence in favor of the mechanism of chain elongation by addition of amino-acyl tRNAs at the C-terminal end of the growing chain by displacement reactions. *Chloramphenicol* and *cycloheximide* (Actidione) (Fig. 18-11) appear to interfere with peptide

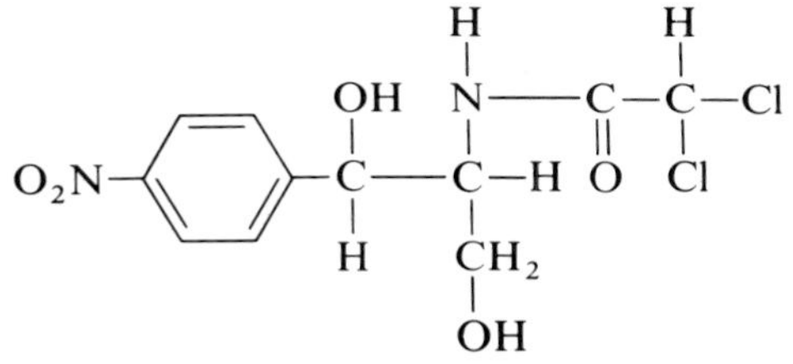

Chloramphenicol

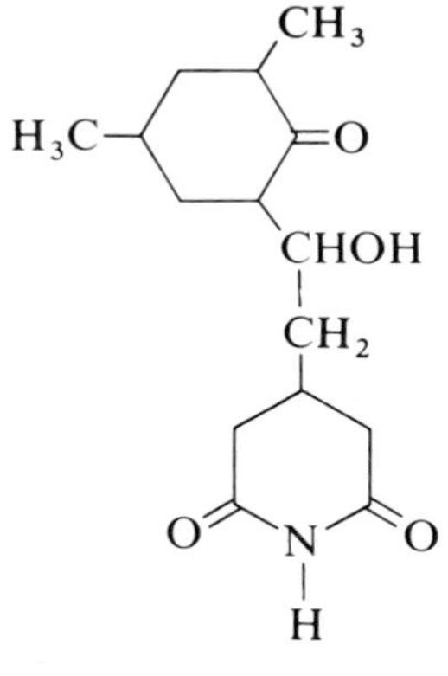

Cycloheximide
(actidione)

Figure 18-11   Structures of chloramphenicol and cycloheximide.

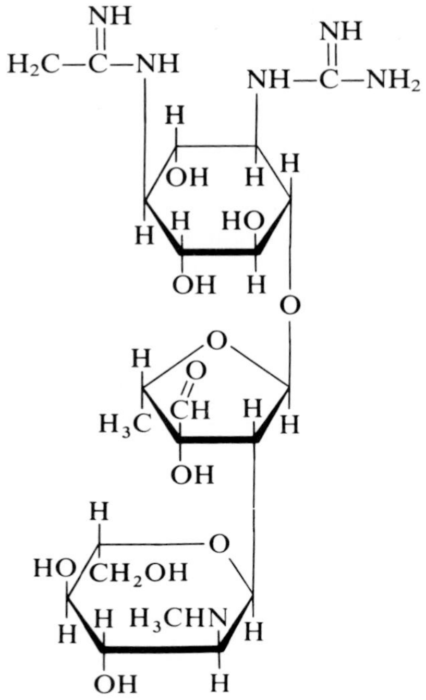

Figure 18-12   Structure of streptomycin.

bond formation. Cloramphenicol inhibits the protein synthesis of procaryotic (70S) ribosomes and ribosomes of eucaryotic mitochondria, but it has no effect on cytoplasmic eucaryotic (80S) ribosomes. It binds to the 50S subunit. Cyclo-heximide has the opposite effect and inhibits protein synthesis in eucaryotic cells, but not in procaryotes. *Streptomycin* (Fig. 18-12) not only inhibits protein synthesis, but also causes incorrect reading of the genetic code. It binds to the 30S subunit of ribosomes and possibly changes the three-dimensional con-formation of the subunit, so that aminoacyl tRNAs are not bound as firmly to codons and hence are less specific.

THE "CENTRAL DOGMA"

In the genetic determination and control of protein structure and biosynthesis, there is evidently a unidirectional flow of genetic information; the DNA nucleo-tide sequence determines the nucleotide sequence of mRNA (transcription), and the mRNA nucleotide sequence determines the amino acid sequence of poly-peptides (translation). The central dogma of molecular biology is the name given to the doctrine that this informational flow is one-way and irreversible. It is summarized in Fig. 18-13. (The curved arrow indicates that DNA provides the template for its own replication.)

Although this would seem to be generally true, there appear to be some exceptions (i.e., some of the arrows in Fig. 18-13 should be pointing in both directions). The first indication that the irreversibility of information flow implied by the central dogma may not be true in every single case was provided by H. Temins's finding in the 1960s that animal cancer induced by RNA-containing viruses was accompanied by alteration of the host DNA code. At about the same time, S. Spiegelman demonstrated an enzyme capable of synthesizing DNA on RNA templates in six different cancer-inducing viruses. That this was a genuine "reverse transcriptase," as it is called, for obvious reasons, was firmly established by work published soon afterward. Leucocytes from leukemia patients and

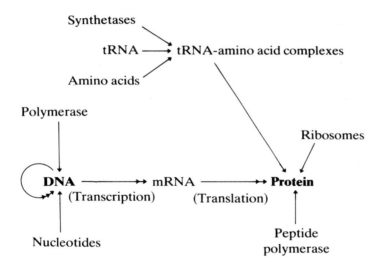

→ Indicates that an agent is necessary for formation of a product but does not contribute structural specificity

→» Indicates that agent contributes to specificity of product

→»» Indicates that specificity is determined solely from identical preceding molecule, which is therefore self-replicating.

Figure 18-13 The "central dogma." This implies that information transfer is linear and unidirectional and that specificity originates in DNA only.

cancerous tissues were found to contain high levels of reverse transcriptase in investigations published during the early 1970s. The corresponding normal cells did not. The first arrow in the central dogma should therefore be pointing in both directions.

GENETIC CONTROL OF PROTEIN BIOSYNTHESIS

The processes of transcription and translation are subjected to controlling influences, as is demonstrated by the phenomenon of differentiation during the growth by mitotic cell division of a higher animal from the fertilized egg or zygote. The genome of the zygote contains all the genetic information necessary to code for all the proteins that go to make up the adult organism yet, as is obvious, it is not all being expressed in every cell of the differentiated adult. The precursors of erythrocytes, for example, produce essentially only one protein—hemoglobin—almost exclusively, and yet their nuclei contain the coding information for thousands of other proteins. Likewise, adult exocrine pancreas cells synthesize various digestive enzymes, but no hemoglobin, myosin, or collagen. Apparently, the process of growth and differentiation from the zygote to the adult involves a progressive and selective irreversible inactivation or "switching off" of large segments of the genetic information contained within the differentiating cells.

As has often been the case in other problems of cell and molecular biology, an insight into how gene action may be regulated in higher organisms was sought in studies of gene regulation in the much simpler cells of microorganisms, especially the bacterium *E. coli.*

(i) Genetic regulation of enzyme biosynthesis in Escherichia coli

The first established and best-known example of gene regulation in this bacterium was discovered by F. Jacob and J. Monod and published in 1961; Jacob and Monod were awarded the Nobel Prize in Medicine in 1965 for their discovery. It concerns the genetic mechanism controlling the biosynthesis of three enzymes involved in lactose metabolism. These are β-galactosidase that catalyzes hydrolysis of lactose to galactose and glucose, galactoside permease that facilitates the penetration of the disaccharide into the cell, and galactoside acetylase.

Wild-type *E. coli* cells grown in culture containing glucose whether lactose or other galactosides are present or not, synthesize hardly any β-galactosidase but, if glucose is absent and lactose is added to the *E. coli* culture as the only source of carbon, the bacterial cells respond by synthesizing high levels of the enzymes

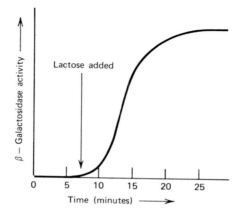

Figure 18-14 Effect of lactose on β-galactosidase levels in *E. coli.*

in a matter of minutes after adding the lactose (Fig. 18-14). When all the lactose is used up, or if it is withdrawn, β-galactosidase falls to its former very low level. This response has been shown to be due to actual *de novo* synthesis of new enzyme and not merely to the activation of enzyme already present. The phenomenon, logically, is called enzyme induction, and lactose is the enzyme inducer in this case.

Certain mutant forms of *E. coli* behave differently toward the presence or absence of lactose from the wild type. One type of mutant does not respond by increased activity of β-galactosidase in the presence of lactose and is therefore said to be noninducible. This may be easily explained by hypothesizing that the gene coding for the enzyme has been changed in such a way that the corresponding altered mRNA codes for a nonfunctional enzyme. This type of mutation is called a structural mutation, because the structure of the gene coding for the enzyme—and also the enzyme itself—has been directly altered.

Another type of mutant *E. coli* continuously produces constant high levels of β-galactosidase, whether or not lactose is present. Such a mutant is said to be constitutive, and it is not so easy to explain.

Jacob and Monod put forward the idea that in addition to the structural gene that determines the structure of the enzyme, there was also a regulator gene—not necessarily close to the structural gene—that controls the activity of the structural gene. They proposed that the regulator gene produces a repressor substance that normally combines with and inhibits or switches off the structural gene in the absence of lactose, so that no β-galactosidase is produced. In the presence of lactose the repressor substance combines with lactose, it is thereby inactivated and is no longer capable of combining with and inhibiting the structural gene. The latter is therefore switched on, and enzyme is produced (Fig. 18-15).

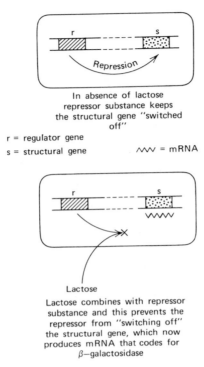

In absence of lactose
repressor substance keeps
the structural gene "switched
off"

r = regulator gene
s = structural gene   ∿∿ = mRNA

Lactose

Lactose combines with repressor
substance and this prevents the
repressor from "switching off"
the structural gene, which now
produces mRNA that codes for
β–galactosidase

Figure 18-15   Control of the β-galactosidase structural gene by the regulator gene. (Redrawn after C. U. M. Smith.)

Presumably, mutations of the regulator gene exist that are no longer capable of synthesizing the repressor substance so that the structural gene would be permanently switched on and β-galactosidase would be continuously produced (i.e., a constitutively mutant cell would be the result). Jacob and Monod postulated that the repressor substance does not act directly on the structural gene, but with another type of gene adjacent to the structural gene called an operator gene, which switches on the transcriptional activity of the structural gene. The repressor inactivates the function of the operator gene so that the structural gene is switched off. Absence of the repressor causes the operator to become active, and the structural gene is switched on and stimulated to transcribe its mRNA (Fig. 18-16). This hypothesis predicts, therefore, that there should be two genetically different types of constitutive mutant. In one type with a mutant regulator gene, the repressor substance is not produced or is inactive, so that the operator is permanently switched on and the structural gene causes enzyme to be continuously produced. The second type would have a mutated operator gene that, no longer being susceptible to inactivation by the repressor, again causes

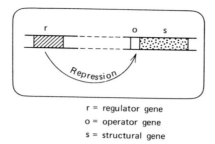

r = regulator gene
o = operator gene
s = structural gene

Figure 18-16 Relation between regulator, operator, and structural genes of the
β-galactosidase synthesizing system in *E. coli.* (Redrawn after C. U. M. Smith.)

the structural gene to be switched on, with the resulting continuous production
of β-galactosidase. The question that arises is, "Is it possible, in practice, to
distinguish between two such genetically different but phenotypically identical
mutations?"

The distinction between the two types of constitutive mutants predicted by the
hypothesis of the existence of operator genes was demonstrated in experiments
carried out by Jacob and Monod in which they mated wild-type *E. coli* with
constitutive *E. coli* cells. *E. coli* can be made to undergo a primitive type of
sexual process in which a part of the chromosome of one cell (the "male") enters
a "female" cell, so that the latter is temporarily diploid for that particular part of
the chromosome. If the chromosomal segment contributed by the male included
the structural and regulator genes involved in β-galactosidase synthesis, then the
behavior of the temporary "diploid" cells toward lactose, as shown in many
experiments, was of two types. One type of diploid continued to behave like its
constitutive "parent," but another type had the wild-type characteristic of producing no enzyme in the absence of lactose, but of doing so in response to the
presence of lactose. The operator and regulator gene hypothesis predicts that
two phenotypically different types of diploid cells like these will, in fact, be the
result of chromosomal transfer between a wild-type cell and the two types of
constitutive mutant. One type of diploid will have one normal and one mutated
regulator gene plus two normal operator and normal structural genes (Fig.
18-17*a*). The one functional regulator produces repressor substance that can act
on and control both operators, so that this diploid will behave phenotypically
like the inducible wild-type cell. The inducible phenotype is therefore dominant
to the constitutive mutant type resulting from a mutant regulator. The other type
of diploid will possess one mutant and one normal operator gene, two normal
structural genes, and two normal regulators (Fig. 18-17*b*). The normal operator
will be repressible by the regulator in the absence of lactose but not the mutated
operator. This second type of diploid will therefore behave as a constitutive cell,

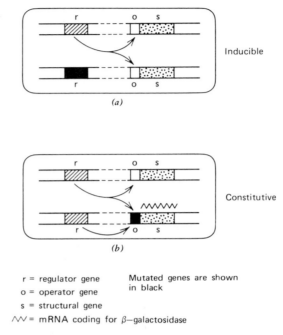

r = regulator gene
o = operator gene
s = structural gene
⋀⋁ = mRNA coding for β–galactosidase

Mutated genes are shown
in black

Figure 18-17   Results of mating two types of *E. coli* constitutive mutants with wild-type *E. coli.* (Redrawn after C. U. M. Smith.)

continuously producing enzyme whether or not lactose is present. The wild type is therefore recessive to the mutated operator type of constitutive mutant.

That such phenotypically different diploids are actually produced by "mating" wild-type *E. coli* cells with constitutive cells, as shown by Jacob and Monod, is striking confirmation of their theory of operator and regulator genes.

As well as structural genes for determining the β-galactosidase enzyme, there are structural genes for the galactoside permease and acetylase enzymes adjacent to it on the bacterial chromosome, and all three are under the control of the same operator and regulator genes. The operator switches the three structural genes on and off simultaneously. Such a "cluster" of functionally related structural, regulator, and operator genes is called an operon, and the system just described is named the lactose operon, or "lac" operon for short. Adjacent to the operator gene is a "promoter site." This binds the enzyme RNA polymerase (transcriptase) as the first step in the transcription of the DNA structural genes onto the corresponding mRNA. The binding of the repressor substance at the operator site prevents the RNA polymerase from transcribing the structural genes. The regulator gene also has a promoter site.

The regulator gene probably codes for a mRNA that directs the synthesis of the repressor, which has been shown to be a protein. Binding of β-galactosides might alter the three-dimensional configuration of the repressor protein so that it can no longer bind to the operator. The molecular weight of the repressor protein is about 150,000 to 200,000, and this is in the range of allosteric protein molecular weights. The structure of the lac operon is shown diagramatically in Fig. 18-18. The lac operon provides an example of negative regulation of gene activity, because it is failure of the regulatory protein–inducer complex to bind to the operator that results in activity of the genes. The enzymes of arabinose metabolism in *E. coli* are controlled by three structural genes and a gene remote from them that codes for a permease. The regulator gene that controls this system is called C, and the control mechanism appears to be positive instead of negative because binding of the arabinose-regulator protein complex to the operator occurs and turns on the structural genes. In this case, mutants of the C gene or operator are uninducible. This is quite different from the lac operon

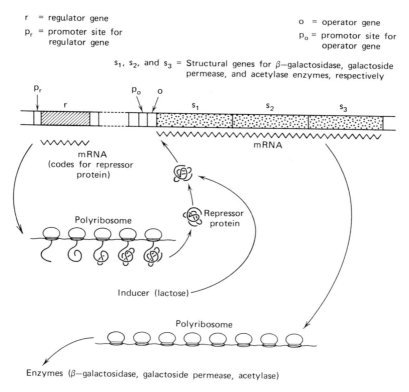

Figure 18-18   The Lac operon (diagrammatic).

situation, which is under negative control of the regulator and where mutated regulator or operator genes are constitutive.

Catabolic enzymes such as β-galactosidase and the enzymes of arabinose metabolism are the sort of enzymes that are inducible, and they are induced by their substrate. On the other hand, anabolic enzymes (i.e., enzymes involved in biosynthesis) exhibit the phenomenon of repression and are repressed by their end products. Enzyme repression is exemplified by the histidine biosynthetic pathway in *E. coli*. The enzyme histidinol dehydrogenase catalyzes the formation of histidine from histidinol, which is the immediate precursor of histidine. When histidine is added to an *E. coli* culture, this enzyme is repressed. The repression can be accounted for by postulating that the repressor produced by the regulator gene is inactive by itself, but combines with histidine to give an active complex that binds to the operator for the structural gene and stops synthesis of the enzyme. The end product (histidine) of this metabolic pathway is therefore called a corepressor. This repressor substance, like the lac repressor, presumably has two binding sites, one of them specific for the operator. We can therefore distinguish two categories of repressors, one involved in the control of inducible enzymes and another involved in the control of biosynthetic enzymes by end-product inhibition.

(ii) Genetic regulation of protein biosynthesis in higher organisms

Do the mechanisms of control of gene function in *E. coli* and other micro-organisms give any insights into gene control and cell differentiation in higher organisms? Not too much, apparently, because coordinate regulation of poly-cistronic systems like the lac operon is uncommon in eucaryotes and is not too common even in bacteria. The operon concept would therefore seem to be of quite limited relevance in the understanding of gene regulation in general. There is increasing evidence for reversible enzyme induction and repression in the cells of higher organisms, but the responses to the repressors and inducers are much slower and less marked than in bacterial cells.

The simple operon theory does not easily account for the following facts regarding higher cells: (1) there is much more DNA in higher cells than in lower cell types, more than can be accounted for by the number of proteins in higher organisms; (2) there is much redundancy of DNA (i.e., repetition of DNA sequences); (3) genes controlling related functions are not "clustered" as they are in bacterial operons and are usually scattered in different parts of the genome in different chromosomes e.g., in *Drosophila melanogaster*, there are two enzymes involved in the phosphogluconic acid oxidation pathway; they are 6-phosphogluconate dehydrogenase and glucose-6-phosphate dehydrogenase. The genes determining them are completely unlinked, and yet they are subject to closely integrated control. Therefore, at least some groups of functionally related genes in eucaryotes must be subject to mechanisms of coordinate control that do not involve physical linkage of genes into clusters, as are found in bacterial operons; and (4) only a small fraction of the total genome is expressed in a given mature cell type.

If differentiation of the cells of a higher organism is a process of progressive and selective repression of most of the genome, as mentioned earlier, a formidable number of different repressors would be required, and this seems improbable. Whatever the mechanism of cellular differentiation is, it certainly is not as easily reversible as bacterial enzyme induction and repression. Repressors and repression must be much more permanent in higher cells, and it may be that the basic proteins of the cell nucleus are involved in inactivation of DNA by forming complexes with it. The possible nature of DNA-histone and protamine complexes is discussed in Chapter Fifteen.

Of considerable significance in this connection is the fact that, as already mentioned in Chapter Sixteen, histone H4 has undergone practically no change in amino acid sequence during evolution. This strongly suggests that H4 functions in suppressing or expressing of genetic information, since it is closely associated with the also highly "conservative" three-dimensional structure of DNA. Evidence that histones mask DNA and make it unavailable for transcription was obtained by V. G. Allfrey and A. E. Mirsky. They removed part of the histone from isolated calf thymus nuclei and found that this resulted in a large increase in mRNA synthesis. Because basic proteins such as these are absent from procaryotic cells that do not exhibit true differentiation and are present in the nuclei of eucaryotic cells suggests that they may be involved in differentiation.

A. E. Mirsky and A. Ris were the first to point out, in 1951, that the size of the eucaryotic genome generally increases with increasing complexity of organization. A more definite correlation is that the minimum amounts of DNA per haploid genome in the different phyla and classes of organisms increase as the complexity of each group increases. The minimal DNA in each group can therefore be regarded as the amount necessary to achieve the degree of complexity of that group. In mammals there seems to be very much more DNA than is needed to code for all the known mammalian proteins, and it does not seem likely that this extra DNA can be accounted for by supposing that it codes for as yet unrecognized proteins. Some but not all of the much larger amount of DNA found in mammals as compared with that in *E. coli* cells can be accounted for by redundancy. The main difference between a sponge and a mammal would seem to be in the much greater integration and regulation of cellular function in the latter, and so the greater DNA content of mammalian cells would probably consist of greater numbers of regulator instead of structural genes. R. J. Britten and E. H. Davidson have put forward a theory, published in a paper in 1969, that attempts to explain mechanisms of gene regulation in higher cells. They define several functionally different entities as follows, and consider that these are the minimal needed to explain regulatory phenomena in higher cells.

1. *Producer genes*, which are analogous to the structural genes of bacteria. They can code for specific polypeptides or for transfer RNA molecules, among other things.

2. *Receptor genes*, which are linked to producer genes and stimulate transcription when molecules of special types of RNAs called activator RNAs form sequence-specific complexes with receptor gene DNA.

3. *Activator RNA*, which forms sequence-specific complexes with receptor genes. The same function might also be carried out by proteins coded for by these RNAs.

4. *Integrator genes*, which synthesize activator RNA. The name underlines the function of these genes in stimulating the coordinated activity of several producer genes through activator RNA. Several linked integrator genes can therefore be simultaneously activated by a single specific event that leads to simultaneous activity of many producer genes that do not share the same receptor gene sequence.

5. *Sensor genes* function to bind substances (e.g., hormones) that initiate "specific patterns of activity" in the genome. The integrator gene linked to a given sensor gene is activated by such specific binding. Specific proteins must usually first bind with, say, an activating hormone, and the resulting complex is what binds to the sensor gene DNA.

6. *Battery of genes*, which is the group of producer genes that becomes activated when the corresponding group of integrator genes is activated by its specific sensor gene. Any given state of the cell is visualized as resulting from the activity of several gene batteries.

Cell differentiation is seen as involving the following sequence.

1. A chemical, such as a hormone, binds to a sensor gene.

2. The integrator genes linked to the sensor gene are stimulated to transcribe activator RNA.

3. Receptor genes bind the activator RNA.

4. Producer genes are activated by the binding of the activator RNA to their receptor genes.

Patterns of activity and interaction far more complex than those observed in simple operons are therefore possible.

An important characteristic of Britten and Davidson's model is its integrative function. They consider that the different states of cell differentiation are caused by concerted activation of one or several batteries of producer genes, and they show how in systems of integrator, receptor, and producer genes of varying complexity the coordinated synthesis of enzyme systems such as the enzymes of urea synthesis in the liver might be achieved.

Changes of the magnitude observed in cell and tissue differentiation clearly must involve coordinated activity of several batteries of producer genes. Accomplishment of this is visualized as being brought about by sensor genes that respond to the products of integrator genes in other integrative groups.

Development requires that gene activations occur in sequences, and this could result from the responses of some of the sensor genes to producer gene products.

Redundant DNA sequences are attributable to receptor or integrator genes. In eucaryotes, as is known, batteries of producer genes may not be physically linked as they are in operons; this model provides for their simultaneous activation. The production of a number of events triggered by, say, a hormone, is also explainable in terms of this model.

There is experimental evidence that is consistent with the model. For example, there exists RNA molecules confined to the cell nucleus that bind to DNA, and this would seem to correspond closely to the postulated activator RNA. Again, for many years it has been known that administration of corticosteroid hormones to higher animals results in *de novo* synthesis of several hepatic enzymes. In 1975, A. J. Eisenfeld et al. discovered that in the cytosol of adult female mammalian liver, there are estradiol-binding proteins that translocate to the nucleus and bind to chromatin. They consider these proteins to be estrogen receptors involved in modulation of biosynthesis of various plasma proteins. Such protein-bound hormones could be the sequence-specific complexes that bind to and activate sensor gene DNA. Regulator genes have been discovered, and mutations would be expected to have pleiotropic effects; in *Drosophila*, the so-called notch series of deficiencies that affect all three germ layers are just what would be expected from a mutated regulator gene.

In this model, established sets of useful producer genes would tend to stay functionally integrated, but there is also the possibility of the formation of already existing producer genes into new integrative combinations, probably as a result of translation.

The clonal selection theory of the immune response described in Chapter Eight provides a good experimental system for investigating cell differentiation. Each plasma cell clone differs from others in the specific antibody protein that it is "committed" to synthesize in response to an antigenic stimulus, and this property is inherited. A single difference such as this is more amenable to experimental study and analysis than the complex biochemical differences between diverse types of cells found in higher organisms.

D. Kabat's chromosomal "looping out-excision" model, also described in Chapter Eight, which was advanced to explain mutually exclusive gene activation, is relevant in this context, because it may be one of the mechanisms of gene selection and activation in differentiating cells.

PART FIVE

CELL ULTRASTRUCTURE: UNITY AMID SEEMING DIVERSITY

CHAPTER NINETEEN
Organization of procaryotic cells

In the preceding chapters we have "dissected" the living cell into its component macromolecular constituents and have studied their interactions and how they are organized into supramolecular complexes such as biomembranes, chromatin fibers, or multienzyme complexes. So far, the approach has been analytic; we will now attempt a synthesis and see how these molecular and supramolecular complexes are organized into functioning living cells—but what is a cell?

BIOLOGICAL ENTITIES

In what we perceive as the world of "biological objects," we can recognize two types of entities, the noncellular and the cellular. The first group is exemplified by the viruses, nucleoprotein complexes that have some of the attributes of what we recognize as "living things," such as the ability to replicate. However, they cannot do so outside of living cells. In this respect viruses are analogous to the chromosomes of eucaryotic cells. Chromosomes are not normally regarded as being "alive," but they replicate in a suitable cellular environment (e.g., the chromosomes within the free-swimming sperm cell undergo no further replication unless they enter a suitable cell and meet similar chromosomes, as occurs when the sperm fertilizes the egg). Perhaps it is best to regard viruses noncommittally as "biological entities" or "infectious nucleoproteins" instead of as "living" things.

423

Cellular entities, on the other hand, are characterized primarily by the possession of a lipoprotein plasma membrane that separates them from the outside world while, at the same time, permitting interaction with it. Perhaps more than anything else, it is the possession of the plasma membrane that makes the cell alive; cells can continue to survive if their nuclei are removed but, if the plasma membrane is damaged too extensively, the cell quickly dies. A second feature that differentiates cells from viruses is metabolism (absorbing substances from the environment, processing them, and excreting waste), in particular, their ability to extract energy from carbon compounds absorbed from the exterior so as to sustain their vital activities.

THE DEFINITION OF A CELL

Perhaps it is foolish to try to define a cell, but the following attempt seems at least to exclude viruses and nonliving matter and to include, so far as can be determined, all entities that are recognizable as cellular.

> *A living cell is a self-sustaining organized assemblage (usually of microscopic dimensions) of many different kinds of interacting molecules and biopolymers and their products of interaction, in a state of dynamic exchange with the environment and able to extract energy from materials absorbed from it, but bound from the environment by a liproprotein membrane and usually containing sufficient encoded genetic information at some time in its existence to ensure its accurate replication if and when this occurs.*

The cell is the material unit of what we call life.

CELLULAR COMPONENTS

In talking about cells and the multitude of organelles and macromolecular assemblages entering into their structure, some sort of classification of cellular components and functions is desirable. Although somewhat arbitrary, the following breakdown of cellular components into three major classes with their associated functions may be practical and useful.

1. *Lamellar (membranous) structures.* Associated functions are compartmentalization of the cell and spatial organization of its components and involvement in metabolic reactions.
   (i) Plasma membrane.
   (ii) Endoplasmic reticulum.
   (iii) Nuclear membrane.
   (iv) Golgi body.
   (v) Lysosomes.
   (vi) Peroxisomes.

    (vii) Glyoxisomes.
  (viii) Mitochondria.
    (ix) Chloroplasts.
     (x) Plastids.
    (xi) Vacuoles.
2. *Fibrillar structures.* These are associated with cellular motion, replication, and information storage.
    (i) Nuclear chromatin fibers. Chromosomes. (Chapter Fifteen.)
   (ii) Protein fibrillar structures.
3. *Granular structures.* These are concerned with protein synthesis.
    (i) Ribosomes.
   (ii) Zymogen granules, secretion granules.

THE SIMPLEST CELLS—PROCARYOTES

We may next ask what is the simplest type of organized biological structure that will satisfy the requirements of the above definition of a cell (i.e., how would a "minimal cell" be organized, and do such minimal cells exist today?). There are, in fact, a few groups of cellular organisms whose structure can be regarded as that of a minimal cell, as defined above, and they are represented by three groups of organisms: bacteria, blue-green algae, and pleuropneumonialike organisms (PPLO). The group name of "procaryote" derives from the fact that these simple cells do not possess definite compact nuclei or well-differentiated mitotic apparatus with chromosomes, centrioles (except in eucaryotic plant cells), and spindle fibers, whereas the second great class of cells—the eucaryotes—do possess these structures, or "organelles," as they are called.

    Of the procaryotes, the PPLOs are undoubtedly the simplest and are, indeed, minimal cells.

(i) Pleuropneumonialike organisms

These are so-called because they resemble the organism that causes bovine pleuropneumonia. Some cause respiratory and other diseases in humans and in animals. The best-known genus is *Mycoplasma.*

    The cells of PPLOs are the smallest known (about 1000 Å) and are smaller than some of the larger viruses such as the vaccinia virus, which is more than 3000 Å in diameter. The PPLO is bounded by a lipoprotein plasma membrane 75 Å thick; contained within this is little more than a "soup" of various macromolecules and metabolites and possibly a few vacuoles. In PPLOs, as in other procaryotes, there is no definite nucleus. The genetic material is distributed diffusely as a double helical strand of DNA that may be circular. In the smallest PPLOs it consists of roughly 500,000 base pairs. In addition, PPLOs

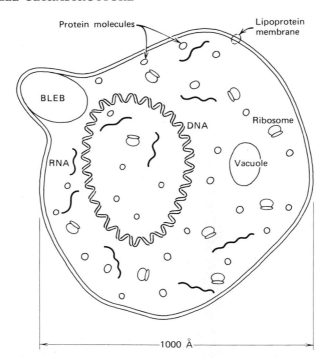

Figure 19-1   Structure of a pleuropneumonialike organism (PPLO).

possess mRNA and ribosomes and are able to synthesize their necessary proteins and to replicate themselves. They contain enzymes that are required for
replication of DNA and for protein synthesis and for breaking down of glucose
to pyruvate so that some of the energy of the glucose molecule can be extracted.
PPLOs are therefore capable of independent existence. Since they can extract
energy from materials in the environment, can synthesize macromolecular
substances from simple precursors and, can replicate themselves, they qualify as
genuine cells, despite their extremely small size. They respresent the bare
minimum of structural organization required for autonomous behavior.

The organization of a PPLO cell is shown schematically in Fig. 19-1.

(ii) Bacteria

Although bacterial cells are much larger than those of PPLOs (e.g., 8000 Å ×
20,000 Å for *E. coli*), they are hardly any more complex (Fig. 19-2). The main
structural difference is that in addition to the plasma membrane, there is a cell
wall external to this composed of protein, lipid, and carbohydrate (Fig. 19-3).

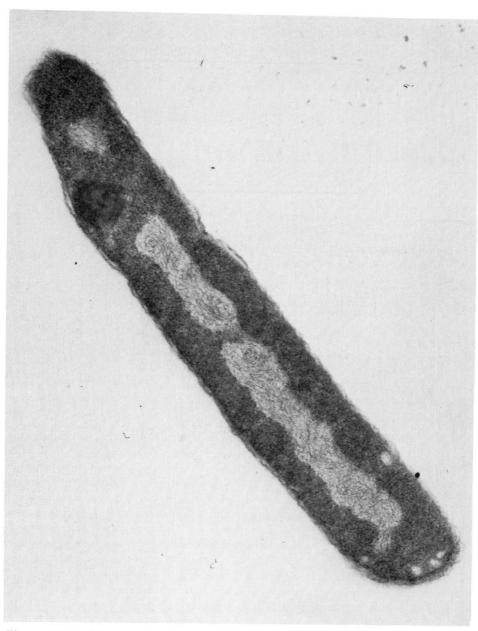

Figure 19-2 Electron micrograph of a thin-sectioned *E. coli* cell. The light-colored electron-transparent structure in the center of the cell is the nucleoid. (Courtesy of Dr E. Kellenberger. *J. Biophys. Biochem. Cytol.*, *4*(2), plate 331, 1958.)

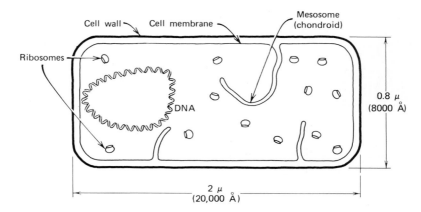

Figure 19-3   Organization of a simple bacterial cell (*E. coli*).

The genetic material is in the form of a circular, double helical DNA molecule attached at one point to the cell membrane. As in the PPLO, it is diffuse and not compacted into a nucleus. There may be two to four copies of the DNA molecule per cell under conditions of optimal growth, and it contains sufficient genetic information to code for several thousands of different proteins. There are no histones associated with the *E. coli* "chromosome," and there are no chromosomes like those of eucaryotes, but polyamines are possibly bound to some of the DNA phosphate groups.

For many years, bacteriologists have employed the Gram staining reaction to differentiate bacteria into two major groups as a preliminary step in the identification of an unknown bacterium in clinical and other investigations. Bacteria are classified as Gram-positive or Gram-negative, according as to whether or not they resist decolorization by an alcohol-acetone mixture after they have been stained with the dye crystal violet in conjunction with Gram's iodine. The reaction of bacteria to the Gram staining procedures has since been found to be closely correlated with certain structural and chemical differences between the two classes. *E. coli*, for example, is Gram-negative and, like other Gram-negative bacteria, its internal organization is very simple and essentially similar to that of the PPLO. On the other hand, in Gram-positive bacteria (e.g., *Bacillus subtilis*), the genetic material, instead of being diffusely arranged, is compacted into a more or less well-defined structure called a nucleoid. This, however, is not a true nucleus, since it is free in the cytoplasm and is not bounded from it by a nuclear membrane. They also possess a primitive type of membranous ultrastructure that is completely lacking in PPLOs and Gram-negative bacteria.

In *B. subtilis* the electron microscope has revealed a whorl of intracellular membranes known as the mesosome or chrondroid, which A. W. Sedar and R.

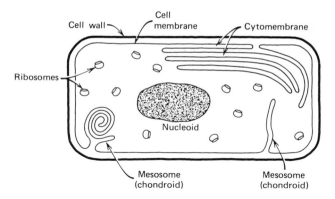

Figure 19-4   Organization of an advanced bacterial cell (*Bacillus subtilis*).

M. Bunde showed to be the site of redox activity, so it may be functionally analogous to a primitive mitochondrion (Fig. 19-4).

The chemical differences between Gram-positive and Gram-negative bacteria are found in the composition of the cell wall and in the effect of antibiotics. Gram-positive bacteria have cell walls composed of polysaccharides and muco-peptides linked together to form a molecular network. Gram-negative bacteria cell walls have a similar network on the outside of which there is, in addition, a layer consisting of lipids complexed with protein and polysaccharides. As a rule, Gram-positive bacteria are more sensitive to penicillin, which disrupts the formation of the cell wall and reduces its rigidity so that the bacterial cell is susceptible to rupture by osmotic effects. The cell walls of Gram-positive bacteria are attacked more readily by the enzyme lysozyme than those of Gram-negative bacteria. Possibly, the lipid in the wall of Gram-negative cells hinders penetration of the enzyme into the cell.

Bacteria often possess hairlike motile organs called flagella. Bacterial flagella are much simpler in structural organization than eucaryotic flagella or cilia. They consist entirely of a protein called flagellin and give an X-ray diffraction pattern indicating the presence of α-helical secondary structure. Examination with the electron microscope at high resolution reveals that the structural units of the flagellum are globular subunits about 50 Å in diameter, arranged in a helical pattern. In the flagellum of *Salmonella typhimurium*, there are four intertwined helices of globular subunits, and the pitch of each helix is about 200 Å (Fig. 19-5).

Bacterial flagella exhibit the same type of motion as eucaryotic flagella; it is screwlike (Fig. 19-6), but how this type of motion is achieved is still a mystery. The procaryotic flagellum is the most primitive type of cellular motile organ, and because of its great simplicity of structure, Astbury regarded it as a "mono-molecular muscle."

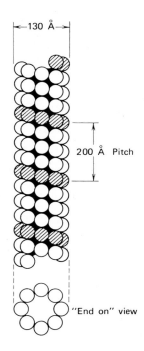

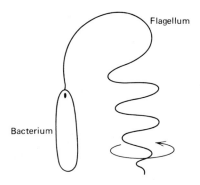

Figure 19-5 Arrangement of spherical protein subunits in the flagellum of *Salmonella typhimurium*. There are four helical strands of globular subunits, one of which is shaded in black in this diagram. (After Lowy and Hanson.)

Figure 19-6 Flagellar motion.

Some bacteria are photosynthetic, and the light-trapping pigments and enzymes are associated with internal membranous tubules or vesicles.

(iii) Blue-green algae

Of the 2000 species of blue-green algae that are known, some grow as single cells, but most form long chains or filaments of cells. They are photosynthetic and are the most primitive plant cells that perform oxygen-yielding photosyn-

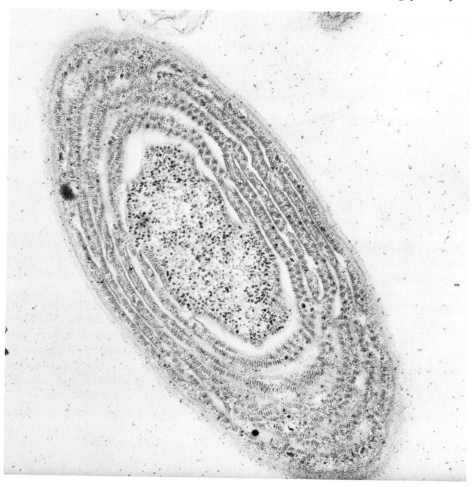

Figure 19-7 Electron micrograph of a thin-sectioned blue-green algal cell (*Synechococcus*). (Courtesy of Dr M. R. Edwards. *J. Cell Biol.*, *50*(3), page 897, 1971). Reprinted by permission of Rockefeller University Press.)

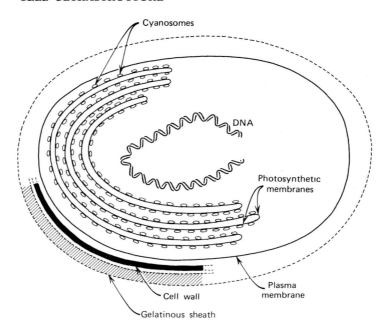

Figure 19-8   Organization of cell of a blue-green alga.

thesis. The organization of the cell is generally similar to that of bacteria. The photosynthetic apparatus of blue-green algae is more elaborate than that of photosynthetic bacteria. There are many extensive parallel layers of chlorophyll-containing membranes (photosynthetic lamellae) within the cell (Fig. 19-7).

The cell wall is covered on the outside with a gelatinous sheath (Fig. 19-8). The system of photosynthetic lamellae in blue-green algae and the mesosomes of Gram-positive bacteria are examples of the unexpected membranous ultra-structure that has been discovered in procaryotic cells in recent years. Even so, these membranes and their associated photosynthetic and redox functions are not contained within well-defined, membrane-bound organelles (e.g., chloro-plasts and mitochondria), as they are in eucaryotic cells.

Locomotor organs such as flagella or cilia are not found in blue-green algae, although the cells are capable of rotary or gliding movement, the mechanism of which is still a puzzle.

SUMMARY

Procaryotic cells represent a very simple level of cellular organization, the simplest of which are the pleuropneumonialike organisms, which may be regarded as "minimal cells." Procaryotic cells lack many of the structural characteristics of eucaryotic cells such as membranous cytoplasmic organelles, a membrane-bound compact nucleus, the endoplasmic reticulum, mitotic apparatus, and complex cilia and flagella. They do not exhibit amoeboid motion or cytoplasmic streaming. In fact, procaryotic cells are more often noted for what they lack than for what they possess when compared with eucaryotic cells.

CHAPTER TWENTY

Organization of eucaryotic cells

The feature common to all eucaryotic cells that differentiates them sharply from procaryotes and that gives them their name is the possession of a compact, well-defined nucleus enclosed within a double membranous envelope that delimits it from the cytoplasm. The appearance of distinct chromosomes in cell division also sharply demarcates eucaryotic from procaryotic cells. Another pronounced and characteristic feature of the eucaryotic cell is the possession of well-defined membranous organelles within the cytoplasm. All three of these features are absent from procaryotic cells. There is never any doubt, therefore, as to whether a cell is procaryotic or eucaryotic. There are no intermediate forms that are difficult to classify as one or the other type.

In eucaryotic cells we find a considerable increase in complexity as compared with even the most advanced procaryotic cells, especially in the eucaryotic cells of certain protozoa. Although at first sight there is relatively little structural specialization in the one-celled body of an *amoeba* (Fig. 20-1), the cells of certain ciliated protozoa, such as *Epidinium*, are quite complex (Fig. 20-2). In *Epidinium* there is a definite mouth whose position is fixed, and all food enters here. Around the mouth and esophagus are retractable fibers like miniature muscles that move the mouth and esophagus; these fibers are controlled and coordinated by fibers that function like nerves. The indigestible remains of food are voided through a well-differentiated rectum and anus. The movements of complex membranellae (locomotor organs) are controlled by neurofibrils, which

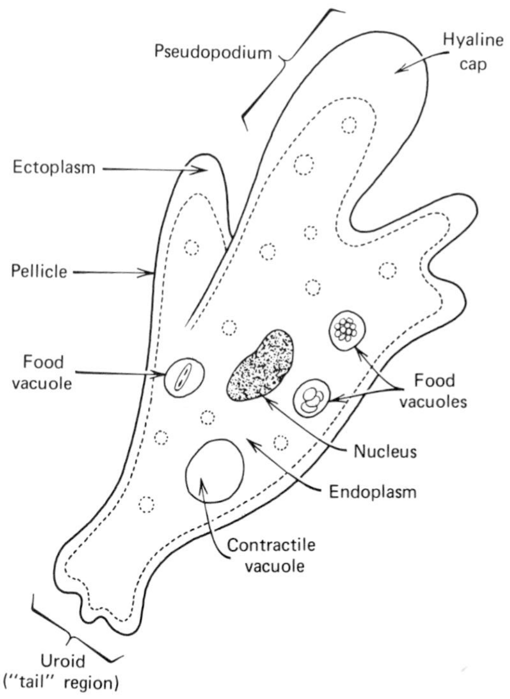

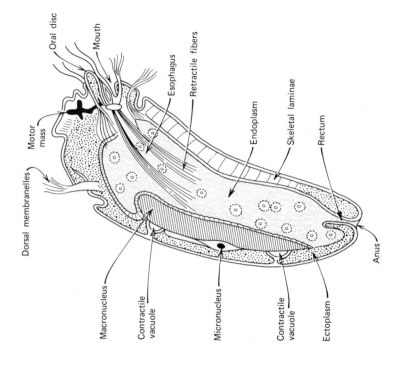

Dorsal membranelles

Motor mass

Oral disc

Mouth

Esophagus

Retractile fibers

Endoplasm

Skeletal laminae

Rectum

Anus

Macronucleus

Contractile vacuole

Micronucleus

Contractile vacuole

Ectoplasm

Figure 20-2 *Epidinium*, a complex, single-celled animal. (Redrawn from *Life: An Introduction to Biology*, Second Edition by George Gaylord Simpson and William S. Beck. Copyright 1965 by Harcourt Brace Javanovich, Inc. by permission of the publishers.)

Hyaline cap

Food vacuoles

Nucleus

Endoplasm

Contractile vacuole

Pseudopodium

Ectoplasm

Pellicle

Food vacuole

Uroid ("tail" region)

Figure 20-1 *Amoeba proteus.*

communicate with a structure called a "motor mass" that resembles a primitive brain. The shape of the cell is due partly to a definite "skeleton."

MEMBRANOUS STRUCTURES
(i) The plasma membrane

In eucaryotic cells the plasma membrane may be specialized in various ways; for instance, in Chapter Eleven, we noted that the myelin sheaths of nerves are constructed from spiral whorls of unit membranes, which are specialized developments of the plasma membrane of the Schwann cells. The membrane stacks of thylakoids of photoreceptor cells are derived from and are specializations of the plasma membrane.

For many years histologists have been familiar with the so-called "brush borders" seen in the epithelial cells lining the inner surfaces of kidney tubules and the intestinal villi. In the light microscope this has a brushlike appearance of faint striping (Fig. 20-3), which electron microscopy of ultrathin sectioned material has shown to consist of myriad fingerlike projections of the plasma membrane (Fig. 20-4). Because of their resemblance to the much larger intestinal villi, they are called microvilli. They are found in cells where active absorption or secretion processes are occurring, and so they probably play an important role in wholesale transport of materials into and out of cells.

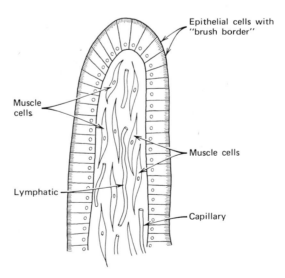

Figure 20-3   Light microscope appearance of epithelial cells with a "brush border" (intestinal villus).

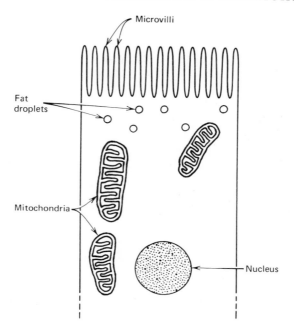

Figure 20-4 "Brush border" (*microvilli*) as seen in ultrathin sectioned cell with the electron microscope.

(ii) The endoplasmic reticulum

The complex system of interconnecting tubular or parallel sheets of lipoprotein membranes in the cytoplasm of eucaryotic cells is called the endoplasmic reticulum (E.R.) (Fig. 20-5). Topologically, J. Robertson considered it to be continuous throughout the cytoplasm and also with the nuclear envelope, but it is not often seen to be continuous with the plasma membrane (Fig. 20-6), or at least the evidence is ambiguous (see later). One hypothesis suggests that the E.R. may have originated in evolution by the outward folding of the plasma membrane of a procaryotic type of cell (Fig. 20-7, first mechanism) or by the invagination of the plasma membrane (Fig. 20-7, second mechanism).

The E.R. is not an inert structural framework, but is constantly changing, being broken down, and resynthesized. It is especially prominent and complex in cells that are actively synthesizing proteins. In such cells polyribosomes (Chapter Eighteen) are seen to be attached to one of the two surfaces of the E.R. membranes; this gives the membrane a rough appearance in electron micrographs and, therefore, it is called "rough E.R." "Smooth E.R." does not have ribosomes attached to it. Because the ribosomes are attached to only one surface

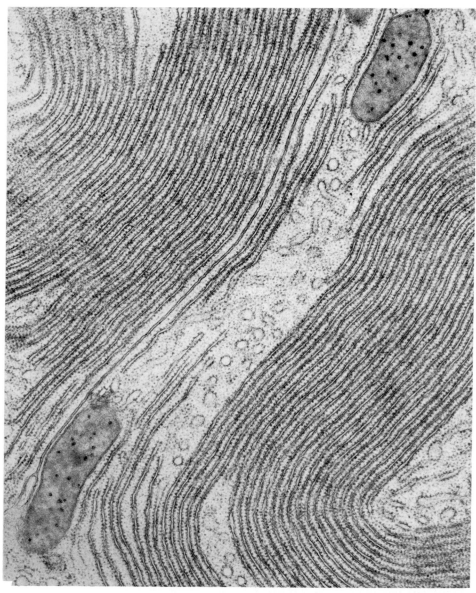

Figure 20-5 Electron micrograph of the endoplasmic reticulum. (Courtesy of Dr D. W. Fawcett. From *An Atlas of Fine Structure* by D. W. Fawcett. W. B. Saunders Company, 1966. Reprinted by permission.)

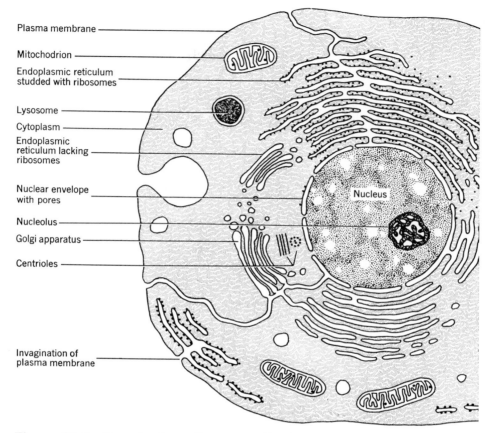

Plasma membrane

Mitochodrion

Endoplasmic reticulum
studded with ribosomes

Lysosome

Cytoplasm

Endoplasmic
reticulum lacking
ribosomes

Nuclear envelope
with pores

Nucleolus

Golgi apparatus

Centrioles

Invagination of
plasma membrane

Nucleus

Figure 20-6 Organization of a generalized metazoan eucaryotic cell as revealed by the electron microscope.

of an E.R. membrane, the cytoplasm is therefore compartmentalized; the ribosomes are contained within one compartment, and their biosynthetic products (proteins) are discharged into the other (Fig. 20-8). Polyribosomes are attached to E. R. membranes only when their protein products are to be transported and extruded from the cell. Polyribosomes that synsezize proteins for local use (sedentary proteins) are not attached to the E.R. membranes. This is seen, for example, in immature red blood cells in which hemoglobin is being synthesized.

(iii) The Golgi complex

The Golgi complex is also prominent in protein-secreting cells and is located near to the surface from which the protein is discharged. It consists of a series of

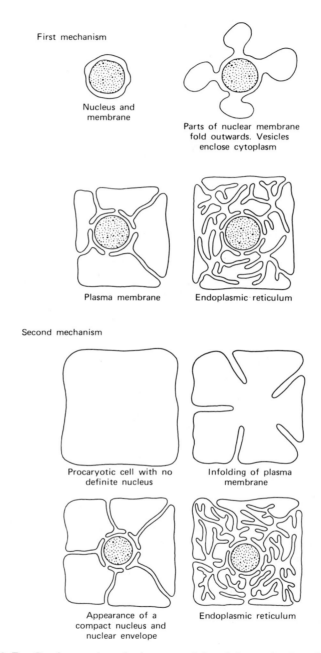

First mechanism

Nucleus and
membrane

Parts of nuclear membrane
fold outwards. Vesicles
enclose cytoplasm

Plasma membrane

Endoplasmic reticulum

Second mechanism

Procaryotic cell with no
definite nucleus

Infolding of plasma
membrane

Appearance of a
compact nucleus and
nuclear envelope

Endoplasmic reticulum

Figure 20-7 Conjectural evolutionary origin of the endoplasmic reticulum (two mechanisms). (Redrawn after W. T. Keeton.)

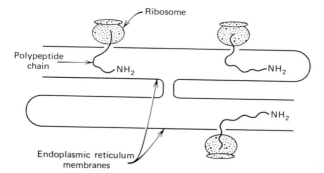

Figure 20-8   Endoplasmic reticulum membranes and ribosomes in protein synthesis.

parallel-stacked, smooth membranes (Fig. 20-9) and appears to be continuous with the E.R., because transitional membrane regions between rough E.R. and Golgi membranes have been observed with the electron microscope in ultrathin-sectioned cells.

Newly synthesized protein is conducted to the Golgi complex, where the protein is "packaged" in membrane vesicles called secretion granules (or zymogen granules, if the protein is an enzyme). The protein packages migrate to the secretory surface of the cell, and the contents are discharged by coalescence of the plasma and vesicle membranes, followed by a process resembling reverse pinocytosis. (Pinocytosis is illustrated in Fig. 20-15.) The role of cytoplasmic membranes and organelles in protein synthesis is summarized diagrammatically in Fig. 20-10.

The discharge of acetycholine from nerve endings may also be achieved by a process of fusion of vesicles with the plasma membrane. Vesicles have been seen in electron micrographs of the cytoplasm of nerve axons close to their terminal neuromuscular or synaptic junctions. That acetylcholine is released in packages has been established from neuromuscular physiology, and it may be that the vesicles seen in nerve axons contain acetylcholine.

The Golgi complex is also thought to be involved in biosynthesis of new E.R. membranes.

(iv) The endoplasmic reticulum in muscle cells

In the cytoplasm between the myofibrils of vertebrate striated muscle (Chapter Five) there are smooth endoplasmic reticulum membranes (or sarcoplasmic reticulum, as it is called in this tissue) in the form of vesicles and flattened tubes (cisternae) that are arranged in a complex regular pattern with respect to the sarcomere A and I band structure and the actomyosin elements. This membrane

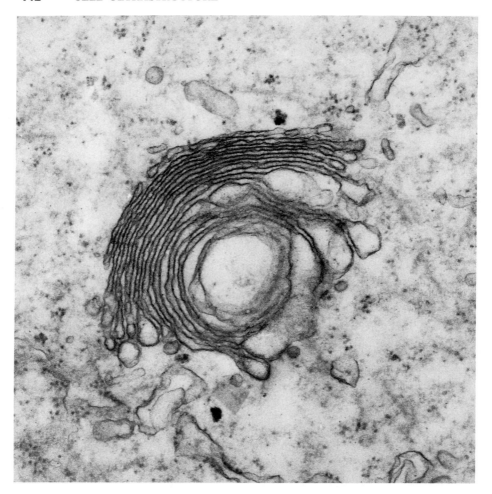

Figure 20-9 Electron micrograph of the Golgi complex. (Courtesy of Dr C. J. Flickinger. *J. Cell Biol.*, *49*(1), page 224, 1971. Reprinted by permission of Rockefeller University Press.)

complex consists of two distinct and unconnected membrane systems, one of which is the true sarcoplasmic reticulum. In the skeletal muscle of amphibians, the sarcoplasmic reticulum proper is made up of a network of parallel tubules wrapped around the sarcomeres and oriented parallel to the long axes of the myofibrils. In the regions of the Z lines, these longitudinal tubules connect with circular transverse tubules that encircle adjacent I bands and overlie the Z region and connect with each other, so that the transverse system of tubules penetrate the thickness of the whole muscle cell. The other system (known as the "transverse" or "T" tubular system of tubules) is formed from invaginations of

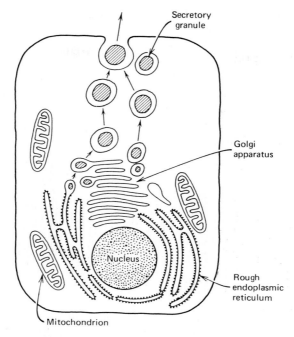

Figure 20-10 Cellular protein synthesis and secretion. (See text for explanation.)

the sarcolemma. These tubules also penetrate the thickness of the muscle cell transversely and run parallel with and closely adjacent to the transverse tubules of the first system. Between them, small spherical vesicles are sandwiched. Hence, in a longitudinal section of muscle, "triads" are seen in the I zones made up of the (larger) transverse components of the sarcoplasmic reticulum (cisternae, spherical vesicles) and the T system (smaller and in between the cisternae) (Fig. 20-11). Notice that the transverse elements are discontinuous at the Z lines.

The sarcoplasmic reticulum and the T system of tubules are deeply involved in the contractile mechanism of muscle. The action potential, which stimulates contraction, travels in the sarcolemma and is transmitted to the deeper lying myofibrils by the invaginated tubular membranes of the T system. It is thought that when the muscle is at rest, the transverse elements (outer vesicles) of the sarcoplasmic reticulum contain most of the free calcium ions necessary for sarcomere contraction. Some kind of signal is conducted along the T system of tubules when the action potential passes over the sarcolemma, which causes the permeability of the sarcoplasmic reticulum membranes to change. The calcium ions within the outer vesicles are therefore released and pass into the sarco-

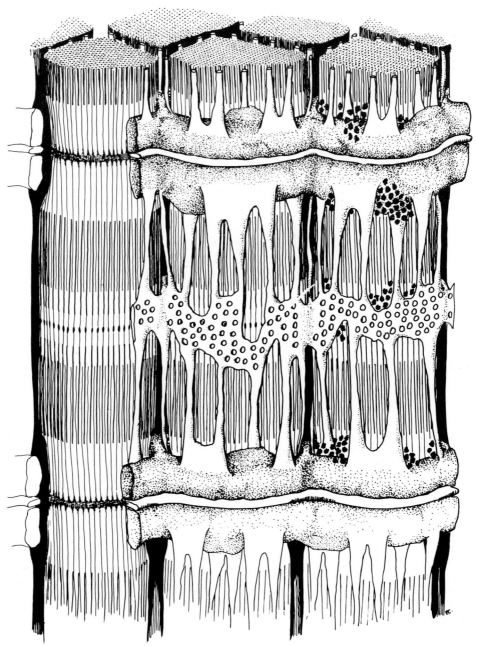

Figure 20-11 Structure of the sarcoplasmic reticulum. (Courtesy of Dr L. D. Peachey. *J. Cell Biol.*, *25*(3), page 222, 1965. Reprinted by permission of Rockefeller University Press.)

plasm. The hydrolysis of ATP and contraction are thus initiated (see Chapter Five). The calcium ions are pumped back into the outer vesicles, and the sarcomere relaxes if there is not another action potential.

Clearly, the intracellular membranous system of eucaryotic cells is deeply involved in vital processes such as the isolation, transport, and secretion of the biosynthetic end products of ribosomes and in the contractile machinery of muscle, to name only two.

(v) The nuclear envelope

This is a double layer of unit membrane material that is continuous with the endoplasmic reticulum. The nuclear material is therefore enclosed by what is really a flattened sac instead of a simple membrane, so it is more usual to refer to it as the nuclear envelope instead of the nuclear membrane (Fig. 20-12). The cytoplasmic surface of the outer membrane may be studded with ribosomes, but they are not seen on the nuclear surface of the inner membrane.

Earlier it was mentioned that whereas the nuclear envelope membranes are continuous with the E.R., there is some doubt about whether the plasma membrane and the E.R. are continuous. If this were so, then this would have important biological consequences, even if such continuity was only intermittent, because it would mean that the nuclear compartment of the cell is in contact with the outside, albeit by a very devious route. Such channels of communication between nuclear compartments and the extracellular environment would play important roles in intra- and extracellular transport of different substances. The question has bearing on virus transmission, because some kinds of freshly synthesized virus find their way into the space around the nucleus in both plant and animal cells.

A direct electron microscope observation of ultrathin sections of cells of *Blasia pusilla* lends qualified support to the idea that the outer nuclear envelope membrane is continuous with the plasma membrane. Z. B. Carothers showed that where the large spherical nuclei of late-stage androgonial cells of *Blasia pusilla* abut against the cell membrane, the outer membrane of the nuclear envelope was continuous with the cell membrane, and the inner membrane was intact and completely surrounds the nucleus (Fig. 20-13). Admittedly, the nucleus is "broadly exposed" in this instance, whereas J. Robertson presumed that the contact between nucleus and cell exterior was by narrow channels. Nevertheless, this observation can be interpreted as supporting the concept of continuity between nuclear and cell (plasma) membranes.

The nuclear envelope is perforated by small apertures (nuclear pores) (Fig. 20-12), where the inner and outer membranes of the nuclear envelope appear to fuse and are bordered by raised "annuli," so that each pore has the appearance of a doughnut or crater (see also Chapter Sixteen). The pore itself is not always

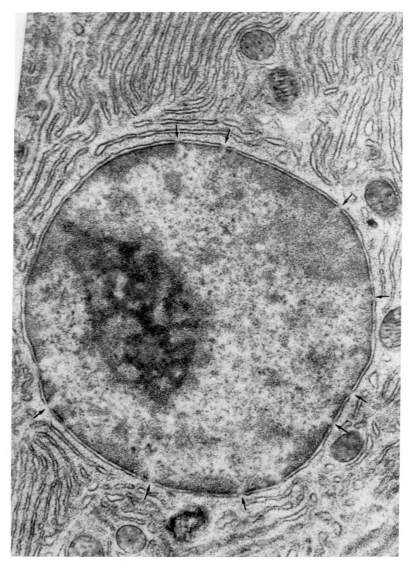

Figure 20-12 Electron micrograph showing the nuclear envelope and nuclear pores in a thin-sectioned cell. (Courtesy of Dr D. W. Fawcett. From *An Atlas of Fine Structure* by D. W. Fawcett. W. B. Saunders Company, 1966. Reprinted by permission.)

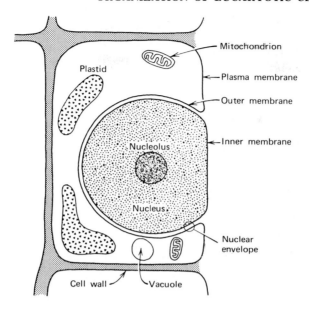

Figure 20-13 Continuity of the plasma membrane and outer nuclear membrane in late stage of androgonial cells of *Blasia pusilla*. (Based on an electron micrograph by Z. B. Carothers.)

open and is sometimes seen covered with a membrane. Passage of materials into and out of the nuclear compartment through these pores would therefore be restricted. The pore membranes probably consist of basic proteins, at least in embryonic honeybee cells, according to DuPraw. The raised rim of the pore has eightfold symmetry and apparently consists of eight symmetrically arranged "spheroids." Electron microscope studies of nuclear envelopes from different animal species show that annuli in the same species may vary in size, although they retain the circular shape, so that they may be contractile and therefore might function like minute sphincters.

The nuclear envelope does not only serve to partition the nuclear chromatin from the cytoplasm; it also functions as an attachment site for many cell structures. As already mentioned in Chapter Sixteen, the interphase chromatin fibers are strongly and specifically attached to the rims of the nuclear pores. The "Barr body" (i.e., the heterochromatic X chromosome of human females) is usually found lying against the inner surface of the nuclear envelope. During meiotic prophase, the pairing chromosomes in many plant and animal species are attached at various points to the nuclear envelope, and both ends of the synaptinemal complex in each pair of homologous chromosomes are also attached to it.

MEMBRANOUS ORGANELLES
(i) Mitochondria

Motichondria were described in detail along with chloroplasts in Chapter Thirteen. They show how lipoprotein membranes are intimately involved in vital activities and that the reactive molecular assemblies may actually be an integral part of the membranes of the organelles. This is again illustrated by the structure and organization of the thylakoid membranes of the outer segments of vertebrate retinal rod cells (Chapter Thirteen). The compartmentalization function of the lipoprotein membranes is also illustrated by the internal membrane systems of mitochondria, chloroplasts, and retinal rod cells.

(ii) Lysosomes

The membrane-bound organelles called lysosomes are about the same size as mitochondria. In 1955, C. de Duve showed that mitochondria and lysosomes were distinctly different organelles. Instead of the enzymes and coenzymes of respiration, lysosomes were found to contain several different hydrolytic enzymes that digest proteins, nucleoproteins, polysaccharides, and other materials. With very few exceptions, such as mammalian erythrocytes, all animal cells contain lysosomes, and there is mounting evidence that they also occur in plant cells. The membrane of the lysosome that seals these enzymes off from the rest of the cell must have remarkable properties, because it does not seem to be attacked by the enzymes. Yet if the lysosomal enzymes escape into the cytoplasm, dissolution and death of the cell follow. In fact, the name of the organelle is derived from its property of causing dissolution or lysis of various substances.

Lysosomes are especially numerous and prominent in cells that ingest and digest particulate matter. Such cells are called phagocytes. Phagocytes (e.g., certain mammalian white blood cells) ingest minute solid particles, such as bacteria, by the cytoplasm flowing around and ultimately engulfing the particle in a membranous bag (phagosome) "nipped off" from the plasma membrane. The process is called phagocytosis, which means literally "cell eating" (Fig. 20-14). The phagosome then fuses with a lysosome by breakdown of the intervening membranes, and the ingested material comes into contact with the lysosomal enzymes and is digested within this digestion vacuole, or "secondary lysosome." (The freshly formed lysosome containing the enzymes only is a primary lysosome.) Undigested materials may be voided from the cell in a "reverse phagocytosis" process (defecation), or the materials may simply remain and accumulate, giving rise to residual bodies; this process is believed to be related to aging.

Cells ingest materials of colloidal or molecular dimensions by the process of pinocytosis (Fig. 20-15). This superficially resembles phagocytosis, but it in-

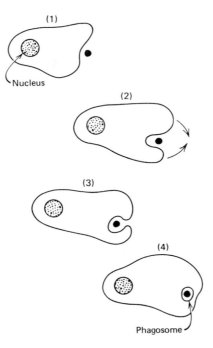

Figure 20-14   Phagocytosis.

volves only the nipping off of plasma membrane vesicles without cytoplasmic flowing and engulfing, as seen in phagocytosis. The resulting pinocytosis vesicles may then unite with primary lysosomes, and their contents are digested. Both phagocytosis and pinocytosis are important feeding mechanisms in unicellular animals such as amoeba.

Bacteria and other foreign particles that enter cells are examples of exogenous materials that are dealt with by lysosomal enzymes. Endogenous materials (i.e., materials from within the cell itself or from cell components) may also become digested by lysosomal enzymes in the process called autophagy ("self-eating"). This occurs, for example, in the liver cells of starving animals, and it appears to be an attempt by the cell to make use of its own substance to supply itself with energy, without causing irreversible damage.

The absorption of the tail during the metamorphosis of the tadpole and the removal of dead cells where renewal is constantly taking place, as in epithelia, are also the result of activity of lysosomal enzymes and phagocytes. When cells die, lysosomal enzymes are released into the cytoplasm and bring about self-digestion or "autolysis" of the cells. Escape of lysosomal enzymes into cytoplasm is believed to occur in certain pathological states, as in vitamin A intoxication.

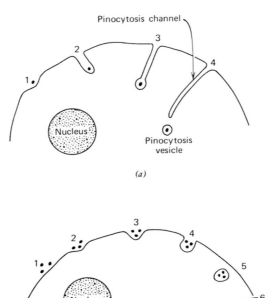

Figure 20-15   Pinocytosis. Two possible mechanisms. (*a*) Formation of pinocytosis channels with formation and detachment of pinocytosis vesicle. (*b*) Direct formation of pinocytosis vesicles at the cell surface.

Vitamin A appears to labilize the lysosomal membrane and to promote leakage of the enzymes. It also occurs when tissues are deprived of oxygen. Lysosomal enzymes may be discharged *outside* the cell and cause lytic effects. This happens with osteoclasts, the cells that resorb bone during the growth and development of the skeleton. This involves both resorption and laying down of new bone tissue. The various possible functions of lysosomes are summarized in Fig. 20-16.

(iii) Peroxisomes

These organelles are approximately spherical, about 0.3 to 1.5 μ in diameter, and are bounded by a single unit membrane. They occur in both plant and animal cells. As seen in the electron microscope, they contain finely granular material, and in several tissues, peroxisomes are seen to contain cores, or nucleoids, which sometimes show a paracrystalline arrangement of tubular

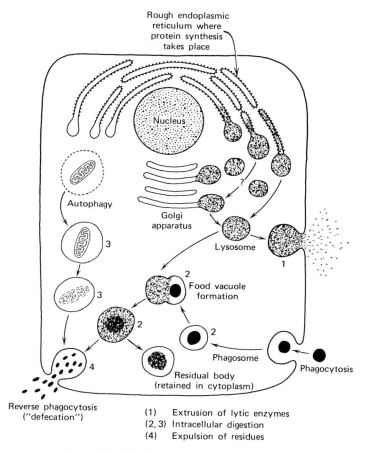

Figure 20-16 Functions of lysosomes.

subunits. The cores seem to be correlated with the presence of the enzyme urate oxidase. In addition to this enzyme, peroxisomes, like lysosomes, contain several other metabolically related enzymes that are involved in the metabolism of hydrogen peroxide. Rat liver and kidney peroxisomes contain three enzymes that produce hydrogen peroxide (urate oxidase, D-amino acid oxidase, and α-hydroxy acid oxidase) and one that destroys it (catalase). Hydrogen peroxide is toxic, so that catalase probably plays an important role in protecting the cell from its harmful effects. Peroxisomes have been found to contain glyoxalate cycle enzymes in germinating castor oil seedlings and other cells in which fat is being converted to carbohydrate (gluconeogenesis). Hence, these are called glyoxysomes. The green leaf cells of plants that contain peroxisomes are able to perform photorespiration, a light-dependent process that requires oxygen and low carbon dioxide levels. Glycolate is produced in chloroplasts during photorespiration, and this passes out into the cytoplasm and is oxidized in peroxisomes. The close spatial relationship observed between chloroplasts and peroxisomes is possibly related to this. A similar relationship of peroxisomes and mitochondria has been observed that may also be related to the metabolic interactions of these two organelles.

(iv) Plastids

This term was first used by Schimper in 1883 to denote certain cytoplasmic organelles that are found in plant cells but not in animal cells. They may contain pigmented materials, in which case they are called chromoplasts; those containing colorless materials are termed leucoplasts. The best-known chromoplasts are the chloroplasts, because they are the sites of photosynthesis. Their detailed structure and function are described in Chapter Thirteen.

Chromoplasts may contain yellow or orange pigments, which occur in the fruits and petals of some higher plants. Red chromoplasts containing the carotenoid lycopene are found in the ripe fruit of the tomato, and chromoplasts containing the pigments phycoerythrin and phycocyanin occur in algae.

Leucoplasts have the ability to synthesize and store food reserves such as starch (amyloplasts), protein (proteinoplasts), and fat (elaioplasts).

(v) Vacuoles

This is a vague term generally applied to roughly spherical, membrane-bound intracellular spaces, or "bubbles," in the cytoplasm. Small vacuoles are called vesicles. Most of the space within a mature plant cell is occupied by a single large vacuole, so that the nucleus is pushed to one side of the cell and the cytoplasm is in the form of a relatively thin peripheral layer. The vacuole contains liquid (cell sap) under pressure that maintains the turgidity of the cell and keeps it pressed tightly against the cell wall. The adult vacuole is formed from several small vacuoles in the immature plant cell that coalesce when the adult stage is reached.

In freshwater protozoa such as amoeba, contractile vacuoles are important in osmoregulation. These are interesting because of their dynamic nature and cyclic activity. Water absorbed by the cell is taken up by many small vacuoles, which may then fuse to form a single large vacuole. This migrates to the cell surface, where its membrane coalesces with the plasma membrane. The membranes break down, and the vacuole contracts and discharges its contents to the exterior. Contractile vacuoles may also be filled from several cytoplasmic channels that converge on a central main vacuole as in *paramecium*.

PROTEIN FIBRILLAR STRUCTURES

(i) Microfilaments

Microfilaments are delicate, cytoplasmic protein threads about 40 to 60 Å in diameter (Fig. 20-16). There are apparently two types. The L or "lattice" type are composed of short lengths and form a reticulum or network near the inner surface of the plasma membrane. The S or "sheath" type are elongated and are found as sheaths of parallel filaments under the plasma membrane.

Microfilaments are found abundantly in cleavage furrows when cells are dividing and in the pseudopodia of amoeboid cells. They seem to be associated with cell movement and, in this connection, it may be significant that the actin filaments of vertebrate sarcomeres are strikingly similar to microfilaments. Perhaps the sliding of these protein filaments over one another in the cell cytoplasm may be the mechanism of cell movement, just as it is in the case of sarcomere contraction.

Structurally, microfilaments are single-stranded polymers of small globular protein subunits.

(ii) Microtubules

Microtubules are more complex than microfilaments and are very fine, straight, cytoplasmic protein tubules that may be up to several microns in length (Fig. 20-17). They have only been thoroughly studied since 1963, when glutaraldehyde was introduced as a cytological fixative in electron microscopy of cells. Microtubules do not survive osmium tetroxide fixation too well. They are composed of a globular protein called tubulin, which has a molecular weight of 110,000. The molecule is dimeric. There are at least two very similar proteins in tubulin preparations, each with a molecular weight of about 55,000. Attempts at *in vitro* reconstitution of tubulin preparations into microtubules were not successful until R. C. Weisenberg showed in 1972 that calcium ions strongly inhibit reconstitution. If calcium ions are rigidly excluded (e.g., by adding a strong calcium chelating agent to the mixture), repolymerization proceeds readily. Magnesium ions and ATP and GTP are also required.

Microtubules are universally present in the cytoplasm of eucaryotic cells and are remarkably uniform in diameter (250 Å). They have never been found to branch. Transverse sections reveal that they have a wall 45 to 70 Å thick that surrounds a less dense central cavity. The microtubule wall appears to be made up of repeating spherical subunits about 50 Å in diameter; transverse sections viewed in the electron microscope have the appearance of being a ring of 12 or 13 of these subunits. Longitudinal views of microtubules have the appearance of consisting of 12 or 13 longitudinally oriented "protofilaments" that have a beaded appearance. Based on these observations, a model of the microtubule

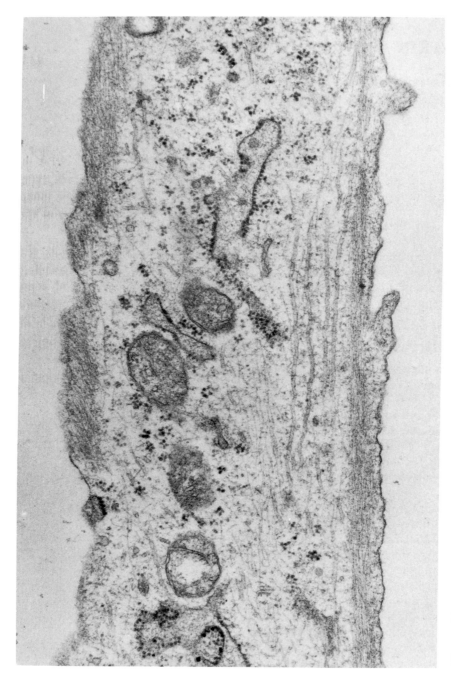

Figure 20-17 Electron micrograph of microfilaments and microtubules. (Courtesy of Dr R. D. Goldman. *J. Cell Biol.*, *51*(3), page 754, 1971. Reprinted by permission of Rockefeller University Press.)

has been presented in which the wall is made up of spherical tubulin subunits helically arranged; these also form the longitudinal rows, which are probably the protofilaments that make up the walls. Microtubule organization is therefore similar to that of the procaryotic flagellum (see diagram of the *Salmonella typhimurium* flagellum, Fig. 19-5, in the previous chapter).

Functionally, microtubules appear to be involved in determining cell shape and in cell movement, both of the cell as a whole and of the movement of materials and structures within the cell. Bundles of microtubules function as a "cytoskeleton" in maintaining the form of structures such as the long, rigid, permanent psueudopodia (axopodia) of the protozoan *Actinosphaerium*. Electron microscopy of thin-sectioned axopodia reveal longitudinal bundles of parallel microtubules running the length of the axopodium. The distinctive biconcave discoidal shape of the mammalian erythrocyte is also thought to be maintained by microtubules.

A suggestion has been made that microtubules are directly responsible for cytoplasmic streaming, but it seems more likely that microfilaments are responsible, and that microtubules may orient or direct the flow. The dramatic cytoplasmic streaming in the formation of the pseudopodia of amoeba or in phagocytic cells is associated with and seems to depend on the presence of microtubules. Amoebae withdraw pseudopodia and microtubules disappear from the cytoplasm if the outside pressure is raised or if the temperature is lowered. Both microtubules and pseudopodia reappear when normal conditions are restored.

(iii) Cilia and flagella

The simplest type of cellular motile organ is the bacterial flagellum, whose structure has already been described (Chapter Nineteen). There is a striking resemblance between bacterial flagella and microtubules, both in the shape and size of the globular subunits and in their arrangement in the organized tubular structure built from them. The flagella and cilia of eucaryotic cells are much larger and more complex. Cilia are 2 to 10 μ long, and flagella may be 100 to 200 μ long. Their diameters are about 0.5 μ. In transverse section eucaryotic cilia and flagella are seen to have a structure consisting of nine peripheral fibers and two axial fibers, the so-called "9+2" arrangement (Figs. 20-18 and 20-19). These fibers are embedded in a structureless matrix. Much earlier (Chapter Five) we noted that the protofibrils of the protein keratin also have a 9+2 arrangement inside the microfibril. Eucaryotic cilia and flagella, unlike bacterial flagella, are invested by extensions of the plasma membrane. The peripheral fibers frequently have a double structure, giving them a "figure of eight" appearance, and there are appendages, as shown in Fig. 20-19. In some cells the two components of the peripheral fibers are comparable to bacterial flagella, both in

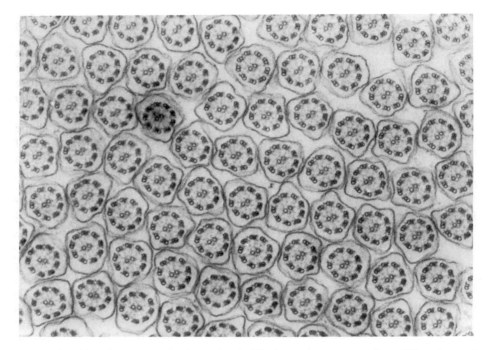

Figure 20-18 Electron micrograph of cilia in transverse section. (Courtesy of Dr P. Satir.)

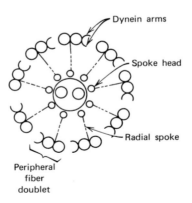

Figure 20-19 Cross section of a eucaryotic cilium.

has been presented in which the wall is made up of spherical tubulin subunits helically arranged; these also form the longitudinal rows, which are probably the protofilaments that make up the walls. Microtubule organization is therefore similar to that of the procaryotic flagellum (see diagram of the *Salmonella typhimurium* flagellum, Fig. 19-5, in the previous chapter).

Functionally, microtubules appear to be involved in determining cell shape and in cell movement, both of the cell as a whole and of the movement of materials and structures within the cell. Bundles of microtubules function as a "cytoskeleton" in maintaining the form of structures such as the long, rigid, permanent psueudopodia (axopodia) of the protozoan *Actinosphaerium*. Electron microscopy of thin-sectioned axopodia reveal longitudinal bundles of parallel microtubules running the length of the axopodium. The distinctive biconcave discoidal shape of the mammalian erythrocyte is also thought to be maintained by microtubules.

A suggestion has been made that microtubules are directly responsible for cytoplasmic streaming, but it seems more likely that microfilaments are responsible, and that microtubules may orient or direct the flow. The dramatic cytoplasmic streaming in the formation of the pseudopodia of amoeba or in phagocytic cells is associated with and seems to depend on the presence of microtubules. Amoebae withdraw pseudopodia and microtubules disappear from the cytoplasm if the outside pressure is raised or if the temperature is lowered. Both microtubules and pseudopodia reappear when normal conditions are restored.

(iii) Cilia and flagella

The simplest type of cellular motile organ is the bacterial flagellum, whose structure has already been described (Chapter Nineteen). There is a striking resemblance between bacterial flagella and microtubules, both in the shape and size of the globular subunits and in their arrangement in the organized tubular structure built from them. The flagella and cilia of eucaryotic cells are much larger and more complex. Cilia are 2 to 10 μ long, and flagella may be 100 to 200 μ long. Their diameters are about 0.5 μ. In transverse section eucaryotic cilia and flagella are seen to have a structure consisting of nine peripheral fibers and two axial fibers, the so-called "9+2" arrangement (Figs. 20-18 and 20-19). These fibers are embedded in a structureless matrix. Much earlier (Chapter Five) we noted that the protofibrils of the protein keratin also have a 9+2 arrangement inside the microfibril. Eucaryotic cilia and flagella, unlike bacterial flagella, are invested by extensions of the plasma membrane. The peripheral fibers frequently have a double structure, giving them a "figure of eight" appearance, and there are appendages, as shown in Fig. 20-19. In some cells the two components of the peripheral fibers are comparable to bacterial flagella, both in

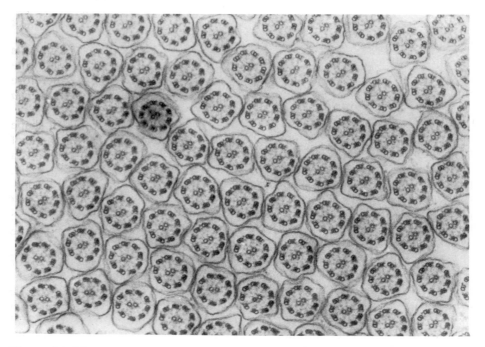

Figure 20-18 Electron micrograph of cilia in transverse section. (Courtesy of Dr P. Satir.)

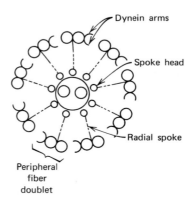

Dynein arms

Spoke head

Radial spoke

Peripheral
fiber
doublet

Figure 20-19 Cross section of a eucaryotic cilium.

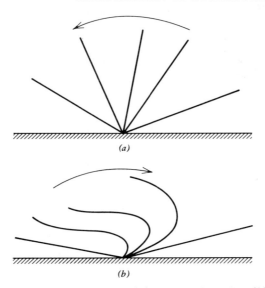

Figure 20-20   Ciliary movement. (*a*) Forward stroke. (*b*) Recovery.

dimensions and in that they are hollow tubes constructed from globular protein subunits. Eucaryotic cells might therefore be regarded as possessing "compound" cilia and flagella that are composed of tubular fibers closely similar to procaryotic flagella. How the three-dimensional spiral motion of the simpler flagellum is converted into the essentially one-dimensional lashing movement of cilia (Fig. 20-20) is still not certainly known.

Ciliary and flagellar motion were once thought to involve the coordinated differential contraction of the microtubules. This could come about by simultaneous rearrangement of the subunits to give a wider and shorter microtubule, a process analogous to the contraction of the bacteriophage tail. However, opposing this idea were electron microscope observations of mollusc gill cilia fixed in various position of bending, by P. Satir. Satir observed that the filaments did not change their lengths during the bending cycle, but that bending was associated with sliding of the filaments over one another. Several other observations by different investigators support an interfilament sliding mechanism.

Isolated flagella continue to beat normally, so that the control process that generates the regular undulations would appear to be within the flagellum itself and probably involves local feedback mechanisms.

That flagellar and ciliary motion might depend on a sliding filament mechanism like that in muscle fibers is an attractive idea, especially since lateral "appendages" are seen on the peripheral filaments (Fig. 20-18). These might permit interaction of some kind between the filaments. I. Gibbons found that these appendages are a protein, which he called dynein; this is an adenosine triphosphatase. The analogy with the protuberances of heavy meromyosin filaments is therefore quite striking. A three-dimensional view of the ultrastructure of a flagellum is shown in Fig. 20-21.

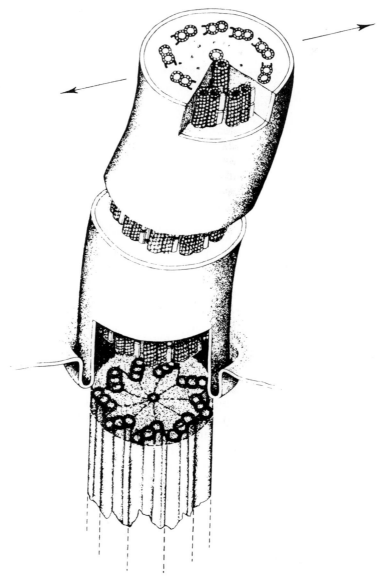

Figure 20-21   Flagellum ultrastructure. (From *Cell and Molecular Biology* by E. J. DuPraw. Academic Press, 1968. Reprinted by permission.)

ATP added to isolated flagella that lack the outer membrane causes them to beat normally. There is a protein that connects the peripheral fibers together and, when ATP is added to preparations in which this protein has been removed enzymatically, sliding movements of the filaments are observed. Gibbons believes that the mechanism of flagellar motion involves interaction of adjacent filamentous doublets, and that it is analogous to the sliding filament mechanism of sarcomere contraction with the energy supplied by ATP. The overall bending of the flagellum must be the result of coordinated movement of several sliding filament systems mutually oriented at different angles. Computer simulation of the motion of a flagellum based on this sliding filament mechanism, and other studies by C. J. Brokaw, show that undulating flagellar motion receives a satisfactory explanation in terms of a sliding filament mechanism.

Not all cilia and flagella are motile; as pointed out in Chapter Thirteen, many cellular sensory receptors, such as the outer segments of retinal rod cells, originate from nonmotile cilia. The grade of organization represented by cilia and flagella in cells therefore seems to be a very fundamental one. This will be even more evident when we examine the ultrastructure and functions of basal granules and centrioles.

(iv) Basal granules

At the base of cilia and flagella, the light microscope reveals a basal granule; in the electron microscope the basal granule is seen to have a structure very similar to cilia and flagella. Basal granules are cylindrical bodies with an internal arrangement of nine peripheral fibers, but there are no central fibers, as seen in cilia and flagella. Basal granules therefore have a "9 + 0" fiber organization (Fig. 20-22). Their function appears to be synthesis of cilia and flagella.

(v) Centrioles

With the light microscope a tiny granule known as the centrosome is seen within a lighter zone—the centrosphere—situated close to the interphase nucleus of

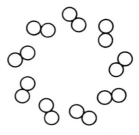

Figure 20-22   Cross section of a basal granule.

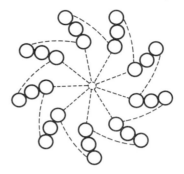

Figure 20-23   Cross section of a centriole.

cells. The electron microscope has shown that the centrosome consists of two smaller bodies, the centrioles, each of which is a short cylinder about 0.15 μ in diameter and 0.3 to 0.5 μ long; both are usually arranged at right angles to each other. The internal organization is similar to that of basal granules; again there is a 9 + 0 arrangement of fibers within them. In centrioles the peripheral fibers have a triplet structure (Fig. 20-23). Like basal granules, the centrioles are involved in production of protein fibers concerned with motion (i.e., the spindle fibers of the mitotic apparatus that are essential for the orderly movement of chromosomes in cell division). Spindle fibers are apparently parallel arrays of microtubules. After cell division, the centrioles duplicate themselves, since only one centriole passes into each daughter cell.

The structural similarity of basal granules and centrioles confirms the belief held since the work of Henneguy and Lenhossek in 1897 that they may be different forms of the same organelle and therefore homologous. In support of this is the observation that in some maturing sperm cells a centriole is associated with growth of a flagellum during division, and this same centriole is in the usual position that it occupies in a dividing cell. Here, then, is the same organelle performing the role of both a centriole and a basal granule. Also, in the differentiated cells of certain higher organisms, the centrioles may undergo many duplications and give rise to basal granules that migrate to the cell surface, where each produces a cilium or flagellum.

(vi) Chromosome movements and spindle fibers

Inoue proposed that mitotic spindle fiber formation and dissolution were the result of polymerization and depolymerization of the subunits that supposedly are always in equilibrium with formed cytoplasmic microtubules. During spindle formation, addition of tubulin subunits outweighs loss, and the fibers grow. They shorten by loss of subunits when loss outweighs addition.

There are orienting centers close to the centrioles; the polar fibers that push the centrioles apart begin at these centers and lengthen by net addition of microtubule subunits. The chromosomal centromeres are the sites of origin of the spindle fibers extending from the chromosomes to the poles. Fiber elongation resulting from tubulin polymerization first pushes chromosomes to the center of the cell, and their subsequent shortening (depolymerization) pulls them apart at anaphase.

Arms extend from the spindle fibers and form connecting bridges between them. Many investigators have suggested that the movements of chromosomes in mitosis are caused by sliding movements of the spindle fibers, and the bridges between them make such interaction possible in a manner reminiscent of the sliding filaments in sarcomere contraction.

CONCLUSIONS

This brief survey of procaryotic and eucaryotic cell structure has shown that among the bewildering variety of cells there are so many common features of structure and function that we are justified in including them all under the one designation of "cells."

Modern ultrastructural research has shown that the simple "amorphous jelly" concept of a cell's cytoplasm, which stemmed from light microscopy, is totally erroneous. Far from being amorphous, the cytoplasm of eucaryotes is almost incredibly complex and structured, and the notion that "simple" unicellular animals like *Ameba proteus* are mere "blobs of jelly" is obsolete; they, too, are really complex and highly organized.

Although there is a marked discontinuity in size and complexity between procaryotic and eucaryotic cells at the molecular level and at the organizational level, represented by macromolecular assemblies, there are many similarities. All cells, for example, possess a plasma membrane that in procaryotic cells simply delimits the cell, but that in eucaryotic cells is developed into the intricate compartmentalized structure of the endoplasmic reticulum and nuclear membrane, both of which may be topologically continuous with the plasma membrane.

Advanced bacterial cells possess internal membranes on which respiratory reactions occur, and blue-green algae possess fairly well-developed internal chlorophyll-containing photosynthetic membranes. In essence, these appear to be similar to eucaryotic respiratory and photosynthetic membranes which, in addition, have become elaborated into the discrete organelles called mitochondria and chloroplasts, respectively, in the higher types of cells. All cells possess the nucleoprotein assemblies called ribosomes, but only in eucaryotic cells are ribosomes found associated with lipoprotein membranes to give the "rough endoplasmic reticulum."

The most primitive type of cellular motile organ is the bacterial flagellum. In eucaryotic cells there are found superficially similar flagella and cilia which, in reality, are much more complex. In fact, they seem to be compounded from fibers organized in a definite $9+2$ pattern, and the fibers appear to be analogous to procaryotic flagella. The constituent fibers of eucaryotic and procaryotic flagella bear a striking resemblance to cytoplasmic microtubules, and microtubules apparently play a major role in amoeboid movement, which superficially is very dissimilar to flagellar or ciliary motion. These two dissimilar motor structures—flagella and pseudopodia—thus have an essential component in common. Also, organelles like basal granules and centrioles that seem to be functionally unrelated are basically similar in structure and are involved in the formation of microtubules either in flagella (basal granules) or in spindle fibers, which are concerned with chromosome movement. Therefore, a wide range of apparently unrelated cell structures have microtubules as common structural components, and all are concerned with movement of one sort or another.

The elaborate chromosomal apparatus of eucaryotic cells may simply represent local "condensations" of what is a single very large circular chromosome (Chapter Sixteen), of which the procaryotic circular DNA molecule is a much smaller and simpler version. The more complex organization of eucaryotic DNA, together with various basic proteins into distinctive chromosomal structures, may be a necessary condition for the specialization that is so prominent in eucaryotic cells. The compacting of the eucaryotic DNA into a membrane-bound nucleus is the most basic and obvious structural difference between procaryotic and eucaryotic cells.

To summarize, all cells are composed of a relatively few basic structural components such as lipoprotein membranes, protein microtubules, and nucleoprotein fibers and granules (ribosomes). In procaryotic cells these are found in simple uncomplicated form, but are found elaborated into more complex diversified membrane systems, fibrous structures, and organelles in eucaryotic cells. Furthermore, as hinted at earlier, we find that the three basic types of macromolecular assembly are associated with certain major "themes"; lipoprotein membranes are associated with cell compartmentalization and vital metabolic activities, protein fibers with movement and skeletal support, and nucleoprotein assemblages with genetic information storage and transmission and its ultimate expression as protein structure.

THE ORIGIN OF EUCARYOTIC CELLS

The basis for the pronounced separation of cells into procaryotic and eucaryotic can be related to the marked difference in size between them. In very small cells such as those of PPLOs and bacteria, intracellular materials, including macromolecular substances, can probably be transported around the cell with sufficient speed by the random process of diffusion. In the much larger cells of eucaryotes, diffusion would not be sufficiently rapid, and it is in these cells that we find cytoplasmic streaming and the highly organized cytoplasmic compartmentalization (both of which are associated with intracellular transport) so well developed.

The surface area of a cell increases as the square of the linear dimensions, but the volume increases as the cube (i.e., volume increase rapidly outruns surface area increase as cells become larger). In the very small cells of PPLOs, the surface area: volume ratio is apparently large enough for the many membrane-associated metabolic reactions to occur on a scale sufficient for cellular needs. In the larger cells of bacteria, there is some development of folded internal membranes to provide extra surface area for metabolic reactions (e.g., the respiratory membranes of the mesosome, which apparently are developed to keep pace with the energy requirements of the bacterial cell, which is much larger than a PPLO cell). In the much larger cells of eucaryotes, the complex internal membrane system is therefore proportionately much more developed, and respiratory membranes are more abundantly developed within the multitude of mitochondria that are found in aerobic cells. Likewise, the simple photosynthetic membranes of blue-green algae are seen in more elaborate form with greatly increased surface area in the internal thylakoid system of the chloroplasts of higher plants.

Eucaryotic cells are supposed to have been descended from procaryotic ancestors. In 1890, Altmann and Schimper presented the idea that eucaryotic mitochondria are the descendants of what were once free-living, procaryotic organisms that became intracellular symbionts of larger procaryotes. A controversial theory of "serial symbiosis" has now been developed to account for the symbiotic origin of eucaryotic cell organelles; its main proponent is L. Margulis, who points to the "aerobic crisis" of millions of years ago as the initiating stimulus for serial symbiosis. That the primitive terrestrial atmosphere was reducing instead of oxidizing, as it is today, is generally accepted by biologists and geologists. The primeval ocean was a "soup" of abiogenically synthesized organic molecules that provided food for primitive heterotrophic procaryotic cells. Such cells were necessarily anaerobic and obtained their energy by glycolysis. Some types of bacterialike procaryotes developed pigments that enabled them to photosynthesize and hence to become independent of the external food supply that was becoming depleted as a result of two processes—gradual cessation of the electrical storms that presumably supplied the energy for abiogenic syntheses, and the increasing cell population. The oxygen evolved by these photosynthetic bacteria began to accumulate (the first known example of atmospheric pollution, as one author has remarked) and changed the reducing atmosphere to one that was oxidizing. This was the "aerobic crisis." It is called this because oxygen in the atmosphere threatened the existence of the obligate anaerobes, since to them it was toxic. Some bacterialike procaryotes apparently "learned" how to cope with and utilize the oxygen and so became aerobic bacteria. Others did not develop metabolic means to detoxify oxygen and sought out anaerobic niches in which to live. Present-day obligate anaerobic bacteria are believed to be their descendants. Other large, anaerobic, amoebalike cells entered, instead, into symbiotic relationships with aerobic bacteria, which became engulfed into the cytoplasm of the larger cell, but were not destroyed. The smaller aerobic organism now had a constant supply of three-carbon compounds produced by glycolysis within the larger cell's cytoplasm, which they could break down completely to carbon dioxide and water, converting the oxygen harmful to the host cell into harmless substances. Ultimately, some degree of autonomy of the intracellular symbionts was lost, and they became mitochondria.

These large aerobic, heterotrophic, ancestral eucaryotes with mitochondria in their cytoplasm are thought to have undergone a second symbiotic process in which photosynthetic bacteria were ingested and became the ancestors of chloroplasts. The large, aerobic, heterotrophic eucaryotes are considered to be ancestral to protozoa, metazoa, and fungi, and the eucaryotes with chloroplasts are considered to be ancestral to green algae and higher green plants. Thus, to the orthodox evolutionary processes of mutation and selection a third process, serial symbiosis, is added.

The theory of serial symbiosis has many attractive features, only one of which is that, according to it, the ancestors of fungi never did possess chlorophyll. Therefore, it is not necessary to invent ad hoc hypotheses of secondary loss of chlorophyll from a hypothetical, ancestral, photosynthetic phytoflagellate which, according to more orthodox ideas of evolution, is supposed to have been the ancestor of both the plant and animal kingdoms.

Eucaryotic mitochondria and chloroplasts exhibit a certain amount of functional autonomy and possess their own specific DNA, protein-synthesizing apparatus, and specific proteins. The circular DNA of these organelles is reminiscent of bacterial chromosomes, and the infoldings (cristae) of the inner membrane of the mitochondrial membrane, whose function is respiratory, reminds us of the bacterial mesosome and its respiratory function. We have also noted (Chapter Eighteen) that procaryotic ribosomes (*E. coli*) are smaller than eucaryotic ribosomes; eucaryotic mitochondrial ribosomes resemble those of bacteria more closely than those in the cell cytoplasm. When the cell divides, mitochondria and chloroplasts also divide in a precise manner that is coordinated with the division of the cell, so that the daughter cells inherit the same number of mitochondria and chloroplasts that the parent cell had. Other cellular organelles do not exhibit this phenomenon. These and other facts strengthen the idea that mitochondria and chloroplasts are descended from what were once independent, free-living microorganisms. The considerable chemical differences between the inner and outer membranes of mitochondria, the outer one of which resembles the E.R. membranes, further strengthens the idea that the mitochondrion is descended from an intracellular symbiont invested with cytoplasmic membranes.

That intracellular symbiosis occurred in these primitive cells is entirely probable and is a well-established biological phenomenon, many examples of which are known today. Two of the best known are the cells of the photosynthetic algae within the cells of *Hydra viridis*, which coexist in a symbiotic relationship, and the hereditary symbiosis between the photosynthetic alga *Chlorella* within the cytoplasm of *Paramecium bursaria*. Bacterial cells have been artificially induced to live symbiotically within the cytoplasm of *Amoeba discoides*, as was achieved by K. W. Jeon in 1972. He infected amoebae with bacteria and, after 5 years, the bacteria that at first were harmful were now harmless, and the host cell nucleus became dependent on the presence of the bacteria for normal function. This appears to be the first demonstration, during the relatively short time of 5 years, of intracellular parasitic organisms becoming observably transformed into genuine symbionts.

The serial symbiosis theory of the origin of eucaryotic cells is plausible, attractive, and aesthetically pleasing. However, there are objections to it. R. A. Raff and H. R. Mahler argue the case for nonsymbiotic origin of mitochondria and point out three objections to the serial symbiosis theory.

1. The postulated ancestral eucaryotes have a primitive character and metabolic inefficiency in spite of their advanced cellular adaptations. This is difficult to reconcile with their apparent success in competing with advanced and metabolically more efficient procaryotes.
2. It is difficult to envisage how wholesale gene transfer from the endosymbiotic-protomitochondrion to the foreign host cell genome could have occurred.
3. The fossil record of eucaryotes and the biochemistry of eucaryotic cells are inconsistent with the serial symbiosis theory.

Raff and Mahler consider that mitochondria could have arisen in eucaryotes by invagination of respiratory membranes and by their becoming "nipped off" as closed vesicles. For increased efficiency of maintenance of this organelle, a protein-synthesizing system was implanted in the mitochondrion by the incorporation of a stable plasmid carrying the necessary genes for determining the components of the mitochondrial ribosomes. This process could have been analogous to a phenomenon known to occur today in amphibian oogenesis when multiple nucleoli are generated. Many circular copies of ribosomal genes are coded by chromosomal DNA that determines rRNA and are concentrated in the nucleoli that synthesize the abundance of RNA that the eggs need.

The whole matter of the symbiotic origin of eucaryotic cells and organelles must therefore still be regarded as controversial.

MACROMOLECULAR SELF-ASSEMBLY

One of the most interesting properties of biological macromolecules is their ability to assemble into higher structures spontaneously. A simple example is the spontaneous assembly of the tetrameric hemoglobin molecule from its globular subunits. Many others have been noted elsewhere in this textbook: the assemblage of collagen and profibrin molecules into fibers; the self-assembly of globular enzyme subunits such as those of lactate dehydrogenase and pyruvate decarboxylase into complex functional assemblies; and the spontaneous reconstitution of the tobacco mosaic virus from its protein and nucleic acid. In these cases no energy or enzymes are required. Perhaps sometime in the future the conditions and laws governing self-assemblage of cellular organelles from macromolecular components may be discovered to be also the result of thermodynamic forces acting on their own. Can we extrapolate and envision undiscovered laws that determine the assemblage of living cells from supramolecular assemblies as a result of such forces? Apparently not, because beyond the stage of spontaneous assemblage of even several different types of protein structural components such as the reconstitution of the pyruvate decarboxylase complex, it would appear that thermodynamic forces alone are not enough. The T_4 bacteriophage, for instance, which is more complex than TMV but much less complex than the simplest cell, cannot be spontaneously reassembled from its protein and nucleic acid components completely. W. B. Wood and R. S. Edgar point out that more than 40 genes in the T_4 bacteriophage control the morphogenetic process, probably by gene products that may direct the assembly process without actually contributing to the structure. The attachment of the tail fibers requires the presence of a virus-induced enzyme. Apparently, as E. Kellenberger points out, while there is sufficient information within the three-dimensional structure of the TMV protein subunits to direct spontaneous reassembly of virus particles, more is needed for reconstitution of more complex structures such as the head of the T_4 bacteriophage, and this is stored in the viral genes. In addition, the viral head needs an internal "scaffolding" to assist proper assemblage.

In these days of astounding advances in science and technology it is perhaps rash to declare dogmatically that anything such as the artificial synthesis of a living cell is impossible. Yet, on what sort of microloom would a biologist weave the membranes of the endoplasmic reticulum, or with what delicate needles could a biologist fashion the intricacies of the cell nucleus? K. E. Jeon, I. J. Lorch, and J. F. Danielli succeeded in "Frankenstein" experiments in which, by using microsurgical procedures, they "made" viable amoebae by reassembling nuclei, cytoplasm, and plasma membranes from different individuals of the same strain. They were even able to reassemble the three components from individuals of different strains of amoebae; these composite amoebae survived, but did not give rise to viable clones. Although this is a great technical achievement, it comes nowhere near to "creating life" in the usually accepted meaning of that phrase (i.e., making something that is alive from materials that were never alive).

Will it ever be possible for a living cell to be constructed from scratch under controlled laboratory conditions? Perhaps it is pointless to continue with such speculations because there seems to be a step beyond which man cannot go—try as he may. Consider an analogy. We can compare the living cell to a motor car. At present we have learned something about the assemblage of some of the components of a car, such as the gears and the batteries, and we have actually isolated these components and got them to work on their own outside of the car. We have yet to learn how all these components are assembled, which along with a full gas tank and fresh batteries, make a car that will function. We may even be able to start the engine and be ready to "go," but even if we could, the car would not really "come to life" because there must be someone to drive it. What—if anything—corresponds to a driver in the living cell? That is, is there a supreme controlling and coordinating power in the cell that is as much beyond our comprehension and as superior as the driver is to the motor car? Perhaps at this point the analogy is being pushed too far and we are straying into too highly controversial and speculative an area. We are really asking again what makes the cell "alive." We do not think of mitochondria, chromosomes, or ribosomes as being alive, although it has been possible to isolate many types of cellular structures that continue to function; *in vitro* replication of DNA has been achieved.

What we call life appears to be some sort of unifying and coordinating power that causes all macromolecular structures and organelles in the cell to function together and, from this emerges the living cell—something greater than the sum of its parts in that it exhibits the phenomenon we call life, just as new and unexpected novel phenomena emerge from the integration of globular protein molecule subunits into quaternary and quintary structure.

The whole question of what is life is further complicated by the fact that there is cellular life that is distinct from organismal life. A multicellular organism can "die," and yet virtually every cell of its body will be recognizably "alive" immediately after death in the sense of exhibiting vital phenomena, and cells can be removed from a freshly dead organism and cultured indefinitely. What, then, is the nature of the difference between cellular and organismal life and death? Perhaps the gulf between the two of these is even greater than that between a living cell and its nonliving molecular components. Possibly we will never penetrate the twin mysteries of cellular and organismal life and death; perhaps it is just as well.

References

CHAPTER ONE

Articles

1. "Origins of Molecular Biology." Hess, E. L. *Science, 168*:664 (1970).
2. "Phage and the Origins of Molecular Biology." Cairns, J., G. S. Stent, and J. D. Watson (Eds.). *Cold Spring Harbor Laboratory* (1966).
3. "That Was the Molecular Biology that Was." Stent, G. S. *Science, 160*:390 (1968).
4. "Molecular Biology: Gene Insertion into Mammalian Cells." Rabovsky, D. *Science, 174*:933 (1971).

Books

1. *Discovering the Basis of Life.* Roller, A., McGraw-Hill (1974).
2. *The Chemical Foundations of Molecular Biology.* Steiner, R. F., D. Van Nostrand Co. (1965).
3. *Molecular Biology: Genes and the Chemical Control of Living Cells.* Barry, J. M. (Concepts of Modern Biology Series) Prentice-Hall (1964).
4. *Molecular Insights into the Living Process.* Green, D. E., and R. F. Goldberger, Academic Press (1967).
5. *Introduction to Molecular Biology* (2nd edition). Haggis, G. H., D. Michie, A. R. Muir, K. B. Roberts and P. M. B. Walker, Longmans Green & Co. (1973).
6. *Cell and Molecular Biology.* DuPraw, E. J., Academic Press (1968).
7. *Molecular Biology of the Gene* (3rd edition). Watson, J. D., W. A. Benjamin, Inc. (1976).

8. *Molecular Biology. A Structural Approach.* Smith, C. U. M., Faber & Faber (1968).
9. *The Chemical Basis of Life.* (Scientific American) Hanawalt, P. C., and R. H. Haynes (Eds.), W. H. Freeman & Co. (1973).
10. *Structural Chemistry and Molecular Biology.* Rich, A., and N. Davidson (Eds.), W. H. Freeman & Co. (1968).
11. *The Molecular Basis of Evolution.* Anfinsen, C. B., John Wiley (1963).
12. *Chance and Necessity.* Monod, J., (Fontana Books) Collins (1974).
13. *Adventures in Molecular Biology.* Astbury, W. T., Charles C. Thomas, (1952).

CHAPTER TWO

Books

1. *A Biologist's Physical Chemistry.* Morris, J. G., Edward Arnold Ltd. (1968).
2. *Matter, Energy and Life.* Baker, J. W., and G. E. Allen, Addison-Wesley (1965).

CHAPTER THREE

Articles, Research Papers, and Reviews

1. "The Structure of Proteins." Stein, W. H. and S. Moore. *Sci. Am.* (February 1961).
2. "Proteins." Fruton, J. S. *Sci. Am.* (June 1950).
3. "How is a Protein Made?" Linderstrom-Lang, K. U. *Sci. Am.* (September 1953).
4. "Chromatography." Stein, W. H., and S. Moore. *Sci. Am.* (March 1951).
5. "The Insulin Molecule." Thompson, E. O. P. *Sci. Am.* (May 1955).
6. "The Amino Acid Sequence in the Phenylalanyl Chain of Insulin." Sanger, F., and H. Tuppy. *Biochem. J., 49*:463 (1951).
7. "The Amino Acid Sequence in the Glycyl Chain of Insulin." Sanger, F., and E. O. P. Thompson. *Biochem. J. 53*:353 (1953).
8. "The Disulphide Bonds of Insulin." Ryle, A. P., F. Sanger, L. F. Smith, and R. Kitai. *Biochem. J., 60*:541 (1955).
9. "The Structure of Pig and Sheep Insulins." Brown, H., F. Sanger, and R. Kitai. *Biochem. J., 60*:556 (1955).
10. "The ACTH Molecule." Choh Hao Li. *Sci. Am.* (July 1963).
11. "Bee and Wasp Venoms." Haberman, E. *Science, 177*:314 (1972).
12. "The Structure and History of an Ancient Protein." Dickerson, R. E. *Sci. Am.* (April 1972).
13. "Cytochrome c." Margoliash, E., and A. Schejter. *Adv. Prot. Chem., 21*:113 (1966).
14. "Computer Analysis of Protein Evolution." Dayhoff, M. O. *Sci. Am.* (July 1969).
15. "The Newer Concept of Cancer Toxin." Nakahara, W., and Fukuoka, F. *Adv. Cancer Res., 5*:157 (1958).
16. "The Identification and Characterization of Amino Acids and Peptides." Leaf, G. *Biochem. J., 107*:11P (1968).

17. "Determination of the N-terminal Amino Acids in Proteins Using Fluoro-Nitro-pyridines." Signor, A., L. Biondi, A. M. Tamburro, and E. Bordignon. *European J. Biochem.*, *7*:328 (1969).
18. "Interaction of Polypeptide Hormones with Lipid Monolayers." Snart, R. S., and N. N. Sanyal. *Biochem. J.*, *108*:369 (1968).

CHAPTER FOUR
Articles and Research Papers

1. "Proteins." Doty, P. *Sci. Am.* (September 1957).
2. "The Structure of Protein Molecules." Pauling, L., R. B. Corey, and R. Hayward. *Sci. Am.* (July 1954).
3. "The Structure of Proteins: Two Hydrogen-Bonded Helical Configurations of the Polypeptide Chain." Pauling, L., R. B. Corey, and H. R. Branson. *Proc. Nat. Acad. Sci. U.S.A.*, *37*:205 (1951).
4. "The Dependence of the Conformation of Synthetic Polypeptides on Amino Acid Composition." Blout, E. R., C. De Lozé, S. M. Bloom, and G. D. Fasman. *J. Am. Chem. Soc.*, *82*:3787 (1960).
5. "A Correlation Between Amino Acid Composition and Protein Structure." Davies, D. R. *J. Mol. Biol.*, *9*:605 (1964).
6. "The Molecular Structure of Collagen." Rich, A., and F. H. C. Crick. *J. Mol. Biol.*, *3*:483 (1961).
7. "The Pleated Sheet, A New Layer Configuration of Polypeptide Chains." Pauling, L., and R. B. Corey. *Proc. Nat. Acad. Sci. U.S.A.*, *37*:251 (1951).

Book

1. *The Structure and Action of Proteins.* Dickerson, R. E., and I. Geis, Harper & Row (1969).

CHAPTER FIVE
Articles, Research Papers, and Reviews

1. "The Structure Proteins." Seifter, S., and P. M. Gallop. In *The Proteins. IV*, p. 155. (H. Neurath, Ed.) Academic Press, (1966).
2. "Collagen." Gross, J. *Sci. Am.* (May 1961).
3. "The Chemistry and Structure of Collagen." Traub, W., and K. A. Piez. *Adv. Prot. Chem.*, *25*:243 (1971).
4. "The Aging of Collagen." Verzar, F. *Sci. Am.* (April 1963).
5. "Elastic Fibers in the Body." Ross, R., and P. Bornstein, *Sci. Am.* (June 1971).

6. "The Structure and Chemistry of Keratin Fibers." Bradbury, J. H. *Adv. Prot. Chem.*, *27*:111 (1973).
7. "The Chemistry of Keratins." Crewther, W. G., R. D. B. Fraser, F. G. Lennox, and H. Lindley. *Adv. Prot. Chem.*, *20*:191 (1965).
8. "Keratins." Fraser, R. D. B. *Sci. Am.* (August 1969).
9. "The Mechanism of Muscular Contraction." Huxley, H. E. *Sci. Am.* (December 1965).
10. "The Cooperative Action of Muscle Proteins." Murray, J. M., and A. Weber. *Sci. Am.* (February 1974).

CHAPTER SIX

Articles and Research Papers

1. "The Three-Dimensional Structure of a Protein Molecule." Kendrew, J. C. *Sci. Am.* (December 1961).
2. "The Hemoglobin Molecule." Perutz, M. F. *Sci. Am.* (November 1964).
3. "The Three-Dimensional Structure of an Enzyme Molecule." Phillips, D. C. *Sci. Am.* (November 1966).
4. "The Formation and Stabilization of Protein Structure." Anfinsen, C. B. *Biochem. J.*, *128*:737 (1972).
5. "The Use of X-ray Diffraction in the Determination of Protein Structure." Low, B. W. *J. Polymer Sci.*, 49:153 (1961).

Book

1. *The Structure and Action of Proteins.* Dickerson, R. E., and I. Geis, Harper & Row (1969).

CHAPTER SEVEN

Articles and Research Papers

1. "The Active Site and Enzyme Action." Koshland, D. E., Jr. *Adv. Enzymol.*, *22*:45 (1960).
2. "A Family of Protein Cutting Proteins." Stroud, R. M. *Sci. Am.* (July 1974).
3. "Protein Digesting Enzymes." Neurath, H. *Sci. Am.* (December 1964).
4. "The Three-Dimensional Structure of an Enzyme Molecule." Phillips, D. C. *Sci. Am.* (November 1966).
5. "Structure of Hen Egg-White Lysozyme—A Three-Dimensional Fourier Synthesis at 2 Å Resolution." Blake, C. C. F., D. F. Koenig, G. A. Mair, A. C. T. North, D. C. Phillips, and V. R. Sarma, *Nature*, *206*:757 (1965).

6. "Structure of Some Crystalline Lysozyme-Inhibitor Complexes Determined by X-ray Analysis at 6 Å Resolution." Johnson, L. N., and D. C. Phillips, *Nature, 206*:761 (1965).

7. "Tertiary Structure of Ribonuclease." Kartha, G., J. Bello, and D. Harker. *Nature, 213*:862 (1967).

8. "The Structure of Carboxypeptidase A. VI. Some Results at 2 Å Resolution, and the Complex with Glycyl-Tyrosine at 2.8 Å Resolution." Reeke, G. N., J. A. Hartsuck, M. L. Ludwig, F. A. Quiocho, T. A. Steitz, and W. N. Lipscomb. *Proc. Nat. Acad. Sci. U.S.A. 58*:2220 (1967).

9. "Three-Dimensional Structure of Tosyl-α-Chymotrypsin." Matthews, B. W., P. B. Sigler, R. Henderson, and D. W. Blow. *Nature. 214*:652 (1967).

10. "Structure of Subtilisin BPN′ at 2.5 Angstrom Resolution." Wright, C. S., R. A. Alden, and J. Kraut. *Nature, 221*:235 (1969).

11. "Structure of Papain." Drenth, J., J. N. Jansonius, R. Koekoek, H. W. Swen, and B. G. Wolthers, *Nature, 218*:929 (1968).

12. "Acetylcholinesterase VIII. Dissociation Constants of the Active Groups." Wilson, I. B., and F. Bergmann. *J. Biol. Chem., 186*:683 (1950).

13. "The Acetylcholinesterase Surface. VI. Further Studies with Cyclic Isomers as Inhibitors and Substrates." Friess, S. L., and H. D. Baldridge. *J. Am. Chem. Soc., 78*:2482 (1956).

14. "Carbamyl Derivatives of Acetylcholinesterase." Wilson, I. B., M. Harrison, and S. Ginsburg. *J. Biol. Chem., 236*:1498 (1961).

15. "Concerted Displacement Reactions. VIII. Polyfunctional Catalysis." Swain, C. G., and J. F. Brown, Jr. *J. Am. Chem. Soc., 74*:2538 (1952).

16. "The Anatomy of an Enzymatic Catalysis." Bender, M. L., F. J. Kézdy, and C. R. Gunter. *J. Am. Chem. Soc., 86*:3714 (1964).

17. "The Current Status of the α-Chymotrypsin Mechanism." Bender, M. L., and F. J. Kézdy. *J. Am. Chem. Soc., 86*:3704 (1964).

18. "Correlations of Structure and Function in Enzyme Action." Koshland, D. E. *Science, 142*:1533 (1963).

19. "Chemistry and Physiology of the Fibrinogen-Fibrin Conversion." Laki, K., and J. A. Gladner. *Physiol. Rev., 44*:127 (1964).

20. "The Preparation of Subtilisin-Modified Ribonuclease and the Separation of the Peptide and Protein Components." Richards, F. M., and P. J. Vithayathil. *J. Biol. Chem., 234*:1459 (1959).

Books

1. *Enzymes* (2nd edition). Dixon, M., and E. C. Webb, Longman (1964).

2. *The Structure and Function of Enzymes.* Bernhard, S., W. A. Benjamin Inc. (1968).

3. *Catalysis and Enzyme Action.* Bender, M. L., and L. J. Brubacher, McGraw-Hill (1973).

4. *Dynamic Aspects of Biochemistry* (5th edition). Baldwin, E. B., Cambridge University Press (England) (1967).

CHAPTER EIGHT

Articles, Research Papers, and Reviews

1. "The Structure of Antibodies." Porter, R. R. *Sci. Am.* (October 1967).
2. "The Structure and Function of Antibodies." Edelman, G. M. *Sci. Am.* (August 1970).
3. "How Cells Make Antibodies." Nossal, G. J. V. *Sci. Am.* (December 1964).
4. "Electron Microscopy of an Antibody–Hapten Complex." Valentine, R. C., and N. M. Green. *J. Mol. Biol.*, *27*:615 (1967).
5. "Structural Studies of Immunoglobulins." Porter, R. R. *Science, 180*:713 (1973).
6. "Antibody Structure and Molecular Immunology." Edelman, G. E. *Science, 180*:830 (1973).
7. "The Antibody Problem." Edelamn, G. M., and W. E. Gall. *Ann. Rev. Biochem.*, *38*:415 (1969).
8. "Antibody Structure: Now in Three Dimensions." Marx J. L. *Science, 189*:1075 (1975).
9. "Antibody Variability." Smithies, O. *Science, 157*:267 (1967).
10. "Mechanism of Antibody Diversity: Germ Line Basis for Variability." Hood. L., and Talmage, D. W. *Science, 168*:325 (1970).
11. "Gene Selection in Hemoglobin and in Antibody-Synthesizing Cells." Kabat, D. *Science, 175*:134 (1972).
12. "The Complement System." Mayer, M. M. *Sci. Am.* (November 1973).

Book

1. *Structural Concepts in Immunology and Immunochemistry.* Kabat, E. A., Holt, Rinehart & Winston (1968).

CHAPTER NINE

Articles and Research Papers

1. "The Hemoglobin Molecule." Perutz, M. F. *Sci. Am.* (November 1964).
2. "Fetal Hemoglobin." White, J. C., and G. H. Beavan. *Brit. Med. Bull.*, *15*:33 (1959).
3. "Sickle Cell Anemia: A Molecular Disease." Pauling, L., H. A. Itano, S. J. Singer, and I. C. Wells. *Science, 110*:543 (1949).
4. "Abnormal Human Hemoglobins. I. The Comparison of Normal Human and Sickle Cell Hemoglobins by 'Fingerprinting.'" Ingram, V. M. *Biochim. et Biophys. Acta*, *28*:539 (1958).
5. "A Specific Chemical Difference Between the Globins of Normal Human and Sickle-cell Anaemia Hemoglobin." Ingram, V. M. *Nature, 178*:792 (1956).
6. "How Do Genes Act?" Ingram, V. M. *Sci. Am.* (January 1958).

7. "Hemoglobin Gun Hill: Deletion of Five Amino Acid Residues and Impaired Heme-Globin Binding." Bradley, T. B., R. C. Wohl, and R. F. Rieder. *Science*, *157*:1581 (1967).

8. "Recombination of Human Adult and Canine Hemoglobins at Neutral pH." Enoki, Y., and S. Tomita. *J. Mol. Biol.*, *11*:144 (1965).

9. "Properties and Inheritance of Haemoglobin by Asymmetric Recombination." Itano, H. A., and E. Robinson. *Nature, 184*:1468 (1959).

10. "A Function for Hemoglobin A_{Ic}?" Schroeder, W. A., and W. R. Holmquist. In *"Structural Molecular Biology,"* p. 238 (A. Rich, and N. Davidson, Eds.) W. H. Freeman Co. (1968).

11. "Human Embryonic Hemoglobins." Huehns, E. R., N. Dance, G. H. Beavan, S. Hecht, and A. G. Motulsky. *Cold Spring Harbor Symposia Quant. Biol.*, *29*:327 (1964).

12. "Gene Selection in Hemoglobin and in Antibody-Synthesizing Cells." Kabat, D. *Science, 175*:134 (1972).

13. "The Evolution of Hemoglobin." Zuckerkandl, E. *Sci. Am.* (May 1965).

14. "Quaternary Structure of Limnodrilus Hemoglobin." Yamagishi, M., A. Kajita, R. Shukuya, and K. Kaziro. *J. Mol. Biol.*, *21*:467 (1966).

15. "Studies on Chlorocruorin. III. Electron Microscope Observations on Spirographis Chlorocruorin." Guerritore, D., M. L. Bonacci, M. Brunnori, E. Antonini, J. Wyman, and A. Rossi-Fanelli. *J. Mol. Biol.*, *13*:234 (1965).

16. "Macromolecular Organization of Hemocyanins and Apohemocyanins as Revealed by Electron Microscopy." Fernández-Moran, H., E. F. J. Van Bruggen, and M. Ohtsuki. *J. Mol. Biol.*, *16*:191 (1966).

17. "Reassociation of Hemocyanins from Subunit Mixtures." Van Bruggen, E. F. J., and H. Fernández-Moran. *J. Mol. Biol.*, *16*:208 (1966).

18. "Structure and Properties of Hemocyanins. I. Electron Micrographs of Hemocyanin and Apohemocyanin from Helix pomatia at Different pH Values." Van Bruggen, E. F. J., and E. H. Wiebange. *J. Mol. Biol.*, *4*:1 (1962).

19. "Horseshoe Crab Lactate Dehydrogenases: Evidence for Dimeric Structure." Selander, R. K., and S. Y. Young. *Science, 169*:179 (1970).

20. "The Control of Biochemical Reactions." Changeux, J. P. *Sci. Am.* (April 1965).

21. "Allosteric Proteins and Cellular Control Systems." Monod, J., J. P. Changeux, and F. Jacob. *J. Mol. Biol.*, *6*:306 (1963).

22. "On the Nature of Allosteric Transitions: A Plausible Model." Monod, J., J. Wyman, and J. P. Changeux. *J. Mol. Biol.*, *12*:88 (1965).

23. "Comparison of Experimental Binding Data and Theoretical Models in Proteins Containing Subunits." Koshland, D. E., G. Nemethy, and D. Filmer. *Biochem.*, *5*:365 (1966).

24. "The Hemoglobin System. VI. The Oxygen Dissociation Curve of Hemoglobin." Adair, G. S. *J. Biol. Chem.*, *63*:529 (1925).

25. "The Osmotic Pressure of Hemoglobin in the Absence of Salts." Adair, G. S. *Proc. Roy. Soc. London* (Series A) *109*:292 (1925).

26. "Multiple Molecular Forms of Enzymes." *Ann. N.Y. Acad. Sci. 151*, Art 1. Pp. 1–689 Vessell, E. S. (Ed.).

27. "The Multienzyme α-Keto Acid Dehydrogenase Complex." Reed, L. J., and R. M. Oliver. *Brookhaven Symposia in Biology*. No. *21*:397–412 (1968).

CHAPTER TEN

Books

1. *Biochemistry*. Chapter 10. Lehninger, A. L., Worth Publishers Inc. (1970).
2. *Textbook of Biochemistry*. Chapter 9. Mazur, A., and B. Harrow, W. B. Saunders Co. (1971).

CHAPTER ELEVEN

Articles, Research Papers, and Reviews

1. "On Bimolecular Layers of Lipoids on the Chromocytes of the Blood." Gorter, E., and R. Grendel. *J. Exp. Med.*, *41*:439 (1925).
2. "A Contribution to the Theory of Permeability of Thin Films. Danielli, J. F., and H. Davson. *J. Cell. Comp. Physiol.*, *5*:495 (1935).
3. "The Chemistry of Cell Membranes." Hokin, L. E., and M. R. Hokin. *Sci. Am.* (October 1965).
4. "The Membrane of the Living Cell." Robertson, J. D. *Sci. Am.* (April 1962).
5. "The Organization of Cellular Membranes." Robertson, J. D. In *Molecular Organization and Biological Function*, p. 65. Harper and Row (1967).
6. "Pores in the Cell Membrane." Solomon, A. K. *Sci. Am.* (December 1960).
7. "The Structure of Cell Membranes." Fox, C. F. *Sci. Am.* (February 1972).
8. "The Fluid Mosaic Model of the Structure of Cell Membranes." Singer, S. J., and G. L. Nicolson. *Science*, *175*:720 (1972).
9. "A Dynamic Model of Cell Membranes." Capaldi, R. A. *Sci. Am.* (March 1974).
10. "Membrane Structure: Some General Principles." Bretscher, M. S. *Science*, *181*:622 (1973).
11. "Cell Membranes: A New Look at How They Work." Culliton, B. J. *Science*, *175*:1348 (1972).

Books

1. *The Permeability of Natural Membranes* (2nd edition). Davson, H., and J. F. Danielli, Cambridge University Press (England) (1952).
2. *Membrane Molecular Biology*. Fox, C. F., and A. Keith, (Eds.). Sinauer Associates Inc. (1972).

CHAPTER TWELVE

Articles and Research Papers

1. "The Mechanism of Photosynthesis." Levine, R. P. *Sci. Am.* (December 1969).
2. "The Evolution of Photosynthesis." Olson, J. M. *Science*, *168*:438 (1970).

3. "Oxidative and Photosynthetic Phosphorylation Mechanisms." Wang, J. H. *Science*, *167*:25 (1970).

4. "Glycogen Plastids in Müllerian Body Cells of Cecropia peltata—A Higher Green Plant." Rickson, F. R. *Science*, *173*:344 (1971).

Books

1. *Bioenergetics* (2nd edition). Lehninger, A. L., W. A. Benjamin Inc. (1971).

2. *Cells and Energy.* Goldsby, R. A., Macmillan (1967).

CHAPTER THIRTEEN

Articles and Research Papers

1. "How Cells Transform Energy." Lehninger, A. L. *Sci. Am.* (September 1961).

2. "Energy Transduction in Membrane Systems." Green, D. E., and J. H. Young, *Am. Sci.*, *59*:92 (1971).

3. "The Organization and Development of Chloroplasts." Bogorad, L. In *Molecular Organization and Biological Function*, p. 134. Harper & Row (1967).

4. "Subunits in Chloroplast Lamellae." Branton, D., and R. B. Park. *J. Ultrastruct. Res.*, *19*:283 (1967).

5. "Quantasome: Size and Composition." Park, R. B., and J. Biggins. *Science*, *144*:1009 (1964).

6. "Hill Reaction Site in Chloroplast Membranes: Non-Participation of the Quantasome Particle in Photoreduction." Howell, S. H., and E. N. Moudrianakis. *J. Mol. Biol.*, *27*:323 (1967).

7. "Function of the 'Quantasome' in Photosynthesis: Structure and Properties of Membrane-Bound Particle Active in the Dark Reactions of Photophosphorylation." Howell, S. H., and E. N. Moudrianakis. *Proc. Nat. Acad. Sci. U.S.A.*, *58*:1261 (1967).

8. "The Continued Presence of Quantasomes in Ethylenediamine tetraacetate-washed Chloroplast Lamellae." Park, R. B., and A. O. A. Pheifhofer. *Proc. Nat. Acad. Sci. U.S.A.*, *60*:337 (1968).

9. "Molecular Basis of Mitochondrial Structure and Function." Lehninger, A. L. In *Molecular Organization and Biological Function*, p. 107. Harper & Row (1967).

10. "The Membrane of the Mitochondrion." Racker, E. *Sci. Am.* (February 1968).

11. "Molecular Isomers in Vision." Hubbard, R., and A. Kropf, *Sci. Am.* (June 1967).

12. "The Organization of Vertebrate Visual Receptors." Dowling, J. E. In *Molecular Organization and Biological Function*, p. 186. Harper & Row (1967).

13. "Studies on Vision. The Nature of the Retinal-Opsin Linkage." Akhtar, M., P. T. Blosse, and P. B. Dewhurst. *Biochem. J.*, *110*:693 (1968).

14. "Localization of Rhodopsin Antibody in the Retina of the Frog." Dewey, M. M., P. K. Davis, J. K. Blasie, and L. Barr, *J. Mol. Biol.*, *39*:395 (1969).

15. "Molecular Localization of Frog Retinal Receptor Photopigment by Electron Microscopy and Low-Angle X-Ray Diffraction." Blasie, J. K., C. R. Worthington, and M. M. Dewey, *J. Mol. Biol.*, *39*:407 (1969).
16. "Biochemistry of Visual Pigments. I. Purification and Properties of Bovine Rhodopsin." Schichi, H., M. S. Lewis, F. Irreverre, and A. L. Stone. *J. Biol. Chem.*, *244*:529 (1969).

CHAPTER FOURTEEN

Articles and Research Papers

1. "The Discovery of DNA." Mirsky, A. E. *Sci. Am.* (June 1968).
2. "Molecular Shape and Size of Thymonucleic Acid." Signer, R., T. Casperrson, and E. Hammarsten. *Nature 141*:122 (1938).
3. "Structure and Function of Nucleic Acids as Cell Constituents." Chargaff, E. *Fed. Proc.*, *10*:654 (1951).
4. "Molecular Structure of Nucleic Acids: A Structure for Deoxyribose Nucleic Acid." Watson, J. D., and F. H. C. Crick. *Nature*, *171*:737 (1953).
5. "The Structure of DNA." Watson, J. D., and F. H. C. Crick. *Cold Spring Harbor Symp. Quant. Biol.*, *18*:123 (1953).
6. "The Structure of the Hereditary Material." Crick, F. H. C. *Sci. Am.* (October 1954).
7. "Genetical Implications of the Structure of DNA." Watson, J. D., and F. H. C. Crick. *Nature*, *171*:964 (1953).
8. "On the Structure of Nucleic Acids." Furberg, S. *Acta Chem. Scand.*, *6*:634 (1952).
9. "Molecular Configuration in Sodium Thymonucleate." Franklin, R. E., and R. G. Gosling. *Nature*, *171*:740 (1953).
10. "Evidence for Two-Chain Helix in Crystalline Structure of Sodium Deoxyribonucleate." Franklin, R. E., and R. G. Gosling. *Nature*, *172*:156 (1953).
11. "Molecular Structure of Deoxyribose Nucleic Acid and Nucleoprotein." Feughelman, M., R. Landridge, W. E. Seeds, A. R. Stokes, H. R. Wislon, C. W. Hooper, M. F. H. Wilkins, R. K. Barclay, and L. D. Hamilton. *Nature*, *175*:834 (1955).
12. "Nucleic Acids." Crick, F. H. C. *Sci. Am.* (September 1957).
13. "The Molecular Configuration of Nucleic Acids." Wilkins, M. F. H. In *Nobel Lectures, Physiology and Medicine, 1942–62*, p. 754. (Lecture given in 1962.) Elsevier Publ. Co. (1964).
14. "DNA Sequencing: A New Era in Molecular Biology." Kolata, G. B. *Science*, *192*:645 (1976).
15. "The Structure of Small Viruses." Klug, A., and D. L. D. Caspar. *Adv. Virus. Res.*, *7*:225 (1960).
16. "The Replication of DNA in Escherichia coli." Meselson, M., and F. W. Stahl. *Proc. Nat. Acad. Sci. U.S.A.*, *44*:671 (1958).
17. "Enzymic Synthesis of Deoxyribonucleic Acid." Kornberg, A., I. R. Lehman, M. J. Bessman, and E. S. Simms. *Biochim. et Biophys. Acta*, *21*:197 (1956).
18. "Biologic Synthesis of Deoxyribonucleic Acid." Kornberg, A. *Science*, *131*:1503 (1960).

19. "Physical and Chemical Properties of Protamine from the Sperm of Salmon (*Onchorhynchus tschawytscha*). I. Preparation and Characterisation." Callanan, M. J., W. R. Carroll, and E. R. Mitchell. *J. Biol. Chem.*, *229*:279 (1957).
20. "Polarized Infra-Red Studies of Nucleoproteins. I. Nucleoprotamine." Bradbury, E. M., W. C. Price, G. R. Wilkinson, and G. Zubay. *J. Mol. Biol.*, *4*:39, 50 (1962).
21. "Protamines and Nucleoprotamines." Felix, K., H. Fischer, and A. Krekels. *Progr. Biophys. Biophys. Chem.*, *6*:1 (1956).
22. "Improved Fractionations of Arginine-Rich Histones from Calf Thymus." Johns, E. W., D. M. P. Phillips, P. Simpson, and J. A. V. Butler. *Biochem. J.*, *77*:631 (1960).
23. "Super-Helical Model for Nucleohistone (DNA)." Pardon, J. F., M. F. H. Wilkins, and B. M. Richards. *Nature*, *215*:508 (1967).
24. "Chromatin Structure." Richards, B. M., J. F. Pardon, and A. S. Pooley. *Biochem. J.*, *124*:39P (1971).

Books

1. *The Double Helix.* Watson, J. D., Atheneum (1968).
2. *Rosalind Franklin and DNA.* Sayre, A., W W. Norton & Co. (1975).
3. *The Path to the Double Helix.* Olby, R., University of Washington Press (Seattle) (1974).
4. *Histones and Other Nuclear Proteins.* Busch, H., Academic Press (1965).

CHAPTER FIFTEEN
Articles and Research Papers

1. "The Composition and Structure of Isolated Chromosomes." Mirsky, A. E., and H. Ris. *J. Gen. Physiol.*, *34*:475 (1950–1951).
2. "The Duplication of Chromosomes." Taylor, J. H. *Sci. Am.* (June 1958).
3. "The Isolation and Properties of Deoxyribonucleoprotein Particles Containing Single Nucleic Acid Molecules." Zubay, G., and P. Doty. *J. Mol. Biol.*, *1*:1 (1959).
4. "The Organization and Duplication of Chromosomes as Revealed by Autoradiographic Studies Using Tritium-Labelled Thymidine." Taylor, J. H., P. S. Woods, and W. L. Hughes. *Proc. Nat. Acad. Sci. U.S.A.*, *43*(1):122 (1957).
5. "Chromosome Puffs." Beermann, W., and U. Clever. *Sci. Am.* (April 1964).
6. "Nuclear Gels and Chromosome Structure." Dounce, A. L. *Am. Sci.*, *59*:74 (1971).
7. "Chromatin Structure." Richards, B. M., J. F. Pardon, and A. S. Pooley. *Biochem. J.*, *124*:39P (1971).
8. "Chromatin Structure: The Supercoil is Superseded." Kelata, G. B. *Science*, *188*:1097 (1975).
9. "Spheroid Chromatin Units (*ν* units)." Olins, A. L., and D. E. Olins. *Science*, *183*:330–334 (1974).

Book

1. *DNA and Chromosomes.* DuPraw, E. J., Holt, Rinehart & Winston (1970).

CHAPTER SIXTEEN
Articles and Research Papers

1. "Experiments on Plant Hybrids." Mendel, G. (translated by E. R. Sherwood.) In *The Origin of Genetics* (C. Stern and E. R. Sherwood, Eds.). W. W. Freeman (1966). Another translation is in *Principles of Genetics* (5th edition). E. W. Sinott, L. C. Dunn, and T. Dobzhansky, (1958). McGraw-Hill (New York). Also in *Mendel's Principles of Heredity*. Bateson, W. Cambridge University Press (England). (1909).
2. "Genetic Control of Biochemical Reactions in Neurospora." Beadle, G. W., and E. L. Tatum, *Proc. Nat. Acad. Sci. U.S.A.*, *27*:499 (1941).
3. "The Genes of Men and Molds." Beadle, G. W. *Sci. Am.* (September 1948).
4. "Fine Structure of a Genetic Region in Bacteriophage." Benzer, S. *Proc. Nat. Acad. Sci. U.S.A.*, *41*:344 (1955).
5. "The Fine Structure of the Gene." Benzer, S. *Sci. Am.* (January 1962).

Books

1. *Mendel's Principles of Heredity*. Bateson, W., Cambridge University Press (England) (1909).
2. *The Physical Basis of Heredity*. Morgan, T. H., Lippincott (1919).
3. *Inborn Errors of Metabolism* (2nd edition). Garrod, A. E., Oxford University Press (England) (1923).
4. *The Mechanism of Mendelian Heredity* (2nd edition). Morgan, T. H., A. H. Sturtevant, H. J. Muller, and G. B. Bridges, Holt, Rinehart & Winston (1922).
5. *The Theory of the Gene* (2nd edition). Morgan, T. H., Yale University Press (1928).
6. *An Introduction to Genetics*. Merrell, D. J., W. W. Norton Co. (1975).
7. *An Introduction to Modern Genetics*. Patt, D. I., and G. R. Patt, Addison-Wesley (1975).
8. *Molecular Biology of the Gene* (3rd edition). Watson, J. D., W. A. Benjamin Inc. (1976).
9. *Genetics*. Levine, R. P., Holt, Rinehart & Winston (1962).
10. *Molecular Genetics*. De Busk, A. G., Macmillan (1968).

CHAPTER SEVENTEEN
Articles and Research Papers

1. "Reconstitution of Active Tobacco Mosaic Virus from its Inactive Protein and Nucleic Acid Components." Fraenkel-Conrat, H., and R. C. Williams, *Proc. Nat. Acad. Sci. U.S.A.*, *41*:690 (1955).
2. "Rebuilding a Virus." Fraenkel-Conrat, H. *Sci. Am.* (June 1956).
3. "The Significance of Pneumococcal Types." Griffith, F. J. *J. Hyg.*, *27*:113 (1928).

4. "Studies on the Chemical Nature of the Substance Inducing Transformation of Pneumococcal Types. Induction of Transformation by a Dexoyribonucleic Acid Fraction Isolated from Pneumococcus Type III." Avery, O. T., C. M. MacLeod, and M. McCarty. *J. Exptl. Med.*, *79*:137 (1944).
5. "Ultraviolet Radiation and Nucleic Acid." Deering, R. A., *Sci. Am.* (December 1962).
6. "An Unstable Intermediate Carrying Information from Genes to Ribosomes for Protein Synthesis." Brenner, S., E. Jacob, and M. Meselson. *Nature*, *190*:576 (1961).
7. "Messenger RNA." Hurwitz, J., and I. J. Furth. *Sci. Am.* (February 1962).
8. "The Genetic Code." Crick, F. H. C. *Sci. Am.* (October 1962).
9. "The Genetic Code: II." Nirenberg, M. W. *Sci. Am.* (March 1963).
10. "The Genetic Code: III." Crick, F. H. C. *Sci. Am.* (October 1966).
11. "On the Colinearity of Gene Structure and Protein Structure." Yanofsky, C., B. C. Carlton, J. R. Guest, D. R. Helsinki, and U. Henning. *Proc. Nat. Acad. Sci. U.S.A.*, *51*:266 (1964).
12. "Evidence for the Universality of the Genetic Code." Abel, P. *Cold Spring Harbor Symp. Quant. Biol.*, *29*:185 (1964).

Book

1. *Molecular Biology of the Gene* (3rd edition). Watson, J. D., W. A. Benjamin Inc. (1976).

CHAPTER EIGHTEEN

Articles and Research Papers

1. "The Structural Basis of Protein Synthesis." Rich, A. In *Molecular Organization and Biological Function*, p. 20. Harper & Rowe (1967).
2. "Involvement of RNA in the Synthesis of Proteins." Watson, J. D. *Science*, *140*:17 (1963).
3. "The Dependence of Cell-Free Protein Synthesis in *E. Coli* upon Naturally Occurring or Synthetic Polyribonucleotides." Nirenberg, M. W., and J. H. Matthaei. *Proc. Nat. Acad. Sci. U.S.A.*, *47*:1588 (1961).
4. "The Nucleotide Sequence of a Nucleic Acid." Holley, R. W. *Sci. Am.* (February 1966).
5. "Ribosomes." Nomura, M. *Sci. Am.* (December 1969).
6. "Ribosomes." (M. Nomura, A. Tissieres, and P. Lengyel Eds.) *Cold Spring Harbor Laboratory* (1974).
7. "Ribosome Structure Determined by Electron Microscopy of Escherichia coli Small Subunits, Large Subunits and Monomeric Ribosomes." Lake, J. R. *J. Mol. Biol.*, *105*:131–159 (1976).
8. "Polyribosomes." Rich, A. *Sci. Am.* December, 1963.

9. "Ribosome Structure and Function Emergent." Kurland, C. G. *Science, 169*:1171 (1970).
10. "Structure and Formation of Ribosome Crystals in Hypothermic Chick Embryo Cells." Byers, B. *J. Mol. Biol., 26*:155 (1967).
11. "The Polarity of Messenger Translation in Protein Synthesis." Thach, R. E., M. A. Cecere, T. A. Sundararajan, and P. Doty. *Proc. Nat. Acad. Sci. U.S.A., 54*:1167 (1965).
12. "Genetic Regulatory Mechanisms in the Synthesis of Proteins." Jacob, F., and J. Monod. *J. Mol. Biol., 3*:318 (1961).
13. "Allosteric Proteins and Cellular Control Systems." Monod, J., J. P. Changeux, and F. Jacob. *J. Mol. Biol., 6*:306 (1963).
14. "Genetic Repressors." Ptashe, M., and W. Gilbert, *Sci. Am.* (June 1970).
15. "Central Dogma of Molecular Biology." Crick, F. H. C. *Nature*, 227:561 (1970).
16. "RNA-Directed DNA Synthesis." Temin, H. M. *Sci. Am.* (January 1972).
17. "Gene Regulation for Higher Cells: A Theory." Britten, R. J., and E. H. Davidson. *Science, 165*:349 (1969).
18. "Estrogen Receptor in the Mammalian Liver." Eisenfeld, A. J., R. Aten, M. Weinberger, G. Haselbacher, and K. Halpern. *Science, 191*:862 (1976).

Book

1. *Molecular Biology of the Gene* (3rd edition). Watson, J. D., W. A. Benjamin Inc. (1976).

CHAPTER NINETEEN

Articles and Research Papers

1. "The Smallest Living Cells." Morowitz, H. J., and M. E. Tourtellotte, *Sci. Am.* (March 1962).
2. "The Bacterial Mesosome." Reusch, V. M., and M. M. Burger, *Biochim. et Biophysica. Acta, 300*:79 (1973).
3. "The Blue–Green Algae." Echlin, P. *Sci. Am.* (June 1966).

Books

1. *Cells and Organelles* (2nd edition). Novikoff, A. B., and E. Holtzman, Holt, Rinehart & Winston (1976).
2. *Cell Biology : A Molecular Approach.* Dyson, R. D., Allyn & Bacon Inc. (1974).
3. *Cell Biology* (6th edition). De Robertis, E. D. P., F. A. Saez, and E. M. F. De Robertis, W. B. Saunders Co. (1975).

CHAPTER TWENTY
Articles and Research Papers

1. "The Living Cell." Brachet, J. *Sci. Am.* (September 1961).
2. "The Organization of Cilia and Flagella." Gibbons, I. R. In *Molecular Organization and Biological Function*, p. 211. Harper & Row (1967).
3. "Cilia." Satir, P. *Sci. Am.* (February 1961).
4. "Flagellar Movement: A Sliding Filament Model." Brokaw, C. J. *Science, 175*:455 (1972).
5. "Amoeboid Movement." Allen, R. D. *Sci. Am.* (February 1962).
6. "The Lysosome." De Duve, C. *Sci. Am.* (May 1963).
7. "Lysosomes and Disease." Allison, A. *Sci. Am.* (November 1967).
8. "The Peroxisome: A New Cytoplasmic Organelle." De Duve C. *Proc. Roy. Soc.* (London) B, *173*:71 (1969).
9. "The Golgi Apparatus." Neutra, M., and C. P. Leblond. *Sci. Am.* (February 1969).
10. "The Sarcoplasmic Reticulum." Porter, K. R., and C. Franzini-Armstrong. *Sci. Am.* (March 1965).
11. "Microfilaments in Cellular and Developmental Processes." Wessels, N. K., B. S. Spooner, J. F. Ash, M. O. Bradley, M. A. Luduena, E. L. Taylor, J. T. Wrenn, and K. M. Yamada. *Science, 171*:135 (1971).
12. "Microtubules: Versatile Organelles." Marx, J. L. *Science, 181*:1236 (1973).
13. "Microtubule Formation in vitro in Solutions Containing Low Calcium Concentrations." *Weisenberg, R. C. Science, 177*:1104 (1972).
14. "How Living Cells Change Shape." Wessels, N. K. *Sci. Am.* (October 1971).
15. "Membrane Continuity between Plasmalemma and Nuclear Envelope in Spermatogenic Cells of Blasia." Carothers. Z. B. *Science, 175*:652 (1972).
16. "The Genetic Activity of Mitochondria and Chloroplasts." Goodenough, U. W. and R. P. Levine. *Sci. Am.* (November 1970).
17. "Are/Were Mitochondria and Chloroplasts Microorganisms?" Cohen, S. S. *Am. Sci., 58*:281 (1970).
18. "Mitochondria and Chloroplasts as Descendants of Prokaryotes." Flavell, R. *Biochem. Genetics, 6*:275 (1972).
19. "A Multiple Origin for Plastids and Mitochondria." Raven, P. H. *Science, 169*:641 (1970).
20. "The Non-Symbiotic Origin of Mitochondria." Raff, R. A., and H. R. Mahler. *Science, 177*:575 (1972).
21. "Development of Cellular Dependence on Infective Organisms: Micrurgical Studies in Amoebas." Jeon, K. W. *Science, 176*:1122 (1972).
22. "The Origin of Plant and Animal Cells." Margulis, L. *Am. Sci., 59*:230 (1971).
23. "Symbiosis and Evolution." Margulis, L. *Sci. Am.* (August 1971).
24. "Cell Biology: Cell Surfaces and the Regulation of Mitosis." Marx, J. L. *Science, 192*:455 (1976).
25. "The Genetic Control of the Shape of a Virus." Kellenberger, E. *Sci. Am.* (December 1966).
26. "Reassembly of Living Cells from Dissociated Components." Jeon, K. W., I. J. Lorch, and J. F. Danielli *Science, 167*:1626 (1970).

Books

1. *Cell Biology* (6th edition) De Robertis, E. D. P., F. A. Sacz, and E. M. F. De Robertis, W. B. Saunders Co. (1975).
2. *Cells and Organelles* (2nd edition). Novikoff, A. B. and E. Holtzman, Holt, Rinehart and Winston (1976).
3. *Cell Biology: A Molecular Approach*. Dyson, R. D., Allyn and Bacon Inc. (1974).
4. *Origin of Eukaryotic Cells*. Margulis, L., Yale University Press (1970).

Index